Microgravity: A New Tool for Space Experiments

Microgravity: A New Tool for Space Experiments

Edited by
Christina Taylor

Willford Press
www.willfordpress.com

Published by Willford Press,
118-35 Queens Blvd., Suite 400,
Forest Hills, NY 11375, USA

Copyright © 2023 Willford Press

This book contains information obtained from authentic and highly regarded sources. Copyright for all individual chapters remain with the respective authors as indicated. All chapters are published with permission under the Creative Commons Attribution License or equivalent. A wide variety of references are listed. Permission and sources are indicated; for detailed attributions, please refer to the permissions page and list of contributors. Reasonable efforts have been made to publish reliable data and information, but the authors, editors and publisher cannot assume any responsibility for the validity of all materials or the consequences of their use.

Trademark Notice: Registered trademark of products or corporate names are used only for explanation and identification without intent to infringe.

ISBN: 978-1-64728-457-2

Cataloging-in-publication Data

Microgravity : a new tool for space experiments / edited by Christina Taylor.
p. cm.
Includes bibliographical references and index.
ISBN 978-1-64728-457-2
1. Reduced gravity environments. 2. Outer space--Exploration. 3. Astronautics--Experiments.
4. Research in reduced gravity environments. I. Taylor, Christina.
TA357.5.R44 M53 2023
620.106--dc23

For information on all Willford Press publications
visit our website at www.willfordpress.com

Willford Press

Contents

Preface

Microgravity is the condition in which people or objects appear to be weightless. The microgravity lab consisting of only-in-space conditions facilitates a wide range of investigations in various disciplines. This lab provides a platform with greatly reduced non-gravitational sources of noise, access to the vacuum conditions of space, extreme heat and cold, large distances, and high velocity that is beneficial for physics research. The concept of microgravity is used in advanced materials science research, because it changes the processes such as crystal growth, fluid mixing, separation between gases and liquids, heat transfer, solidification, and combustion. Access to microgravity is beneficial for biology and human health. Data about how different organisms react to a microgravity environment and the ways various biological functions change in the absence of gravity increase the understanding related to organisms. This book is a compilation of chapters that discuss the most vital concepts and emerging trends on the subject of microgravity. It aims to serve as a resource guide for students and researchers alike.

This book is a result of research of several months to collate the most relevant data in the field.

When I was approached with the idea of this book and the proposal to edit it, I was overwhelmed. It gave me an opportunity to reach out to all those who share a common interest with me in this field. I had 3 main parameters for editing this text:

1. Accuracy – The data and information provided in this book should be up-to-date and valuable to the readers.

2. Structure – The data must be presented in a structured format for easy understanding and better grasping of the readers.

3. Universal Approach – This book not only targets students but also experts and innovators in the field, thus my aim was to present topics which are of use to all.

Thus, it took me a couple of months to finish the editing of this book.

I would like to make a special mention of my publisher who considered me worthy of this opportunity and also supported me throughout the editing process. I would also like to thank the editing team at the back-end who extended their help whenever required.

Editor

1

Experiment Preparation and Performance for the Electromagnetic Levitator (EML) Onboard the International Space Station

Angelika Diefenbach, Stephan Schneider and Thomas Volkmann

Abstract

The electromagnetic levitation (EML) facility on board the ISS is a powerful tool for investigation of solidification phenomena of metallic melts and precise measurement of thermophysical properties of the liquid. Containerless processing enables deep undercoolings prior to solidification and the analysis of crystal nucleation and growth phenomena. The microgravity environment allows studying these processes under reduced fluid flow and moreover under different levels of melt convection by systematic variation of electromagnetic stirring. Material properties like density, specific heat, surface tension, viscosity, thermal and electrical conductivity of liquid metals and semiconductors are determined in the absence of disturbances caused by container walls and gravity forces. Scientists are supported by facility and mission specialists for preparation and performance, which is decisive for successful operation on orbit. User support comprises the determination of material data, development of experiment procedure, parameter sets and their validation in the ground model as well as the conduction of space experiments by real-time monitoring and control. The comprehensive support program for the entire life cycle of science project from experiment definition to its operation ensures high-quality data and an optimum of scientific results.

Keywords: electromagnetic levitation, containerless processing, International Space Station, undercooled melts, ground support program

1. Introduction

Since more than three decades, electromagnetic levitation in ground-based research is an established technique for thermal experiment processing (heating, melting, and solidification) of electrically conductive material samples, without any contact of the sample to a containment. The samples are free-floating in an ultraclean environment, typically high-purity noble gases, in order to preserve intrinsic behavior during heating up, melting, in liquid state, and during solidification behavior. This unique feature enables the scientific investigation and measurement of material properties and characteristics that are otherwise not possible due to the interaction of the liquid samples with the container walls and any undesired contamination adhering to it. For example, liquid metallic materials are typically highly reactive with even smallest amounts of oxygen. Contact with any solid particle or element can trigger the immediate nucleation and forming of crystal structures while cooling down through the solidification point. The absence of any such disturbance on the other hand allows observation of the pure material behavior. In particular it enables the possibility to reach an undercooled state, in which the material stays liquid even below the equilibrium freezing temperature. This is achieved by levitating and heating the material sample through high-frequency electromagnetic fields generated by alternating currents in a coil. The field induces eddy currents in the sample, which is heated by Ohm's losses of the currents, while their interaction with the electromagnetic field leads to forces toward the center of the coil system. The technique can be applied both on ground and in reduced or zero gravity, but the latter provide much better experiment conditions. To levitate a sample under normal gravity requires strong levitation fields, which induces convection electromagnetically and an unwanted heating bias. Furthermore, the liquid sample is getting deformed by the strong electromag-netic force needed to compensate gravity. In zero-g the sample floats by itself and requires only very small positioning forces to stay in the center of the coil system with a perfectly spherical shape, which is beneficial for a variety of techniques to measure thermophysical properties of melts. The heating can be controlled almost independently of the positioning, allowing processing of, e.g., highly reactive material, thermophysical measurements with much higher accuracy, and solidification experiments, with varying stirring conditions. Therefore, electromagnetic levitation under microgravity provides unique opportunities for the investigation of liquid metals, alloys, and semiconductors, both above and below their melting temperatures, i.e., also including the undercooled regime. Besides fundamental scientific interests, the research is also oriented to industrial applications where reliable data for accurate modeling of industrial processes are difficult or impossible to be obtained on ground.

The payload "electromagnetic levitator" (EML) is the realization of that concept for the International Space Station (ISS) and was jointly developed by the European Space Agency (ESA) and the *Deutsches Zentrum für Luft- und Raumfahrt* (DLR) space administration [1]. It was built by Airbus DS and installed in the European research lab COLUMBUS on the ISS by the ESA astronaut Alexander Gerst during the Blue Dot Mission in fall 2014. Containerless processing of metallic alloys and semiconductors with the electromagnetic levitator has been performed since April 2015. Using sophisticated methods of initiating and analyzing temperature modulations and liquid surface oscillations and their decay, thermophysical properties can be investigated. Measurements include the specific heat capacity, the surface tension and viscosity, as well as the electrical conductivity in the stable liquid and undercooled metastable liquid states. Entire sets of high precision data of thermophysical properties in the liquid and undercooled liquid regime have thus been obtained over extended periods of time using the ISS-EML device.

The scientific experiments in EML are grouped into batches. One batch is related to a specific sample chamber with 18 dedicated samples. Each sample can be melted many times; hence one batch consists in the range of about thousand individual melt cycles. Since its successful on-orbit commissioning onboard the ISS beginning of 2015, EML has been under quasi-continuous operation, and two complete batches of experiments have been conducted since, processing 36 samples with a total of more than 2000 individual melt cycles [2].

Before the realization of the EML facility for the ISS, the scientific community had gained substantial experience with ground-based research by means of electromagnetic levitation technique since the late 1980s. Driven by its shortcomings under terrestrial conditions, a first approach for a levitator to be used under reduced gravity conditions was developed in the Spacelab era in the 1990s. The technology development for the positioning and heating technique as well as the selection of suitable noninvasive diagnostics has been closely supported by the research teams of the former DLR Institute of Space Simulation (today "*Institut für Materialphysik im Weltraum*") which had established ground-based levitation facilities in their labs. When it was decided to adapt the levitation technique for application under reduced gravity, the DLR technical and scientific expertise was used in the proof of concept during early parabolic and sounding rocket flights of a precursor model of EML, which in the end led to the development of a Spacelab payload TEMPUS which was flown on three missions in the 1990s.

The authors are convinced that the success of the current EML payload onboard the ISS in its fifth year of on-orbit operation with currently approximately 2000 individual melt cycles performed on 36 pure metals, alloys, and semiconductors is the result of a longstanding facility and experiment development phase where a substantial built-up in technical, operational, and scientific expertise could be achieved. This will be substantiated by zooming in on preparatory work and performance of EML experiments. We will detail the evolution of the "EML type" payloads for use under microgravity and describe the user support provided for the preparation and conduct of the EML experiments for the ISS. Moreover, an outlook to near-term EML program activities and further future enhancements is given increasing the facility's capabilities for the coming years and sound scientific experiment batches.

2. Motivation and scientific background

It is well established that modern technologies can work properly only if suitable materials with tailor-made properties are utilized. Metallic materials play an important role in nearly all fields of daily human life and are used for many applications, e.g., in automobile or aerospace industries. The material properties (strength, hardness, corrosion behavior, and magnetic and electrical properties) are not only determined by the chemical composition but also sensitively depend on the microstructure. Nearly all metallic alloys are produced from the liquid state by solidification processes. The conditions during solidification of a metallic melt control the evolution of the microstructure and therefore the properties of the as-solidified material. Therefore, efforts of researchers are directed to a detailed understanding of the physical mechanism involved in the transition from the liquid to the solid state, which is mandatory for computer-assisted modeling of solidification processes and the formation of the microstructure. Computer simulations are increasingly demanded in industry since it enables the optimization of production routes in order to save costs and energy during manufacturing. The production of metallic materials belongs to the largest industrial sectors worldwide. For instance, in the European Union, there were more than 400,000 enterprises with more than 5 million employees in 2006 [3]. Therefore, large economic gains can be achieved even by small progresses in metal processing.

Solidification of melts is a very complex process. It is initiated by the formation of tiny crystals on an atomic scale (the so-called crystal nucleation process) and their subsequent growth into macroscopic dimensions. For the formation of the microstructure, the morphology of the moving solid-liquid interface plays a decisive role. Among a broad variety of pattern formation, the major growth mode in metallic systems is dendritic growth (see, e.g., review articles [4, 5]). In many cases an initially planar interface between the growing solid and liquid is not stable. Local fluctuations in growth rate lead to the formation of protrusions, and finally a tree-like dendritic structure with a primary stem and secondary branches is established as illustrated in **Figure 1**. Branches or dendrite arms are growing preferentially in certain directions with respect to the crystallographic structure of the solid. The cubic crystal structure of the solid phase in $Fe_{99}B_1$ alloys is reflected by the fourfold symmetry of the dendritic grain where growth directions of the primary dendrite and secondary arms are perpendicular to each other (**Figure 1**).

However, the microstructure is then composed of many dendritic grains originating from different crystal nuclei. Key parameters such as size of grains and their orientation to each other (texture), thickness and spacing between dendrite arms, and the distribution of solute atoms across the sample determine the material properties of the final product. These parameters can be controlled externally by the cooling rate or the temperature gradient during solidification, but they also sensitively depend on the liquid phase properties for the transport of heat and mass. As illustrated in **Figure 2**, solidification is accompanied with the release of latent heat of fusion, which needs to be removed from the solid-liquid interface. Generally, in alloys the solid phase and the melt exhibit different chemical compositions so that growth of the crystal requires a rejection of solute atoms ahead the interface. Transport of heat and solute atoms is not only ruled by thermal conduction and atomic diffusion, respectively, but also by fluid flow (see, e.g., [4]). Therefore, melt convection is considered as an additional process parameter controlling the evolution of the microstructure. Under the conditions of microgravity, a major source of natural convection, i.e., buoyancy-driven flow, is completely avoided, thus enabling the study of solidification processes under near-diffusive conditions. Moreover, external triggering of fluid flow, e.g., by applying electromagnetic fields, allows analyzing the influence of fluid flow systematically without superposition by gravity-driven effects.

Another important process parameter is the level of undercooling of the melt prior to solidification. When a melt is cooled down, the phenomenon of undercooling below the equilibrium freezing temperature always occurs. At the melting point, nucleation does not set in and the sample remains in the liquid state. Any level of undercooling, defined by $\Delta T = T_m - T$ (T_m: melting temperature, T: temperature of

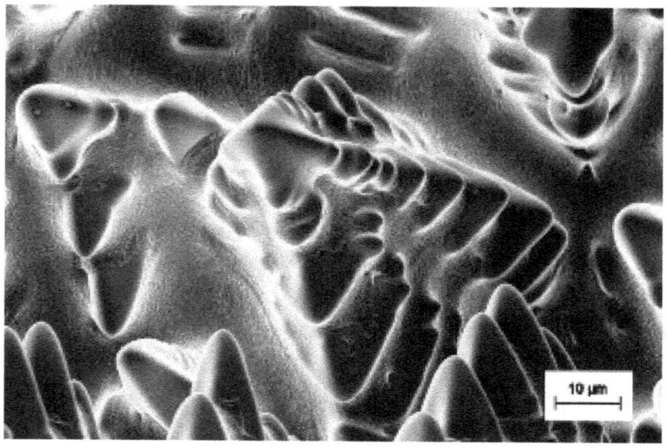

Figure 1.
Scanning electron micrograph of the surface of an as-cast $Fe_{99}B_1$ alloy showing the morphology of metallic dendrites. The length scale is indicated by the horizontal bar (M. Kolbe, T. Volkmann, DLR Cologne).

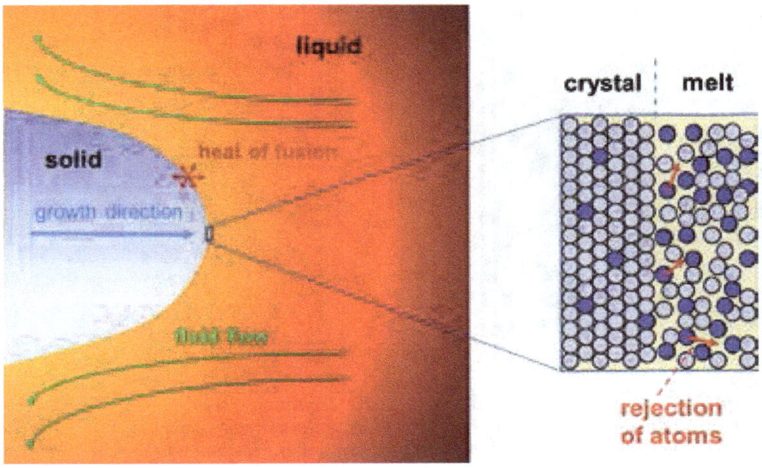

Figure 2.
Left: Illustration of a dendrite tip of the solid phase growing into the liquid. The heat of fusion is released at the solid-liquid interface. Right: The concentration of solute atoms (marked in dark blue) in alloys is different for the solid and the liquid (see text for details).

the melt prior to solidification), is necessary to generate a thermodynamic driving force in order to create nuclei of the solid phase. However, undercooling of melts is of special interest because it gives access to the formation of metastable solid phases and microstructures far from equilibrium [6, 7]. A large variety of metastable materials with new physical properties can be prepared comprising metastable crystalline phases, quasicrystals, metallic glasses, supersaturated solids, and grain-refined materials. In industrial applications, deep levels of undercooling are mainly achieved by rapid quenching of melts such as melt spinning and inert gas atomization of melts into fine droplets (powder fabrication) in order to produce metastable materials.

However, methods of rapid quenching are not suitable for a direct diagnostic of the physical mechanism involved in the selection of alternative solidification pathways into metastable phases. Deep undercoolings that typically amount to about 20% of the melting temperature can also be achieved by the avoidance or reduction of heterogeneous nucleation sites such as solid metal oxides on the sample and container walls. Thus, large undercooling levels can be realized even at slow cooling rates of the order of a few K/s of bulk samples. Containerless processing of melts such as electromagnetic levitation is a powerful technique since heterogeneous nucleation on containers is completely avoided. Moreover, the freely suspended droplet is accessible for a direct monitoring by pyrometers and high-speed video cameras, thus enabling the analysis of solidification processes and the measurement of various thermophysical properties of the liquid.

In electromagnetic levitation technique, the sample is inductively heated and levitated by a radiofrequency (RF) electromagnetic field. The sample is placed in a coil that is connected to a high-frequency generator, which is operated at frequencies of about 400 kHz. An example for a coil design that is used for levitation on ground is shown in **Figure 3**. The conical-shaped coil consists of seven water-cooled copper windings and four counter windings at its top. Such coil geometry produces an inhomogeneous electromagnetic field with a strong field gradient against the gravity vector. The levitation coil together with the sample of 6–7 mm in diameter (about 1 g in mass) is integrated in a vacuum chamber made of stainless steel. In order to avoid oxidation of the sample, the chamber is evacuated to pressures of about 10^{-7} mbar and then backfilled with inert gases like purified He up to about

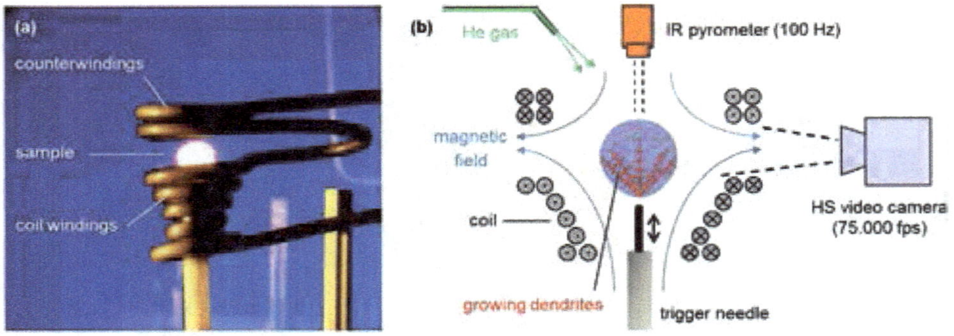

Figure 3.
(a) Coil design used in ground-based electromagnetic levitation facility at DLR Köln, Institut für Materialphysik im Weltraum. (b) Sketch of experimental setup showing a cross section of the levitation coil with an inhomogeneous magnetic field generated by the opposite directions of the RF current in lower coil winding and the upper counterwindings. The temperature is measured by an infrared pyrometer, and rapid solidification is monitored by a high-speed video camera. The nozzle at the top produces a He gas stream to undercool the sample convectively until spontaneous solidification set. Alternatively, nucleation can be triggered at a preselected undercooling by touching the sample with a ceramic needle.

500 mbar before the experiment. During the experiment the temperature is measured contactless by an infrared pyrometer at a rate of 100 Hz, and the sample is observed by a radial high-speed video camera. Changing the temperature of the levitating sample by variation of the electromagnetic field strength is possible only within a limited range because a strong field is needed for compensation of gravity. For many materials like Fe and Ni, the minimum temperature that can be reached by reducing the power is above the melting point. Therefore, for solidification experiments in ground-based levitator and to reach temperatures in the regime of the undercooled melt, the sample must be cooled by applying forced convection in a He gas stream to increase the heat flux to the gas environment. For that purpose, a nozzle system producing a He gas flow is installed in the facility on top of the levitation coil. In particular, the speed of the gas stream can be varied by a tunable valve in order to achieve different cooling rates or to establish a constant temperature. The levitated droplet can be cooled and undercooled below the equilibrium melting temperature until spontaneous nucleation and solidification set in. Alternatively, nucleation can be triggered externally at a preselected undercooling level by touching the sample with a ceramic needle, which acts as a heterogeneous nucleation site. Rapid solidification of the undercooled melt is recorded by the high-speed video camera at frame rates up to 75,000 fps.

An important aspect is that the huge field amplitudes required on Earth in order to generate a levitation force to compensate gravity lead to strong electromagnetically induced turbulent melt convection with fluid flow speeds up to about 0.5 m/s. For processing in microgravity, only slight positioning fields are needed being three orders of magnitude smaller than on the ground. Accordingly, fluid flow is significantly lowered to some centimeter per seconds [8]. After the melting process, the sample can be cooled and solidified without convective gas cooling by setting the electromagnetic field strength to a minimum. Alternatively, small field amplitudes can be applied, still allowing the sample to cool down but generating certain levels of electromagnetic stirring. Thus, the microgravity environment does not only allow the minimization of electromagnetic stirring but also the variation of the field strength in order to study solidification processes systematically under different levels of fluid flow. In parallel, a larger temperature regime is accessible for investigation. While convective gas cooling needed for processing on ground has only a limited efficiency, deeper temperatures can be achieved by reducing the electromagnetic field to a minimum, which is only possible under reduced gravity. This enables to cool and to solidify levitated samples with a low freezing temperature such as glass-forming alloys. Moreover, for many materials, thermal radiation is sufficient to cool the sample down so that it can be processed without a gas atmosphere. The benefit of vacuum conditions with a residual air pressure of typically 10^{-7} mbar in EML on the ISS is the high purity. For comparison, in a He gas atmosphere at 1 bar and with the highest purity of 99.9999%, the residual pressure of impurities (air) equals to 10^{-3} mbar being four orders of magnitude larger than in vacuum. Hence, under vacuum oxidation of extremely reactive materials such as liquid titanium or aluminum based alloys can be reduced to a large extend, which is mandatory to suppress heterogeneous crystal nucleation by solid metal oxides and to achieve deep undercoolings of the melt.

The schematics of a typical temperature-time profile during melting, undercooling, and rapid solidification are shown in **Figure 4**. The solidification process consists of two steps. The blue part of the temperature curve indicates when the sample is solid, while the yellow part shows the sample being liquid. In the undercooled regime, where the sample temperature has fallen below the melting temperature T_m, the rapid growth of dendrites into the undercooled melt leads to a steep temperature rise (recalescence) up to the melting point. The residual liquid in the inter-dendritic region solidifies subsequently under equilibrium conditions during the plateau at T_m. The rapid solidification process can be visualized by high-speed video records taken during recalescence as shown by the example in **Figure 5**. Due to the released heat of fusion at the solid-liquid interface, a steep temperature gradient is established ahead of the solidification front. The sharp contrast in brightness is due to the distinct temperature difference, which makes the solidification front visible when it intersects the surface of the droplet. Monitoring the propagation of the front enables the determination of the growth velocity. Measuring growth velocity as a function of undercooling is an effective method for testing and improving of models for crystal growth in undercooled melts by comparing model predictions of growth kinetics with experimental data.

It must be noted that the solidification front shown in **Figure 5** is macroscopic in dimension and is composed of many microscopic dendrites. The morphology of the front contains valuable information about the growing solid phases and the growth process. In case of deeply undercooled $Fe_{50}Co_{50}$ alloys (**Figure 5**), the recalescence event proceeds in two steps, which originate from the primary formation of a

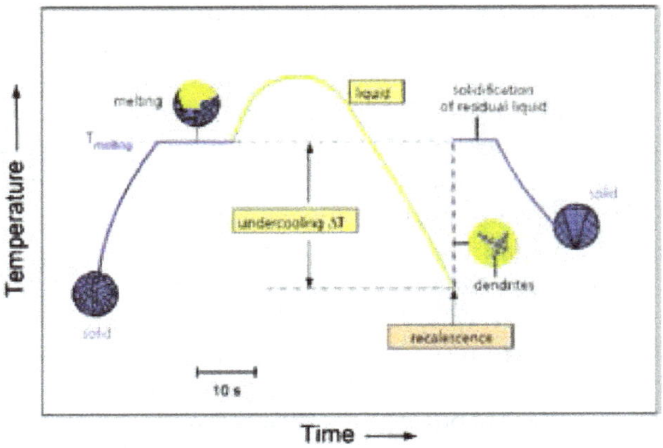

Figure 4.
Schematic temperature-time profile during heating, melting, undercooling, and rapid solidification (recalescence).

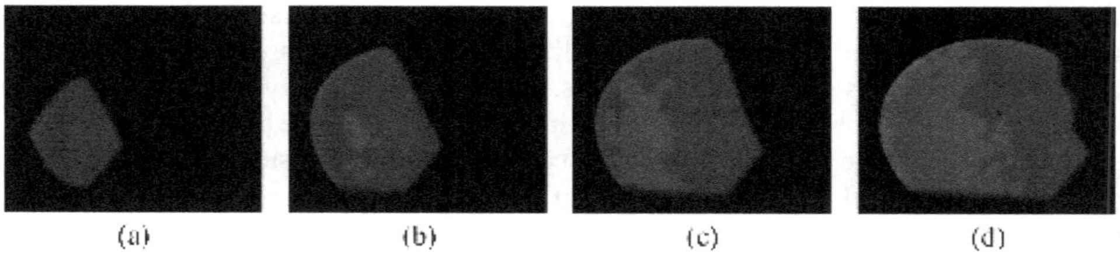

Figure 5.
High-speed video images during rapid solidification of $Fe_{50}Co_{50}$ showing a double recalescence at an undercooling of 245 K. The sample (diameter ~6.5 mm) was processed in a ground-based electromagnetic levitation facility at DLR Cologne. The camera was operated at 75,000 fps, and the figure shows each fifth frame, i.e., the time interval between the frames (a)-(d) is about 0.066 ms (images provided by C. Kreischer, T. Volkmann, DLR).

metastable solid with body-centered cubic crystal structure and subsequent transformation into the stable phase with face-centered cubic structure. The contour of the primary front reveals special features such as a fourfold symmetry and sharp edges and faces. The symmetry is due to dendrite branches originating from a single nucleation point and growing perpendicular to each other, which forms a dendritic grain with a pyramidal-shaped envelope. In that way, containerless processing is a powerful tool for the in situ analysis of rapid solidification processes.

The second research area, which is investigated using electromagnetic levitation techniques under microgravity (TEMPUS, EML), is the thermophysical property of liquid metals and metallic alloys [9]. For the simulation of casting and solidification processes including electromagnetic stirring in the melt and the effect of fluid flow on phase formation, the knowledge of thermophysical parameters of the liquid state, such as density, specific heat, thermal conductivity, and viscosity, are crucial. Contactless processing allows to obtain thermophysical data of highly reactive materials and to extend the temperature range of the liquid into the undercooled metastable state. A broad range of contactless measurement techniques on electromagnetically levitated samples was developed over the years. Many of these properties are difficult to measure in ground-based facilities due to the presence of stronger electromagnetic fields to counter gravitation. The microgravity environment provides optimal conditions for measurements with high precision since fluid flow is minimized and the levitating droplet is nearly free of external forces and exhibits an ideal spherical shape.

Several thermophysical parameters of the liquid state can be obtained from video records of the levitated droplet. By image analysis the volume of a spherical sample can be determined with a high precision and finally the density. Measurement of the density at different temperatures yields the thermal expansion coefficient. Viscosity and surface tension are analyzed using the oscillating drop technique [10–12]. As surface tension acts as the force driving a deformed liquid droplet to its equilibrium shape (sphere), surface tension determines the frequency of surface oscillations. The viscosity originates from the internal friction of liquid motion during oscillation and is therefore related to the damping of droplet vibrations. In microgravity experiments, surface oscillations of the molten sample are induced by short pulses with high amplitudes of the electromagnetic field, while droplet oscillations are recorded by the video camera as shown in **Figure 6** (left). A plot of the sample's radius as function of time reveals a damped oscillation (**Figure 6**, right). A fast Fourier transformation is applied to the radii data to obtain a frequency spectrum from which the oscillation frequencies and thus the surface tension are derived. The viscosity is proportional to the damping constant of the oscillation, which is determined from the decaying oscillation amplitude as shown in the right diagram in **Figure 6**. Under terrestrial conditions the oscillation spectrum is more complicated, and frequencies are shifted due to external forces and the aspherical shape of the droplet, which complicates the quantitative analysis with respect to surface tension. Improved evaluation procedures have to be applied in order to obtain surface tension data with the same precision than in microgravity experiments [12, 13]. However, measurements of viscosity in ground-based experiments could not be performed so far since the electromagnetically induced strong fluid flow is turbulent and the huge magnetic field causes an additional damping.

A contactless method for the measurement of specific heat is the modulation calorimetry [15, 16]. In this method the electromagnetic field is set to a certain level to establish a constant temperature of the sample. Then, the field strength and thus the heating power input are modulated, which lead to a periodical temperature response of the sample (see **Figure 7**). A precise determination of specific heat from the power input and the measured temperature response require a suitable choice of the modulation frequency, which is typically below 1 Hz. Determination of the effective power input requires the knowledge of the heat balance of the sample with the environment. Ideal conditions for calculation of the heat loss are provided under microgravity and by processing in vacuum. Under vacuum (at pressure of typically 10^{-7} mbar), heat flux by conduction to the residual gas atmosphere is negligible; heat loss is thus solely due to thermal radiation. The heat flux is proportional

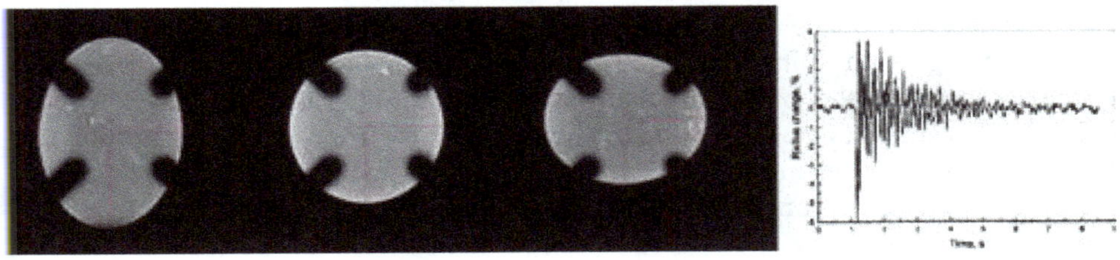

Figure 6.
Oscillating sample at three different times where the radii are marked by red lines (left). The time dependence of the sample diameter reveals a damped oscillation (right) [13].

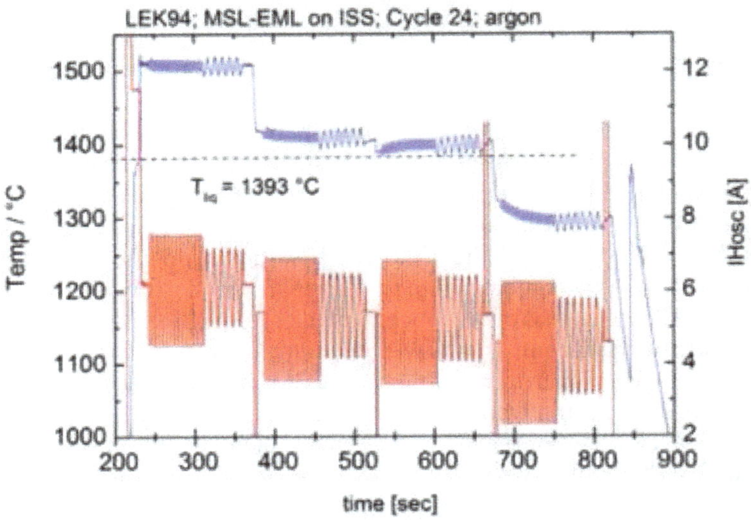

Figure 7.
Specific heat measurement at LEK94—Ni-base superalloy [14].

to the total surface area of the droplet, which has to be obtained from video records of the levitating sample. The highest accuracy is achieved if the sample is an ideal sphere. The modulation calorimetry also allows extracting the thermal conductivity if appropriate values for power input and modulation frequency are chosen.

The electrical conductivity is another important parameter, which can be deduced from electromagnetic levitation experiments by contactless methods. In particular, for liquid metals the electrical and thermal conductivity are based on nearly the same conduction mechanism (transport of free electrons) and are related according to the Wiedemann-Franz law. Hence, thermal conductivity can be derived from measurement of the electrical conductivity, which is less influenced by effects of melt convection. The principle is based on the analysis of the frequency-dependent resistance (impedance) of the electronic circuit consisting of coil including the sample [17]. The impedance is determined by measuring alternating current, voltage, and their phase shift and depends on the electrical resistivity (which is the inverse conductivity) as well as the size and the shape of the droplet. However, precise determination of resistivity from the measured impedance data requires a sophisticated coil design and evaluation methods of the radiofrequency data. A suitable coil system for utilization in microgravity providing measurements with high accuracy is presented in more detail in Section 3.5.

A broad variety of technically important materials such as steels; Ni-, Ti-, Al-, Zr-based alloys; and semiconductors are investigated by an international community using electromagnetic levitation on ground and the ISS-EML facility onboard the International Space Station [18]. The scientific program comprises crystal nucleation, growth kinetics, multiphase solidification (eutectic, peritectic systems), metastable phase formation, liquid-liquid phase separation, and solidification shrinkage phenomena including the measurement of thermophysical properties. The experimental investigations are supported by computer-aided modeling of solidification and the effect of fluid flow.

3. The payloads for use under microgravity and their respective purposes

3.1 Heritage: the spacelab facility TEMPUS

The idea to develop such a facility for the Space Shuttle stems from the Institute of Space Simulation at the German Aerospace Center (DLR) in the 1980s. This facility was built under the acronym *Tiegelfreies elektromagnetisches Prozessieren unter Schwerelosigkeit* (TEMPUS) and means containerless electromagnetic processing under weightlessness. The concept for the TEMPUS facility was tested in several parabolic flights with the KC-135 of NASA from 1988 onwards and one sounding rocket flight in 1989. After this successful proof of concept, it was decided to build a Spacelab facility. This TEMPUS facility was first flown on the Spacelab Mission IML-2 in July 1994. As a bilateral project, TEMPUS was co-funded by DLR Space Agency and NASA. The technical concept of electromagnetic levitation and the specification of the main components were again developed by the DLR Institute of Space Simulation. The scientific program for this mission consisted of 18 samples shared by 8 research teams from Germany and the USA. The scientific resources were equally shared between the US and German scientists. In total, approximately 1000 thermal individual experiment runs were performed within 14 days. The investigated materials comprised pure metals such as Au, Cu, Ni, and Zr but also alloys, such as steels, glass formers, or quasicrystals. During a reflight of TEMPUS on the MSL-1 and MSL-1R Missions, a second set of 18 samples was successfully processed under microgravity conditions, complementing the results of the IML-2 experiments [19].

3.2 Parabolic flights with the TEMPUS spacelab model

In the course of the experiment development for the Spacelab Missions, it became obvious that one major obstacle could not be overcome by ground-based activities alone: the positioning behavior in the electromagnetic fields under reduced gravity was *per se* inaccessible, as well as the coupling efficiency in the molten state. The need for submitting each of the selected samples to a testing during parabolic flights was acknowledged. In the early summer of 1994, shortly before the IML-2 launch on STS-65, the TEMPUS ground model was shipped to Houston for its first parabolic flight campaign. It was operated by a joint team of DLR and Airbus (former Dornier) personnel. All IML-2 science teams had representatives on-site at Ellington Airfield who joined the flight days as observers. With three operators and up to four members from the science community, the TEMPUS team was by far the largest group onboard the aircraft. Only a few minutes after takeoff and still in steep ascent, the teams were allowed to change their seats against positions in front of their respective hardware for setting up the experiments. For our experiments, preparation included running up the facility, selecting the first sample, and establishing vacuum conditions (**Figure 8**).

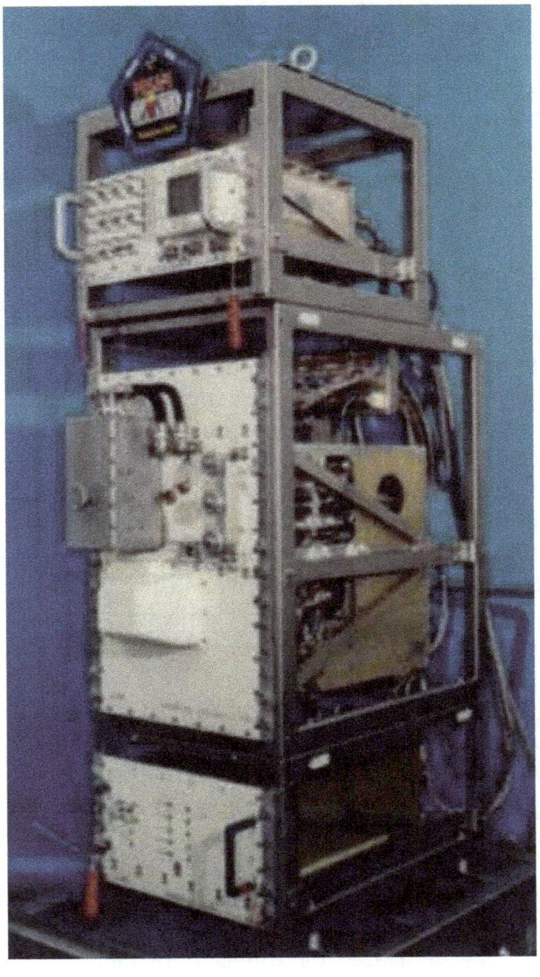

Figure 8.
Engineering model of the TEMPUS facility used during the shuttle missions (Photo Airbus).

The science program had been adjusted to the standard NASA flight profile which, back then, consisted of 48 parabolas per flight day, divided in 6 blocks of 8 parabolas. The eight parabolas in one block were performed back to back with a pause of 3–5 minutes in between the blocks. For TEMPUS flight days, one block was always dedicated to one sample, with several positioning and melting tests. The time between the blocks was spent on level flight and could be used to exchange samples. Apart from the dress rehearsal for the Spacelab experiments, the KC-135 campaign was also used for additional crew training sessions under the most realistic conditions for the Astronauts Rick Hieb, Chiaki Mukai, Don Thomas, and Leroy Chao who were all selected as IML-2 payload specialists. The most important finding from this first scientifically motivated parabolic flight was that all selected samples could be successfully positioned and subsequently melted in the TEMPUS coils system within the 20 seconds of microgravity and that the samples and the experiments were "Go" for launch. However, the learning curve was steep. To maximize the number of successful science runs for future campaigns, there were still some operational challenges to be worked. For example, samples needed to be handed over to the quadrupole field for a stable positioning before heating them by the dipole field well above their respective liquidus temperatures of between 800 and 1850°C. Then, the samples also needed time to cool down and solidify again, all of that in less than 20 seconds. It was planned to preheat the respective sample during the pull-up phase by having it sitting on the sample holder and to hand it over to the field at elevated temperatures at the onset of microgravity, by pulling the holder back. Unfortunately, there is a risk of the sample sticking to the holder at high temperatures, either due to reactions with the holder material or due to a thin layer of evaporated sample material condensing on the holder over the course of repeated melting cycles. It was found that in this case, a smooth transition is difficult and the started experiment has to be aborted and retried in the next parabola, ideally with changed heater parameters.

Usually, the pilots entered the low-gravity phase by slightly "overshooting," i.e., applying negative g levels and then oscillating around the zero line. It was found that due to the coil geometry, the TEMPUS facility was especially susceptible to microgravity disturbances in negative z-direction. In some cases, a free-floating sample was catapulted out of the coil center due to an overshoot or due to weather conditions and was thus lost for further science runs. Interactions with the flight personnel helped to minimize these events. With the parabolas being executed back to back, the time between the reduced gravity periods was in the order of 1 minute only. In case of a sample loss, the sample holder needed to be withdrawn and exchanged which took more time than that. In that case, one or two parabolas would pass unused.

During the first TEMPUS parabolic flight, the ground model of the spacelab facility was used, being identical in form, fit, and function, featuring the original flight software. This model was obviously not conceived for easy mechanical access to subsystems for, e.g., sample exchange, or for flexibility in experiment performance. With the benefits from that campaign, NASA and DLR decided to fund the development of a TEMPUS lab facility, dedicated to parabolic flight campaigns in the future. Between 1996 and 2019, this model and its updates have been used in 17 campaigns. The successful performance of experiments on a parabolic flight campaign today is the precondition for the selection of a sample material for operation in the EML facility onboard Columbus.

3.3 The dedicated parabolic flight facility: testing on a yearly basis

After the shuttle era, the dedicated parabolic flight facility was flown in the two Airbuses operated in Europe during 17 campaigns from 2001 onwards (**Figures 9** and **10**). A typical parabolic flight campaign provides 3 flight days with 31 parabolas each.

The duration of the weightlessness in each parabola is about 20s, and a small break between each parabola enables the operators to optimize the experiment conduction.

Usually, special requirements for the timing of the flight day can be negotiated with the support crew. In total, 2056 parabolas were flown and roughly 200 alloys were measured over the last 18 years. The facility itself consists of a vacuum chamber, which can be evacuated to high vacuum. For optimal cooling conditions, the facility is equipped with He and Ar gas. Typically, the sample is heated in (less cooling) Ar gas and then He is filled in for increased cooling when the maximum temperature is reached, and the sample must be cooled as fast possible, to solidify within the 20s of weightlessness. For temperature measurement, the facility is equipped with a pyrometer and the experiment is recorded with digital video cameras from two sides. One camera can be operated with frequencies up to 40 kHz, allowing the measurement of the fast solidification processes.

The TEMPUS parabolic flights have three aspects. New techniques can be tested under microgravity, scientific projects can measure thermophysical or solidification data under microgravity, and alloys planned for EML can be tested for their levitation behavior.

Figure 9.
Sample magazine of the parabolic flight facility with eight samples in cups and cages.

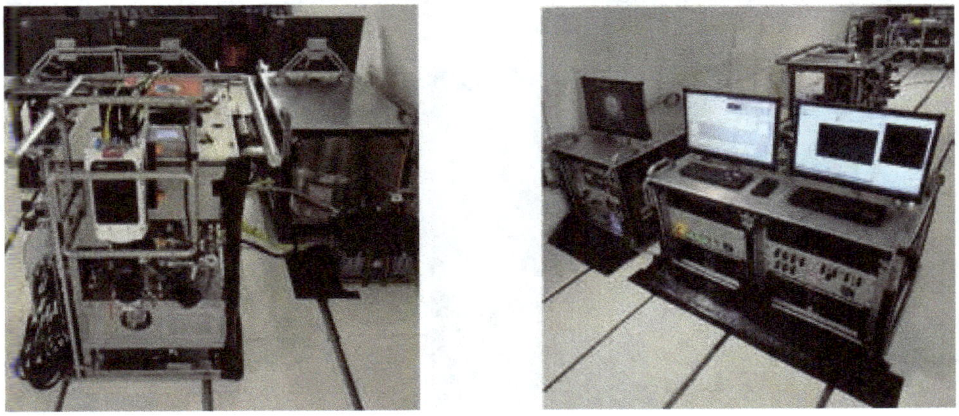

Figure 10.
TEMPUS facility for parabolic flight. The authors share the experience of more than 3000 parabolas both in the NASA KC-135 and in the ESA-CNES-DLR Airbus A300s.

3.4 The TEXUS facility: advanced testing capabilities in preparation for ISS

In addition to the parabolic flight campaigns, five sounding rocket missions between 2005 and 2013 were performed. The TEXUS rockets are launched at Esrange Space Center near Kiruna in northern Sweden (**Figure 11**). One launch provides about 6 minutes of microgravity and two samples can be processed. The TEXUS EML radiofrequency generator and coil system are comparable to the parabolic flight and EML facilities. The TEXUS EML experiments are completely remotely controlled and the housekeeping data and process control video images are downlinked to ground. Commands to optimize the heating or positioning can be uplinked, as necessary. Due to the longer microgravity duration, it is possible to run experiments, which are not possible in the limited time of the parabolic flights. For instance, modulation experiments need a long temperature hold, and also samples with a long solidification interval can be completely solidified in the available microgravity time of a sounding rocket flight. As the facility is available for refurbishment after the experiment, also strongly evaporating samples can be processed.

3.5 EML onboard the ISS

Based on the precursor facilities, the EML for the ISS was developed. The sample chamber can again store 18 samples, which defines the so-called sample batch. The diagnostic is comparable to the parabolic flight facility, with a high-speed camera from the side (with up to 90,000 frames per second) and a second digital camera mounted on top of the sample chamber. The temperature measurement is contactless by a pyrometer integrated in the same housing as the top camera. The technique of the RF generator is the same, and for all current facilities the same SUPOS coil system is used (see **Figure 12**). This coil system was developed at the DLR [20] and is advancement from the TEMPUS coils in the shuttle era where two independent coils for heating and positioning were installed. The so-called SUPOS system consists of a single coil and superimposes a quadrupole field with a strong gradient (but low field strength) for positioning and a (nearly) homogeneous dipole field with high strength for heating the sample as illustrated in **Figure 13**. The benefit is that both fields can be varied independently, i.e., positioning and heating are decoupled in contrast to electromagnetic levitation on ground.

In particular, the dipole heating field serves as the measurement circuit to determine the impedance and finally the electrical resistivity of the sample (compared to Section 2). The measurements are performed during cooling the sample where the

Figure 11.
Launch of the TEXUS sounding rocket at Esrange Space Center in Kiruna, Sweden, during the EML-2 mission (left) and real-time teleoperation of the experiment (right).

Figure 12.
Photo of the EML coil (Photo Airbus).

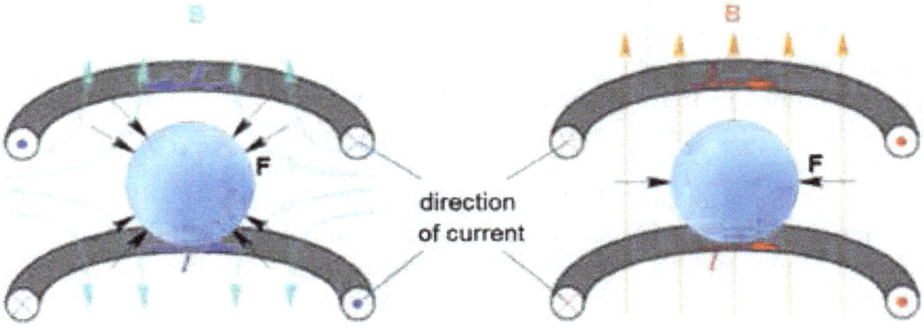

Figure 13.
Sketch of the superpositioning principle (reproduced from Ref. [21]). A current in the opposite direction for upper and lower coil winding creates a magnetic quadrupole field for positioning in the coil's center (left). A current in the same direction creates a magnetic dipole field for heating the sample (right).

heating field is set to a minimum so that the sample is not significantly deformed from its equilibrium shape by the residual weak forces. The homogeneity of the field and the spherical sample shape provide simple and well-defined conditions for evaluation of impedance data, thus enabling the determination of the resistivity with a high precision. For that purpose, the so-called Sample Coupling Electronics (SCE) has been developed [21]. The SCE is specially designed for the SUPOS coil system and measures the RF current and voltage of the heating circuit from which the electrical resistivity of the sample is derived.

EML was installed by the German ESA Astronaut A. Gerst in the European Drawer Rack (EDR) in October 2014, followed by a checkout of all subsystems. After that, two checkout experiments were performed in November 2014 and February 2015. The first scientific runs were successfully performed in spring 2015. The first two batches were already successfully performed in orbit; the third one is currently under preparation. The first sample chamber has already been downloaded, and the samples are being analyzed by the scientists.

4. The microgravity user support center of DLR

Being conceived as a fairly complex multiuser payload, the experiment development for TEMPUS and later EML was centralized at one focal point where the TEMPUS ground model was localized, and the scientific infrastructure to develop the experiments was available: the Microgravity User Support Center (MUSC) of DLR Cologne. Founded in 1986 by the DLR Institutes of Space Simulation and Space Medicine, MUSC has gained broad expertise in the scientific and operational preparation and conduction of space experiments. It started as the national point of contact for scientific users of microgravity facilities. In the first years, the focus was on bilateral scientific collaborations with NASA or Russia, who usually provided the launch vehicles for payloads developed in Germany. In these cases, the experiment resources were usually shared between the two collaborating countries. With the evolving ISS era, Europe has bundled its resources in order to provide a significant contribution to this large endeavor.

Today, MUSC is part of the network of European User Support and Operation Centers who, on behalf of ESA, are in charge of the operation of ESA developed payloads onboard the ISS (**Figure 14**). Over time, hundreds of microgravity experiments were developed in the ground models located at the DLR premises and supported from the control rooms at MUSC. Due to the broad range of scientific topics, payloads, and experiments, merely a small set of examples is listed here. In the 1980s–1990s, the MUSC activities were primarily focused on the preparation for and conduction of Spacelab experiments for the German Missions D-1 (1985) and D-2 (1993) mainly in the fields of materials and life sciences as well as technology development. In addition, experiments on unmanned carriers were supported. The retrievable European carrier EURECA was flown in 1992 and had six payloads onboard, including a protein crystal growth facility and different types of furnaces for solidification experiments. All of them were operated from the MUSC control room. From the early 1990s, first experiments on the Russian MIR Station were supported; the first time was for the Russian-German MIR92 Mission of astronauts Reinhold Ewald and Klaus-Dietrich Flade. During the EuroMIR 1994 and 1995 missions, a German furnace (TITUS) was operated by ESA astronauts Ulf Merbold and Thomas Reiter. Both had received training on the TITUS ground model located at MUSC. The Matroshka experiment from the DLR Institute of Space Medicine located on the outside of the Zvezda Module of the ISS in 2004–2005 marked the start of the new era of international collaboration. In the meantime, MUSC has operated the ISS payloads MSL, Biolab, DOSIS, EXPOSE, EDE, EML, and FASTER. There is a continuing close collaboration with the neighboring scientific institutes and the mutual exchange of staff and expertise, which puts MUSC in the position to support science teams in technical and science operational matters with respect to the optimization of their respective science runs onboard the ISS.

In general, MUSC tasks include the preparation and conduction of experiments in the payloads under its custodianship, including data dissemination and archiving. In order to fulfill these tasks, MUSC operates ground models of the experiment

Figure 14.
The microgravity user support center of DLR.

facilities and provides the ground infrastructure for the real-time on-orbit experiment performance in the control room located at DLR Cologne. For TEMPUS and later EML, even more customized services are provided (see next section).

5. Comprehensive experiment preparation for EML: transcription of scientific goals into facility settings

For TEMPUS and EML, a so-called ground support program was established. It accompanied the TEMPUS scientists during the entire life cycle of their experiments from the proposal to the delivery of the final experiment data. Activities comprised the measurement of sample properties needed for the development of experiment control parameters, their validation in the TEMPUS Ground Model, and the operation of the Spacelab and nowadays ISS experiments. All activities at DLR are performed in close coordination with the science teams and often performed in their respective presence. Preparation activities consist of three major parts.

First of all, the availability of certain physical sample properties, e.g., evaporation rates, coupling to the HF field, and spectral and hemispheric emissivity, is mandatory for the experiment preparation and execution. The knowledge of the amount of sample material evaporating from the sample surface during processing is essential. Therefore, the evaporation rate of each flight sample composition is measured in a dedicated facility on flight sample material provided by the science teams (**Figure 15**). The obtained data are later used in the experiment planning in order to design the temperature-time profiles such that the amount of evaporated material is minimized. Under vacuum conditions, the evaporated sample material condenses on the surfaces of the process chamber, while under gas atmosphere, the particles will agglomerate to fine aerosols or dust particles. It must be assured that the amount of dust remains in the nontoxic regime to protect the astronauts in case of a failure with the EML process chamber sealing.

In order to mitigate both aspects of sample evaporation, a software tool was developed at MUSC which calculates the mass loss of a sample during processing from the incoming temperature data and the known evaporation rates. In a next step, the layer thickness condensing on the coils is derived, and the element specific evaporation is calculated. This tool is used during the on-orbit experiment execution.

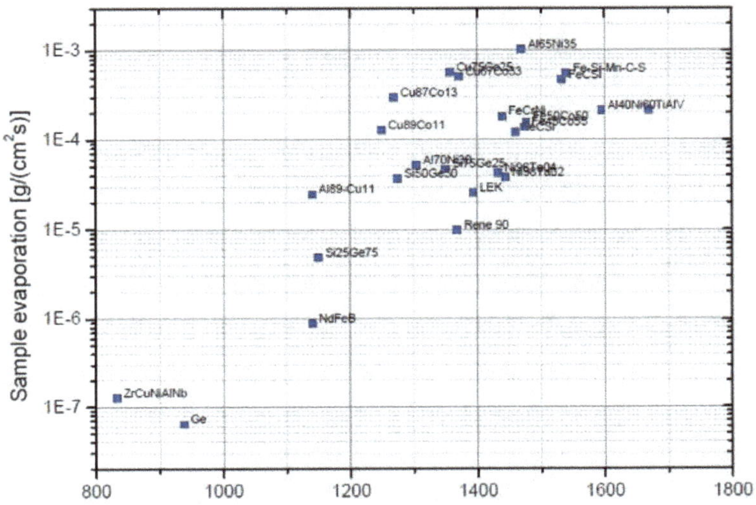

Figure 15.
Measured evaporation rates of batch 1 samples at the corresponding liquidus temperature in °C.

For the detailed planning of the experiments, the coupling of each sample to the RF field has to be known. The coupling describes the temperature reaction of the sample to applied field strength. Since TEMPUS and EML are not designed to levitate samples under terrestrial conditions, all ground-based measurements are limited to the solid state. They are performed on samples suspended in the coils on ceramic rods. In addition, due to the longstanding history of EML, a number of experimental data stemming from parabolic flight and sounding rocket campaigns are available today and will be used in support of future experiment developments. The obtained coupling data are needed to simulate the required temperature-time profile of the liquid sample with a TEMPUS/EML simulator software. This simulator tool is a MUSC development which has undergone refinements by systematic comparison of as-flown profiles with simulation results over the years of operation. For a correct temperature measurement with the pyrometer onboard, the sample emissivity at the melting temperature is measured in the EML ground model. The sample is heated until melting sets in, and by a comparison of the measured and literature melting point, the emissivity can be derived. The optical setup and used pyrometer of the ground model is comparable to the flight facility.

Secondly, the individual operational flow for each experiment is developed, yielding the outline of all nominal and contingency operation. For that purpose, the so-called science protocols are prepared by MUSC on behalf of and in cooperation with the science teams. This document is kept in a narrative form, describing, e.g., key temperatures of the samples to be reached; facility settings for, e.g., camera recording and downlink; aimed pressure; or vacuum values. In addition, it contains a detailed experiment planning with respect to all aspects needed to fully describe the scientific and operational requirements of the respective experiment (**Figure 16**). The assembled information includes the following:

- The required EML configuration at the start of the experiment
- The applied process strategy, planned temperature-time profiles, maximum temperatures, and heating rates
- The planned sequence of cycles within experiments

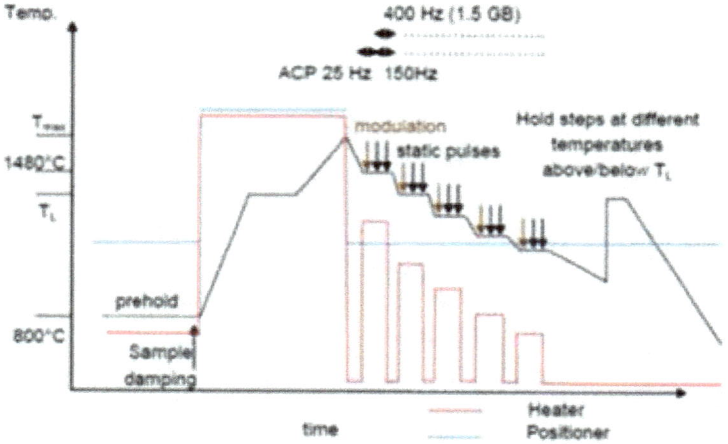

Figure 16.
Example of planned temperature-time profile (black) including the heating (red) and positioning control voltages (blue) as part of the science protocol showing specific heat measurements with the modulation calorimetry at five constant temperatures.

- Any planned modulation settings and heater stimuli settings
- The required diagnostics settings, e.g., camera settings (such as frequency, resolution, allocated memory)
- Strategies in case of unexpected sample behavior such as unsatisfactory sample stability or undercooling

With the determined sample material properties and the individual science protocols, the MUSC team can start the third part of the ground support program, which is the development of the so-called parameter sets. These parameter sets are interpreted by the EML process control software.

Experiments in EML are divided into individual thermal cycles which usually encompass one melting and solidification event. In the liquid phase, the scientific data are obtained either by high-speed video observation of the solidification front or by sample stimulation by modulation or applying pulses of the heater field and observing the effect on the temperature and sample shape. Each individual cycle is divided into several steps, which are described by two sets of parameters. The experiment parameters (EPs) define the experiment flow. With 38 parameters per step, the facility settings can be controlled. Up to 99 steps determine the time flow of one experiment thermal cycle. In addition to the experiment parameters, the so-called limit parameters (LPs) provide an independent guard rail envelope of RF values for each experiment that can never be exceeded. This safety measure was put in place in order to prevent any facility damage in case of a malfunction of the RF generator. Since the Spacelab era, MUSC has been developing and updating a TEMPUS/EML experiment simulator tool, which is used to derive the core experiment and limit parameters. This tool predicts temperature-time profiles from applied RF parameters. The calculation is based on the coupling behavior of the sample to the RF field and takes the planned process atmosphere into account. With this tool, the heater and positioner voltages for heating and melting of the sample and equilibrium temperatures for modulation measurements are determined. These voltages are then transformed into the final EP and LP sets for the facility (**Figure 17**).

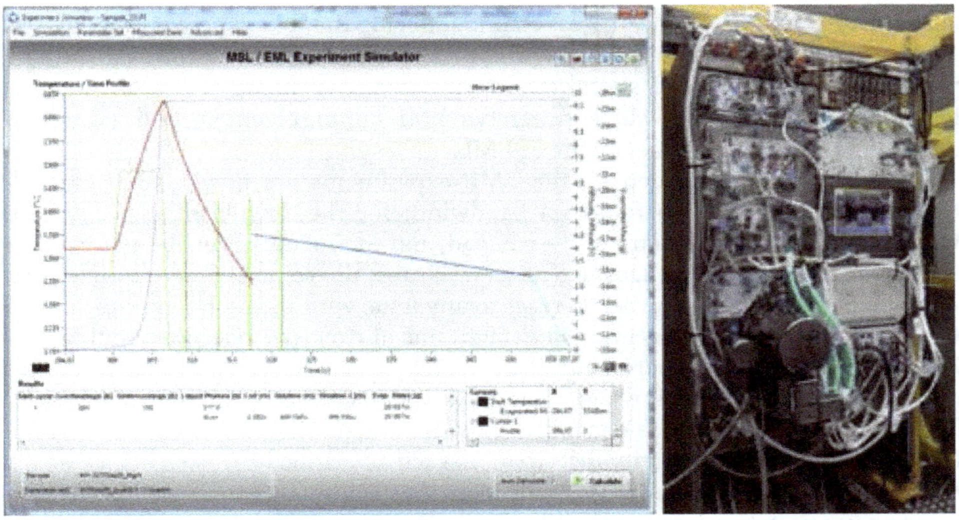

Figure 17.
Left: Thermal profile (red curve: liquid, blue curve: solid) simulated by using parameters developed for the space experiment, here an oscillating drop measurement where surface oscillations are induced by applying short pulses of the heating field (green curve) at different temperatures during cooling. Right: EML ground model used for validation of experiment parameter sets.

After the development of the flight parameter sets, they must be validated in a representative EML ground model. For this purpose, the Operational Model (OM), which also acts as flight spare, is used. During the validation the parameter sets are processed on a suspended high melting sample which remains solid during the validation run. With this approach, it is possible to test that all parameter sets are syntactically correct and that the experiment procedure follows the correct logic in all steps. A successful validation run in the EML OM is a precondition for the experiment performance onboard the ISS.

For the analysis of the video data, DLR MUSC developed a software, which displays the video and housekeeping data simultaneously [22, 23]. This enables the scientists to analyze both data together and investigate the sample behavior, observe the sample surface, etc. For the scientific analysis, image processing routines were developed, which measure the edge of the sample and calculate two perpendicular radii, area, center of mass, and half axes of an ellipsoid fit. On these data, a Fourier analysis can be performed, and the oscillation frequency of the sample surface can be measured. From the oscillation frequency, the surface tension can then be calculated. The decay of the surface oscillations is related to the viscosity. The development of this tool was started for the parabolic flight campaigns and later enhanced to be compatible with the EML data.

6. Operational preparation: how the experiments are implemented in the ISS context

The operational preparation is ongoing in parallel to the scientific preparation described above and is transparent to the scientific community. It consists of (a) resource allocation and coordination, (b) procedure development for ground and on-orbit operations, and (c) ground segment setup [24]. The most restricted resource is crew time. For EML, the experiment itself is performed via ground commanding without crew interaction; nevertheless, an astronaut is needed to change facility settings according to the requirements of the respective experiment, e.g., on the EML high-speed camera and hand gas valve settings regarding Ar or He as the gas atmosphere in the process chamber during the runs. These crew activities are relatively short and recurring with typical durations of 5–10 minutes and can usually be accommodated straightforward. More challenging is the planning of long crew activities like a swap between two EML sample chambers with a duration of 60 minutes or a hardware update of EML.

Further resources needed by the EML experiments are power, access to the Columbus vent line, high rate data bandwidth for the download of the science videos, medium rate telemetry of the payload, and commanding capability for active control of the experiment. During the science runs, the response time from ground needs to be minimized in order to react to any issue with the molten sample in due time. To ensure this, EML is granted a timeframe of exclusive telecommanding during "hot phases" of the experiments. A good microgravity level without disturbances during experiment phases is also mandatory. To achieve optimum microgravity conditions, experiments are deconflicted with all activities on the station that are known to induce microgravity disturbances such as planned thruster firings of the ISS or docking/undocking events. In addition, the experiments are always performed during crew sleep to minimize disturbances induced by crew presence in Columbus.

The experiment execution on the levitated sample is always observed with real-time video of both EML internal cameras. This ensures the proper experiment performance by visually monitoring the sample behavior and by sending safing commands if required. Periods when the onboard crew is not sleeping are used to download the stored science videos obtained during the night. It is coordinated with the flight control team that the maximum available bandwidth is made available for EML data download in order to cope with the high amount of data generated by EML. For operating EML from the MUSC control room located at DLR Cologne, MUSC is connected to the ESA Interconnecting Ground Segment. The EML Ground Segment is based on the commonly used monitoring and control architecture CD-MCS provided by the Columbus Control Center. MUSC receives telemetry (low and medium rate facility status data and video) and has telecommand capabilities to control the payload behavior and experiment performance in real time. In addition, dedicated ground support equipment is available, which transforms the incoming high-rate data streams into readable video data for process control purposes and off-line scientific evaluation.

7. Successful experiment performance on orbit

The EML experiments of a sub-batch are typically grouped into weeks of operation with “24/5” operations. This means that on Monday morning, the facility is switched on, and process conditions are established. The experiments only start late in the evening, when the crew sleep period begins, and finish prior crew wake-up. This yields a period of 8 hours and 30 minutes of science operations four times a week. Right after the last experiment cycle of the night, the conditioning process as described above is started again to optimize processing conditions for the upcoming experiments. In parallel, the download of the stored science video data is initiated. During the download, the operator on console needs to check the completeness of the incoming data, and in case of packet losses, the download has to be restarted. Afterwards the raw video data is processed into CINE file format. This complex task consists of many steps performed in the ground segment utilizing a dedicated software tool. Rundown of EML is usually performed on Friday evening, after the data download is completed.

Figure 18 shows the flow chart of a typical science campaign showing the schedule of samples with allocated number of melting cycles, gas atmosphere, and high-speed camera settings. In the frame of the ground support program, the experiment flows have been defined in detail. For some experiments, it is required to implement

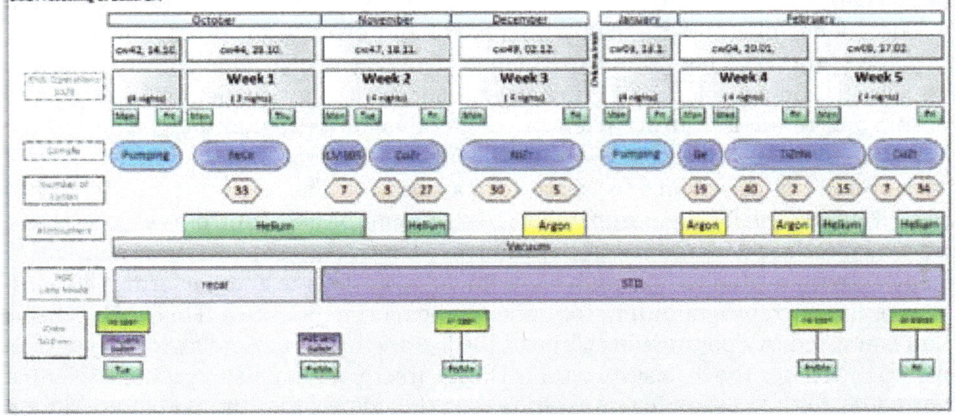

Figure 18.
Flow chart of a typical EML science campaign, in this case sub-batch 2.4 performed in autumn 2019. The Fe-Co sample processed in week 1 is investigated by a science team including the author T. Volkmann.

an intermediate break for a quick look analysis of the science data obtained before continuing. The provided flow chart depicts one specific experiment which requires such a break. During this time experiments on other samples are not excluded.

Once crew sleep has started, the operations team on console initiates the science operations. Prior to the start of cycle 1, EML is commanded to operational mode by selecting and calling a cycle. The cycle starts with the option to reprogram the experiment parameters by telecommand. It must be noted that the limit parameter sets are not accessible via commanding. They can only be changed by ground via file transfer (and only from ground) if the dedicated EML reprogramming cable is mounted by crew. Prior to starting science operations, it is ensured that telemetry and real-time video are available and that MUSC is enabled for commanding. The operators also check that the duration of the signal connection to the Space Station is long enough to perform the thermal cycle. The positioner is switched to ON. Then, the sample is mechanically stabilized in the field center by contact of the sample with the sample holder bottom and preheated according to the parameters in the EP set.

After verification that all prerequisites required to perform the thermal cycle are met, the semiautomated performance of the experiment parameter set is started. At this point in time, the EP Set is synchronized with the Limit Parameter Set (LP) which is independently controlled in the so-called Hazard Control Electronics (HCE). The sample is then heated, molten, and subsequently overheated. Upon reaching the maximum temperature, the heater is switched off or reduced. During cooling, the scientific measurements are performed. This may be any of the following or combinations thereof: (a) heater modulation, (b) heater pulses, (c) external triggering of recalescence, and (d) capturing of the recalescence event with high-speed camera. The cycle is finished once the sample has solidified and subsequently reached thermal equilibrium at a low temperature.

While performing the EP set, the HCE is monitoring the voltages in the oscillating circuits, ready to automatically switch off the heater and positioner in case of a limit violation. For safety reasons, any HCE cycle, that has been partially performed, may not be reused for a later experiment.

The automated cycle performance is closely monitored by the ground operators. One console position is dedicated to the telemetry supervision, and a different console position is responsible for telecommand generation if required. Commanding may be required, e.g., to rapidly cool down a sample in case of unexpected sample movement, if the sample temperature does not reach the expected value or if mechanical damping of sample oscillations is required. Additional reprogramming may be required in order to optimize EP values for an upcoming cycle.

The science operation is highly interactive and often requires ad hoc decisions on how to proceed. Therefore, experiments are always performed with the respective scientist on console (see **Figure 19**) to monitor the experiment on a dedicated console and to advise during scientifically motivated reprogramming. The science representative has also access to the EML video GSE to observe the process control video and to the voice loops for situational awareness. The science representative is assisted by two MUSC operators who actively monitor and control the experiment and who interface with the Columbus Control Center. They support the science team in any activity related to optimum experiment performance and perform the necessary parameter reprogramming for upcoming experiment cycles. This EML console team is assisted by operators in charge of the hosting EDR rack and a ground controller responsible for the data systems at MUSC. Directly after a melt cycle, the scientist is provided with the sample temperature data that allows fine-tuning of experiment parameters for the upcoming run (**Figure 20**). For this task, the EML simulator tool already used in the experiment preparation is available to the console team. It is not uncommon to send as many as 500 single commands to EML during one night.

Figure 19.
Operation of ISS-EML by telecommanding at the console in the control room at MUSC on 30 October 2019. The authors T. Volkmann and S. Schneider performing experiments with a Fe-Co sample as part of batch 2.4.

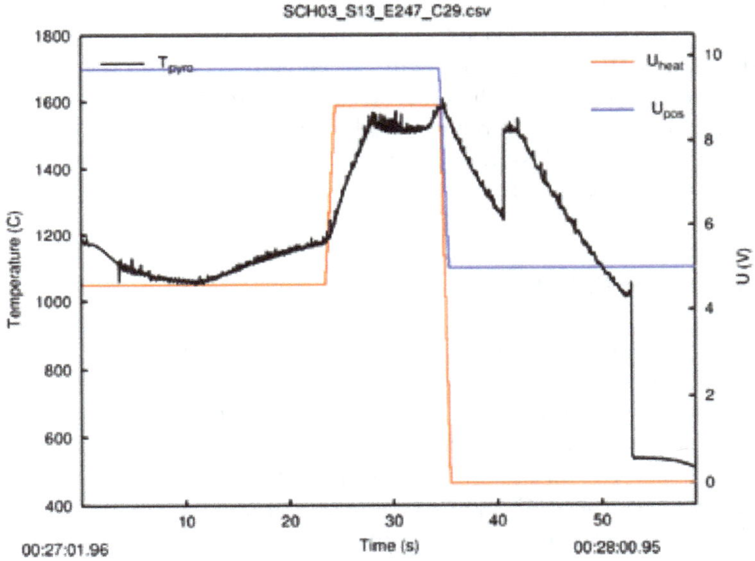

Figure 20.
Temperature-time profile (black), heater (red), and positioner control voltage (blue) during heating, undercooling, and solidification of $Fe_{60}Co_{40}$ processed on October 30, 2019.

After the performance of the first thermal cycle on a given sample, the as-flown sample temperature profile is analyzed by the EML telemetry/telecommand console position with the toxicity tracker software tool. From the temperature profile and the known evaporation rate of the material, the occurred mass loss is determined and compared to the respective limit for this cycle. The analysis is presented to ESA safety in a later stage, if needed, to prove that the result remained below the allowed limit values. This is required as under gas conditions; the evaporated material forms an aerosol, which, depending on its concentration, is considered toxic. The EML experiments are planned in such a way that they always remain in a nontoxic concentration (EP set performance controlled by HCE limit parameter set). In case of a violation of the limit for a specific cycle/experiment, processing of the specific sample would not be continued until a solution has been found. This toxicity tracker additionally keeps track of the evolution of dust the process chamber.

In parallel, it must be ensured that the overall facility degradation by evaporated material at any point within EML remains below a predefined limit value in order to preserve, e.g., the coil and surfaces of the optical components over the planned EML lifetime. Each experiment is provided with predefined maximum values for these layer thicknesses.

The video data of the high-speed camera are stored during a cycle in a ring memory. After performing the cycle, the video data are transferred to the hard disk of the computer controlling the high-speed camera system. The transfer of this video files must be completed before the HSC is reconfigured for recording (series of videos) with different settings for the next science run. Subsequent cycles are then performed according to this operational flow, starting over with the mechanical stabilization of the sample in the coil, if needed. After a night of science operations, the video data are transferred to ground where they are post-processed for later scientific evaluation. For that purpose, the science teams can use the MUSC developed tool, by which video and payload data can be displayed synchronously.

Last step in the experiment support chain is the archiving and distribution of the EML data in a data archive. Already during the parabolic flight era, a data archive basing on the web-based data management platform Hypertest was developed. All TEMPUS/EML data with associated metadata are stored in this archive and can be accessed by the respective data owners. This enables the user to search for metadata such as sample material, responsible scientists, campaign, date of performance, etc. The archive is accessible via the Internet, and thus all involved scientists can easily download their experiment data.

8. Evolution of the EML facility

Since its arrival on the ISS in 2014, two EML subsystems were upgraded. The operating system of the radial high-speed camera was exchanged with a faster one, reducing the waiting time between the cycles for video storage on orbit and thus allowing processing more cycles per night. The upgrade involved the exchange of an electronics board within a submodule of the experiment controller module. The pictures below show the EML operations team supporting the onboard activities performed by NASA Astronaut Andrew Feustel (**Figure 21**).

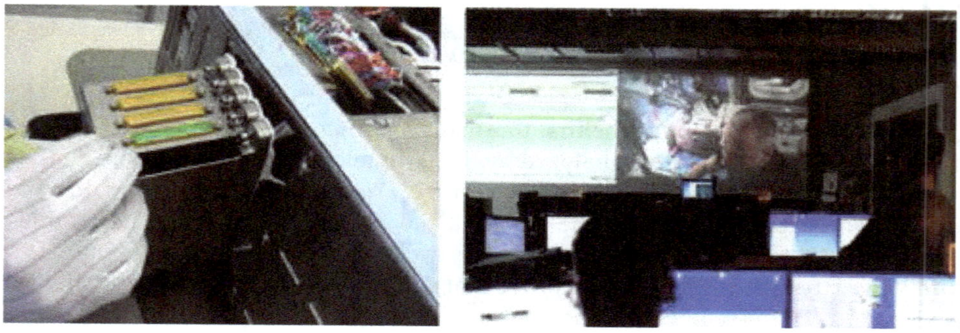

Figure 21.
Upgrade of the high-speed video camera operating system: location of board (left photo: Airbus) and operation team in the control room supporting Andrew Feustel (right).

The second upgrade applied to EML is the Sample Coupling Electronic (SCE) for the measurement of electrical conductivities. For the parabolic flight facility, a comparable SCE is available, which provides a higher measurement rate and can be used for the analysis of surface oscillations. Since its successful installation and on-orbit commissioning beginning of 2017, it has been used by the majority of EML science teams. This system is also a development of the DLR *Institut für Materialphysik im Weltraum.*

In the meantime, the community has opted for a third upgrade, the possibility to monitor and control the oxygen content in the samples. In order to keep the facility up to date on scientific level, Airbus is working on an "oxygen sensing and control system" (OCS), which is an insert that will provide adjustable and defined oxygen partial pressure of the EML processing atmosphere [25]. Oxygen represents a serious and potentially harmful contaminant to many materials at elevated temperatures due to its high chemical reactivity. Especially in containerless materials science experiments in which the processed sample is directly exposed to the process atmosphere and not contained in a cartridge, the presence of oxygen in the atmosphere might lead to a contamination of the sample leading to the formation of an oxide layer on the surface or to dissolution of oxygen into the liquid sample. These occurrences could significantly alter the experimental results and influence the nucleation of the solid phase in the undercooled liquid sample. The technology used for the OCS is founded on ceramic-based oxygen sensors and oxygen pumps and provides adjustable oxygen partial pressure of the processing atmosphere in the range from nominal down to the "parts per billion" range. The Phase B study of EML OCS has established a preliminary design that is compatible with the scientific-, technical-, and safety related requirements and interfaces for the future use in space environment. The functionality and performance were demonstrated with a dedicated prototype. According to present planning, the OCS shall be ready for launch in 2023.

9. Conclusions

The electromagnetic levitator EML onboard the ISS is a multiuser facility for containerless processing and undercooling of melts of metallic systems and semiconductors. The level of undercooling prior to solidification is an important process parameter controlling the evolution of the microstructure during solidification as well as the selection of alternative solidification pathways with metastable phases and therefore the properties of the solidified material. Processing in microgravity in combination with advanced diagnostic devices and evaluation methods enable the investigation of solidification phenomena in undercooled melts as well as the precise measurement of a variety of thermophysical properties of the liquid state as function of temperature including the undercooled regime. While strong levitation fields are needed for electromagnetic processing under normal gravity positioning: heating is decoupled in the ISS-EML. This allows minimizing electromagnetic stirring in the melt, thus enabling to study nucleation, solidification, and phase formation phenomena under (nearly) diffusive conditions. Moreover, under microgravity, electromagnetic stirring can be varied so that these processes can be investigated systematically under different levels of melt convection. Such experimental results provide the basis for verification and refinement of physical models for crystal nucleation and growth kinetics including the influence of fluid flow. Thermophysical properties like density, specific heat, viscosity, and thermal and electrical conductivity are not only fundamental quantities in physics and materials sciences but also serve as input parameters for numerical modeling of casting and solidification processes. Only with a detailed understanding of the physical mechanism involved and reliable input data the simulation of microstructural evolution is a powerful tool for fine-tuning of material properties and optimization of production routes.

Experimentation with the EML in space is very complex and needs a careful planning and preparation. Many tasks cannot be accomplished by the scientists who need a comprehensive support from facility and mission specialists for experiment preparation and performance. In particular, the infrastructure with hardware and software for preparation, conduction of the space experiment, and analysis of flight data is provided in the frame of the ground support program. Moreover, the ISS-EML facility must be constantly monitored and kept ready for operation. The support program is performed in close cooperation with the scientists and comprises the measurement of material properties relevant for processing in EML (evaporation rate, coupling parameters, and emissivity), the development of experiment procedure and parameter sets, and their validation in the EML ground model. Experiment planning also includes scheduling of all samples in a batch or sub-batch, which must be coordinated with the ISS activities concerning allocation of crew time, settings of the high-speed camera, change of gas atmosphere, energy consumption, and data transfer to ground. During space experiments, the operators in the control room actively run the EML facility by telecommanding in attendance of the scientists. Reprogramming of experiment parameters ensures the optimization of experiment cycles. A full support program for the entire science project from experiment definition to its operation on orbit and the management of flight data is decisive in order to obtain high-quality data and an optimum of scientific results.

Acknowledgements

The authors acknowledge the access to the ISS-EML, which is a joint undertaking of the European Space Agency (ESA) and the DLR Space Administration. The experiments on Fe-Co alloys were conducted in the framework of the ESA research project PARSEC (AO-2009-0898) including MAGNEPHAS (AO-1999-101) and NASA-funded project ELFSTONE. T.V. expresses his gratitude to C. Kreischer, O. Shuleshova, and D.M. Matson for their fruitful cooperation within the PARSEC science team and to the team at MUSC for successful EML operation.

Author details

Angelika Diefenbach[1*], Stephan Schneider[2] and Thomas Volkmann[2]

1 Microgravity User Support Center, Deutsches Zentrum für Luft- und Raumfahrt (DLR), Cologne, Germany

2 Institut für Materialphysik im Weltraum, Deutsches Zentrum für Luft- und Raumfahrt (DLR), Cologne, Germany

*Address all correspondence to: angelika.diefenbach@dlr.de

References

[1] Glaubitz B, Kullack K, Dreier W, Seidel A, Soellner W, Diefenbach A, et al. The Electro-Magnetic Levitator (EML) on board the ISS—An overview and outlook. In: Proceedings of Deutscher Luft- und Raumfahrt Kongress 2015, 22--24 September 2015, Rostock, Germany. Bonn: Deutsche Gesellschaft für Luft- und Raumfahrt—Lilienthal-Oberth e.V; 2015

[2] Soellner W. The Electromagnetic Levitator (EML) on-board the ISS: Capabilities, on-orbit performance and recent enhancements. In: Proceedings of the 69th International Astronautical Congress, IAC 2018, 1-5 October 2018, Bremen, Germany. Paris: International Astronautical Federation; 2018

[3] European Commission—Eurostat. Chapter 9: Metals and metal products. In: European Business: Facts & Figures 2009. Available from: https://ec.europa.eu/eurostat/web/products-statistical-books/-/KS-BW-09-001-09

[4] Boettinger WJ, Coriell SR, Greer AL, Karma A, Kurz W, Rappaz M, et al. Solidification microstructures: Recent developments, future directions. Acta Materialia. 2000;**48**(1):43-70

[5] Asta A, Beckermann C, Karma A, Kurz W, Napolitano R, Plapp M, et al. Solidification microstructures and solid-state parallels: Recent developments, future directions. Acta Materialia. 2009;**57**(4):941-971

[6] Herlach DM. Non-equilibrium solidification of undercooled metallic melts. Materials Science and Engineering. 1995;**R12**(1):177-272

[7] Herlach DM, Holland-Moritz D, Galenko PK. Metastable Solids from Undercooled Melts, Pergamon Materials Series. Amsterdam: Elsevier Science; 2006

[8] Hyers RW, Matson DM, Kelton KF, Rogers JR. Convection in containerless processing. Annals New York Academy of Sciences. 2004;**1027**(1):474-494

[9] Egry I, Diefenbach A, Dreier W, Piller J. Containerless processing in space—Thermophysical property measurements using electromagnetic levitation. International Journal of Thermophysics. 2001;**22**(2):569-578

[10] Chandrasekhar S. Hydrodynamic and Hydromagnetic Stability. New York: Dover; 1961

[11] Egry I. Surface tension measurements of liquid metals by the oscillating drop technique. Journal of Materials Science. 1991;**26**(11): 2997-3003

[12] Sauerland S, Eckler K, Egry I. High-precision surface tension measurements on levitated aspherical liquid nickel droplets by digital image processing. Journal of Materials Science Letters. 1992;**11**(6):330-333

[13] Egry I, Lohöfer G, Seyhan I, Schneider S, Feuerbacher B. Viscosity and surface tension measurements in microgravity. International Journal of Thermophysics. 1999;**20**(4):1005-1015

[14] Fecht HJ, Wunderlich R. Fundamentals of liquid processing in low earth orbit: From thermophysical properties to microstructure formation in metallic alloys. Journal of the Minerals Metals & Materials Society. 2017;**69**(8):1261-1268

[15] Fecht HJ, Johnson WL. A conceptual approach for non-contact calorimetry in space. Review of Scientific Instruments. 1991;**62**(5):1299-1303

[16] Wunderlich RK, Diefenbach A, Willnecker R, Fecht HJ. Principles of non-contact A.C. calorimetry. In:

Hofmeister W, editor. Containerless Processing: Techniques and Applications. Warrendale, PA: The Minerals Metals and Materials Society; 1993. pp. 51-56

[17] Richardsen T, Lohöfer G. Contactless electrical conductivity measurement of electromagnetically levitated metallic melts. International Journal of Thermophysics. 1999;**20**(4):1029-1039

[18] Herlach DM, Matson DM, editors. Solidification of Containerless Undercooled Melts. Weinheim: Wiley-VCH; 2011

[19] Team TEMPUS. Containerless processing in space: Recent results. In: Ratke L, Walter H, Feuerbacher B, editors. Lecture Notes in Physics: Materials and Fluids under Low Gravity. Berlin: Springer; 1996. pp. 233-252

[20] Lohöfer G. German patent 3836239 (8 August 1991); US patent 4.993.043 (12 February 1991)

[21] Lohöfer G. High-resolution inductive measurement of electrical resistivity and density of electromagnetically levitated liquid metal droplets. Review of Scientific Instruments. 2018;**89**(12):124709-124710

[22] Willnecker R, Schneider S, Münstermann F, Pütz M, Koerver W. Digital video system for TEMPUS on parabolic flights. Space Technolology. 2006;**26**(2):1-7

[23] Schneider S, Egry I, Wunderlich R, Willnecker R, Pütz M. Evaluation of thermophysical data from electromagnetic levitation experiments with digital image processing. Journal of the Japan Society of Microgravity Application. 2008;**25**(3):387-392

[24] Schumann S, Diefenbach A, Schneider S, Willnecker R. Containerless processing on ISS: Experiment operations in ESA's EML, the Electromagnetic Levitator. In: Proceedings of the 69th International Astronautical Congress, IAC 2018, 1-5 October 2018, Bremen, Germany. Paris: International Astronautical Federation; 2018

[25] Aicher W. Future capabilities of the Electromagnetic Levitator (EML) on-board the ISS: Oxygen sensing and control system (OCS). In: Proceedings of the 69th International Astronautical Congress, IAC 2018, 1-5 October, Bremen, Germany. Paris: International Astronautical Federation; 2018

2

What we Learned from Cavitation Bubbles in Microgravity

Mohamed Farhat

Abstract

The present chapter is about the *Flash and Splash* project, which is dedicated to the study of bubble dynamics in microgravity. The story of this project started in 2004 with a simple curiosity on how a cavitation bubble may behave within a water drop and evolved into an outstanding, internationally renowned science project as well as a wonderful human adventure. So far, we have participated in nine European Space Agency (ESA) parabolic flight campaigns (PFC) and made a significant progress in understanding the cavitation phenomenon. First, we investigated the dynamics of a cavitation bubble within a water drop and learned how the collapse may lead to the formation of a double jet. We discovered the formation of secondary cavitation due to the confinement of shockwaves within the drop. We used this result to propose a new path for erosion due to a high-speed impact of water drops on a solid surface. Then, we addressed the effect of gravity on bubble dynamics and came up with a unified framework to explain and predict key phenomena, such as microjets, shockwaves and luminescence. Parabolic flights gave us the unique opportunity to modulate the gravity-induced pressure gradient, which is crucial for the fate of a collapsing bubble.

Keywords: cavitation, bubble, shockwave, microjet, erosion

1. Introduction

I will always remember this sunny afternoon of September 2004 when I received in my office four talented students, D. Obreschkow, Ph. Kobel, N. Dorsaz and A. De Bosset, who had set themselves the goal of flying in microgravity. What else? The timing was perfect as, at that time, the European Space Agency (ESA) was organizing parabolic flight campaigns reserved for students. Still, to win their boarding pass, the students had to propose an innovative research project related to microgravity, and obviously, the competition was rather tough. I did not know how to help, as I had never been involved in such a crazy adventure. I politely explained that my research activity was about cavitation in hydraulic machines and there was nothing to do about it in a microgravity flight. However, this was not enough to discourage the students. A long and passionate brainstorming followed in search of the most credible project, or should I say the perfect excuse, to realize the dream of floating in microgravity. It was only by chance that, in the course of this rambling discussion, I mentioned an ongoing research project about the interesting dynamics of a cavitation bubble within a liquid jet, performed in collaboration with the European Organization for Nuclear Research (CERN) [1]. Bingo! What if replaced the jet by a drop and proposed to study bubble dynamics within a water drop? Indeed, generating a centimeter spherical drop of water requires microgravity conditions. We could not find a better excuse! The students worked hard on their proposal, which ESA accepted straight away. Then, they worked even harder in building the experimental setup. The *Flash and Splash* project was born.

The beauty of the scientific research is that you never know where it may lead you. I did not know that this meeting of September 2004 was the start of a long and fruitful research activity on bubble dynamics in microgravity, which continues today. Following the first participation in the students flight campaign, we took part in eight research campaigns with the involvement of four PhD students, M. Tinguely [2], O. Supponen [3], A. Sieber and D. Preso with many other master students and technicians. This innovative and exciting research, which received financial supports from the Swiss National Science Foundation, the European Union and many other governmental and private institutions, produced a strong impact within the scientific community and beyond. It received many awards and recognitions as well. *Flash and Splash* is not only a successful research project, it is also a great human adventure made of friendship, generosity and unforgettable memories.

In the following sections, I will give an overview on the cavitation phenomenon, which motivates the actual research. I will then summarize the technical aspects and main results related to bubble dynamics in microgravity.

2. What is cavitation?

Cavitation is the formation and collapse of vapor bubbles within a liquid due to a transient pressure drop. It may occur in a variety of hydraulic systems such as water turbines and pumps, ship propellers and space rocket inducers. In these applications, cavitation is a "nightmare" for design engineers and operators because it may lead to severe erosion, noise and vibration as well as alteration of hydrodynamic performances.

In the particular case of hydropower generation, the occurrence of cavitation in water turbines often requires periodic shutdowns of the power plant to allow for inspections and repairs, which increase significantly the operational cost. **Figure 1** illustrates the different types of cavitation in hydraulic machines as well as a pump impeller severely eroded by cavitation.

Figure 1.
Left: cavitating vortex in a Francis turbine, operating at off-design conditions; middle: cavitation in a ship propeller (courtesy Franc et al. [4]); right: pump impeller heavily eroded by cavitation.

In the course of the past 100 years, the booming development of hydropower and pumped storage power plants around the world and the need for faster and quieter ships and submarines have fuelled an intense and sustained research activity on the cavitation issue. This led to a better understanding of the physics behind the transient growth and collapse of vapor cavities in flowing liquids and the mechanism of induced erosion and noise. It is well known nowadays that cavitation erosion is due to vapor bubbles, which nucleate and grow in low-pressure area and then collapse violently as they move to a higher-pressure zone. Therefore, the cavitation aggressiveness is influenced by the minimum pressure, which governs the size of the vapor cavity, as well as the downstream pressure gradient and the liquid velocity, which together characterize the violence of the collapse. Despite the progress made so far, many issues related to cavitation remain unanswered.

High-speed sailing is another field where cavitation is a challenging issue and a limiting factor. In fact, any submerged body will face cavitation when it travels faster than a threshold speed. Interestingly, this also holds for fishes, whose speed is likely limited by cavitation. The fact that only few predatory fishes, e.g. sailfish and black marlin, can swim faster than 40 knots is due to cavitation occurrence. We have illustrated on **Figure 2** the case of the *Hydroptère* boat, which sat the speed sailing record to 50.17 knots in September 2009 in Hyères (France). Beyond a threshold speed, this "flying boat" uses lifting foils to raise its hull out of the free surface, leading to a significant drag reduction. We have presented on the same figure the experimental evidence of cavitation occurrence on the lifting foil at reduced scale, which we have performed in the high-speed cavitation tunnel of the *Ecole Polytechnique Fédérale de Lausanne* (EPFL). Our investigations reveal that the cavitation occurrence is responsible for a sharp drop of the lift force, which increases the risk of the boat capsizing. We have also learned that it is almost impossible to avoid cavitation when the speed is beyond ~40 knots. Therefore, any attempt to break the actual speed sailing record (55.32 knots established by Vestas Sailrocket team in 2012) requires accommodation of cavitation. This is exactly what an EPFL student's team, among others, is trying to achieve in the frame of an ongoing project (SP80) with the help of supercavitating foils.

An interesting example of cavitation in nature is that of snapping shrimps, which are among the noisiest underwater creatures. Colonies of such crustaceans can produce an intense and wideband noise that interferes with sonars and underwater communication. Snapping shrimps, also called pistol shrimps, use the extraordinary power of cavitation to defend their territory and hunt prey animals. Upon an extremely rapid closure of their large claw, a transient high-speed jet develops and forms vortex rings. The resulting pressure drop within these vortices leads to the growth and collapse of toroidal vapor cavities [5].

From a fundamental viewpoint, the first mathematical model was proposed by Lord Rayleigh in 1917 [6], who used the potential flow theory to provide a

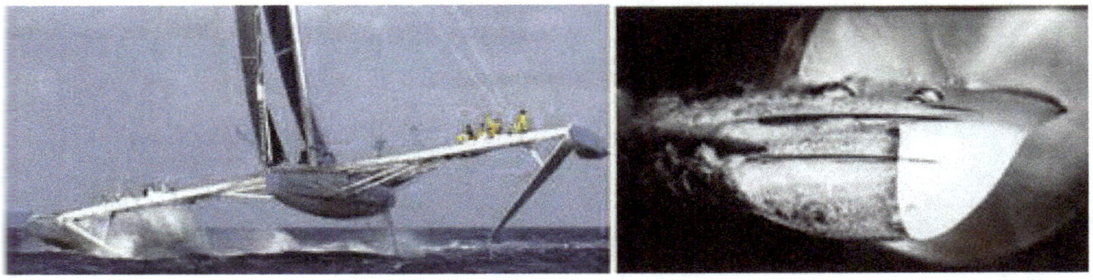

Figure 2.
Left: the Hydroptère sailing boat (courtesy A. Thébaut) and right: evidence of cavitation on a model of the Hydroptère *lifting foil.*

remarkable prediction of the collapse time of an empty spherical cavity in an incompressible fluid. He also predicted that before the cavity closes, the pressure rises very high in the fluid close to the interface. Since then, more sophisticated models have been proposed, taking into account the phase transition, viscosity, surface tension, gas content and compressibility. Besides, a large number of experimental investigations took profit from the development of pulsed lasers to generate spherical bubbles on-demand and unveiled peculiar phenomena. **Figure 3** illustrates a typical growth and collapse of a cavitation bubble induced by a pulsed laser in still water along with the evolution of the bubble radius and the signal of pressure alteration, measured far from the bubble centre. We can observe how the hot plasma induced by the laser leads to an explosive growth of a bubble, filled with gas and water vapor. As the bubble expands, it cools down and reaches a maximum radius of ~4 mm before it collapses violently. The bubble lifetime is less than a millisecond. During the collapse phase, the water vapor condenses while the noncondensable gas is strongly pressurized and heated up before it rebounds and collapses again with a decreasing intensity. Interestingly, the Rayleigh theory predicts an infinite velocity and hence infinite energy densities at the end of the collapse. According to general relativity, such a singularity would generate a black hole—a scary scenario that, fortunately, never happens because a plethora of nonlinear effects, such as liquid compressibility, heating and radiative energy transfer, halt the catastrophic collapse well before relativistic physics becomes noticeable.

Figure 3 illustrates also the emission of intense shockwaves at the initial and final stages of bubble growth and collapse. What makes these shockwaves visible is the sharp density gradient at their front, which alters the refraction index and deviates light. Shockwaves are attributed to the supersonic motion of the bubble interface and the liquid compressibility and may reach Giga Pascal amplitudes.

Another fascinating aspect of collapsing bubbles is their capacity to generate light (luminescence), as illustrated on **Figure 3**. This phenomenon was first observed in the context of bubbles excited by ultrasound in 1933. It is due to a spectacular compression of the noncondensable gas, which heats up to such a degree that light is emitted in the form of a nanosecond flash. Many experimental and theoretical studies have been devoted to luminescence to better understand its mechanism. It is well accepted that the temperature of the gas enclosed in a spherically collapsing bubble may reach 10,000°K, which leads to the generation of free radicals that recombine and radiate light. This result raised a hope that nuclear

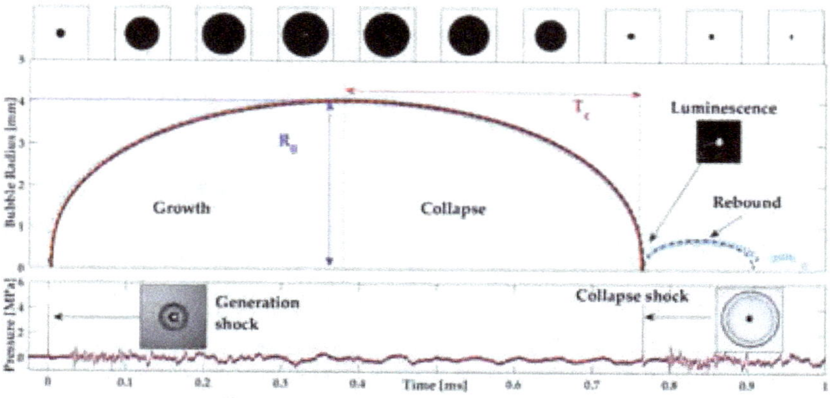

Figure 3.
Top: visualization of a laser-induced bubble dynamics (112 μs inter-frame); middle: measured bubble radius (blue circles), along with theoretical predictions; bottom: pressure signal measured far from the bubble centre (reproduced from [3]).

fusion might occur inside a collapsing bubble, provided that the bubble is large enough and remains spherical. It is hard to meet these requirements on earth, because large bubbles deform significantly in the presence of the gravity-induced pressure gradient. After many attempts to increase the temperature reached in the centre of a collapsing bubble, the objective of nuclear fusion was abandoned amid a large controversy. Yet, research on luminescence continues, motivated by its ability to act as a catalytic host for chemical reactions with promising outcomes for sterilization, nanomaterials, cancer therapy and more.

Moreover, it is also well known that when a bubble is set to grow and collapse near a rigid wall, hydrodynamic instabilities develop at the bubble interface, due to pressure anisotropy. This leads to the generation of a high-speed microjet directed towards the boundary. When the solid surface is replaced by a free surface, the microjet develops in opposite direction, and the bubble moves away from the surface while a counter-jet emerges from the free surface (see **Figure 4**). The cavitation erosion is believed to originate from a combined action of the shockwaves and microjets. The current research is still struggling to further understand and predict erosion due to collapsing bubbles.

What makes cavitation bubbles fascinating is their extraordinary ability to focus energy and produce the powerful phenomena reported earlier. Cavitation remains a hot topic of research because of its multidisciplinary aspect, which involves multiscale fluid and solid mechanics, plasma physics, thermodynamics and chemistry, with a strong interaction between them. While cavitation bubbles were historically associated with negative outcomes, e.g. noise, vibration and erosion in hydraulic systems, more and more efforts are dedicated to take profit from their power in a variety of applications. This includes surface cleaning, material and food processing and water treatment. There is also a growing interest in using cavitation bubbles in medicine, e.g. treatment of musculoskeletal system disease, breakage of kidney stones, drug delivery, cancer cell treatment, etc. All these promising developments require a better knowledge of cavitation bubbles to allow for mastering their effects.

3. Microgravity experiments: what for?

We have initiated 15 years ago a research programme to address the knowledge gaps associated with cavitation bubble dynamics with the help of a state-of-the-art instrumentation. The focus was accorded to the distribution of the bubble's energy into its various collapse phenomena, namely, the microjets, shockwaves, rebound and luminescence. To this end, we built a versatile experimental setup, which produces initially spherical bubbles, first using an electric discharge and then a pulsed laser. Unfortunately, despite all the care given to the experimental setup, spherical bubbles can hardly be created on earth. In fact, the hydrostatic pressure gradient in

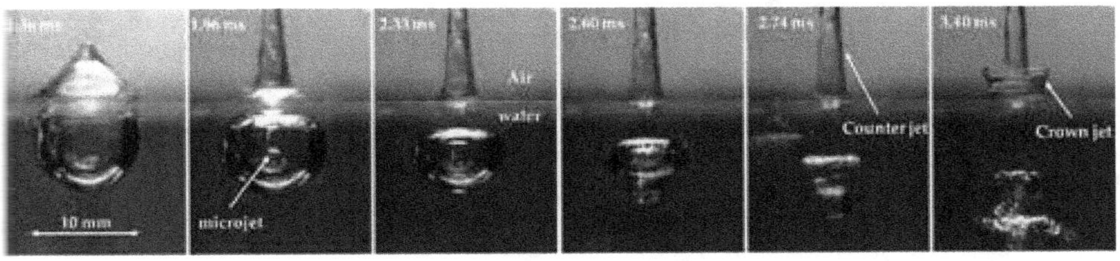

Figure 4.
Growth and collapse of a cavitation bubble near a free surface with the formation of jets inside the bubble and out of the surface.

the liquid caused by gravity will always deform the bubble. To cope with this limitation, one "simply" needs to get rid of gravity effects. This is precisely what parabolic flights can achieve. Running the tests in microgravity allows us to generate bubbles that are more spherical and explore the direct effect of different levels of gravity on their dynamics. So far, we took part in nine parabolic flight campaigns (PFC), eight organized by the ESA and one organized by a Swiss entity, as listed in **Table 1**.

Our experiment flies, with a dozen of other experiments, aboard an Airbus A310 (A300 before 2015), which was adapted to perform parabolic flight maneuvers. The microgravity campaign typically includes three flights, operated on three consecutive days. A typical flight lasts for about 2–4 h and includes 31 parabolas, each offering 20 seconds of microgravity. If requested, a series of steep turns may be also performed to generate stabilized hyper-gravity levels (e.g. 1.2, 1.4, 1.6 and 1.8 g). These maneuvers are illustrated in **Figure 5** with a typical evolution of the gravity level measured by an on-board accelerometer. It should be noted that the accumulated time spent in microgravity is only 10 min per flight! This highlights the importance of the thorough design and preparation of the setup, which should leave no room for failure. In particular, the experiment must be simple and automated as much as possible. During the parabolas, an operator can have difficulties controlling the experiment efficiently. Should operators become sick because of the awkward hyper-gravity phases, the experiment must be able to continue running without their intervention. In addition to the valuable assistance provided by parabolic flights experts, we also had the privilege to collaborate with Claude Nicollier (first Swiss astronaut) to improve these important but nonobvious aspects.

The flight day always starts at 6 a.m. with an ultimate check of the experimental setup using a well-honed procedure. The test chamber is filled with fresh water, and the test sequence is uploaded to the computer to allow for its automatic execution. 8 a.m. is the time for scopolamine injection. This medication is made available to flying participants to overcome the so-called space adaptation syndrome, also referred to as motion sickness. At 9 a.m. sharp, the doors are closed and the plane is ready to leave. About 30 min after take-off, the experimenters are allowed to leave

Year	2005	2006	2011	2012	2013	2014	2015	2015	2017
Campaign ID	SPFC 8	PFC 42	PFC 53	PFC 56	PFC 58	PFC 60	PFC 62	Swiss PFC 1	PFC 67

All flights departed from Bordeaux (France) except the Swiss PFC 1 (from Dübendorf, near Zurich, Switzerland).

Table 1.
List of attended parabolic flight campaigns.

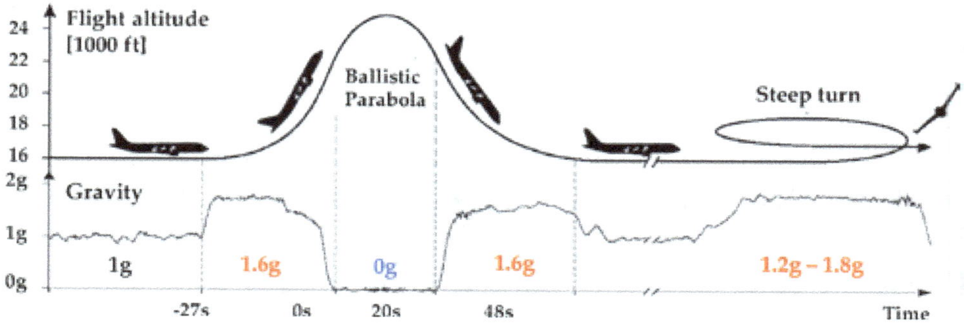

Figure 5.
Typical evolution of the vertical acceleration during parabola and steep turn maneuvers.

their seats to power up their experiments and get ready for the first parabola. This point of no return is probably the most stressful moment of the flight, definitely not suited for nervous fliers. Everyone tries to hide the palpable anxiety with all kinds of distractions until the captain announces in a lovely French accent "5, ... , 3, 2, 1, pull up, ... , 20, 30". The plane rears up to bring its altitude from 16,000 to 30,000 feet in only 20 s. During this unpleasant and long hyper-gravity phase (~1.8 g), everybody remains perfectly still. Fortunately, we suddenly switch to something extraordinary. "... 40, injection". At 45° inclination, the pilots shut down the engines, and the plane is injected into a parabolic arc offering an extraordinary period of free fall with a wonderful and liberating sensation. Everyone in the plane is floating, laughing and even shouting as to replace the noise from the engine, which dropped suddenly. This is the most exciting moment of the flight, which lasts for only 20 brief seconds. Follows another hyper-gravity phase and the plane recovers its horizontal flight. The sequence restarts 2 min later with the same anxiety and happiness, except for those less fortunate who must remain seated at the back of the plane holding a sick bag. After landing, everybody gathers in a debriefing to report issues and prepare the next flight. We use the afternoon to collect data and fix various problems. Once back in our headquarters, we all discover the freshly collected data and debate it for hours. We always appreciated this moment of prolific exchange.

4. Bubbles in drops, from erosion to exploding stars!

4.1 Experimental setup

The first two parabolic campaigns (SPFC 8 and PFC 42) were dedicated to the investigations of cavitation bubble dynamics inside a water drop. Here, the use of microgravity flights is motivated by the possibility to generate a centimeter spherical drop of water, which is hard to achieve on earth. The experimental setup, presented in **Figure 6**, is made of a transparent test chamber, where the bubble and drop are generated, a high-speed camera (up to 120,000 frames per second) and a 100 Joule flashlight. At the start of the microgravity phase, detected by an accelerometer, water is gently expelled through a 1-cm-diameter pipe filled with foam to form a water drop of about 2 cm in diameter. At the end of this phase, which lasts for about 15 s, the drop remains attached to the injector tip. A cavitation bubble is

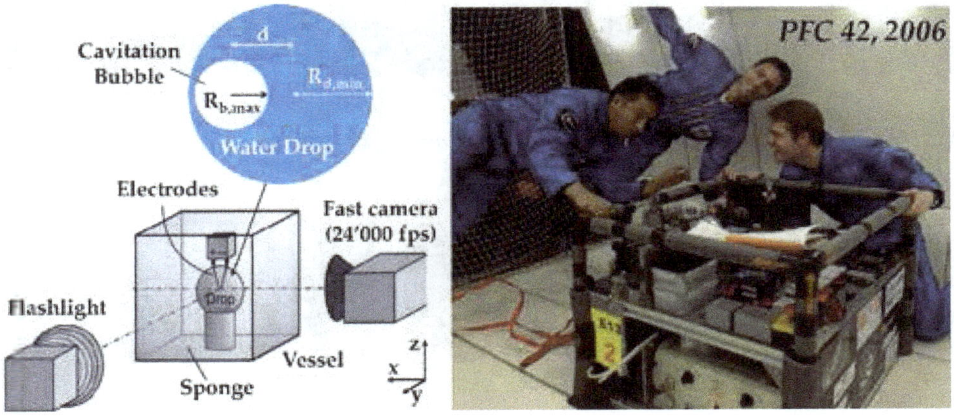

Figure 6.
Schematic of the bubble-in-drop experiment and a photo in microgravity.

then generated with the help of a spark electric discharge between two immersed platinum electrodes. The maximum radius of the bubble is varied by adjusting the discharge energy (up to 1 J). The governing parameters are the initial drop radius ($R_{d,min}$), the maximum bubble radius ($R_{b,max}$) and the eccentricity d.

4.2 Results

A sequence of high-speed visualization of a cavitation bubble within a $\sim$ 2 cm water drop is presented in **Figure** 7. We may observe how the electrical spark, visible on frame N°2, gives birth to an off-centred bubble, which expands to a maximum radius of $\sim$1 cm. The bubble then undergoes several collapses and rebounds, visible on frames N°13, N°21 and N°27. Interestingly, these frames reveal the formation of a myriad of short-lived microbubbles, which we do not observe for bubbles collapsing in a large volume of water. These transient microbubbles are due to confined shockwaves, which reflect on the drop surface and turn into tension waves. With the discovery of this *secondary cavitation*, we could propose a new path for the mechanism of erosion by a high-speed impact of water drops [7, 8], a recurring problem in various fields (e.g. rain erosion of wind turbines and aircrafts, steam turbines erosion, etc.). We argue that upon a high-speed impact of a water drop on a solid surface, the resulting shockwave travels inside the drop and turns into tension wave as it reflects on its boundary, leading to secondary cavitation, which is responsible of the erosion. To validate this hypothesis, we performed ground-based experiments with the precious help of the Swiss army. We visualized the impact of a 9 mm bullet on a $\sim$ 2 cm water jet. The result, illustrated in **Figure 8**,

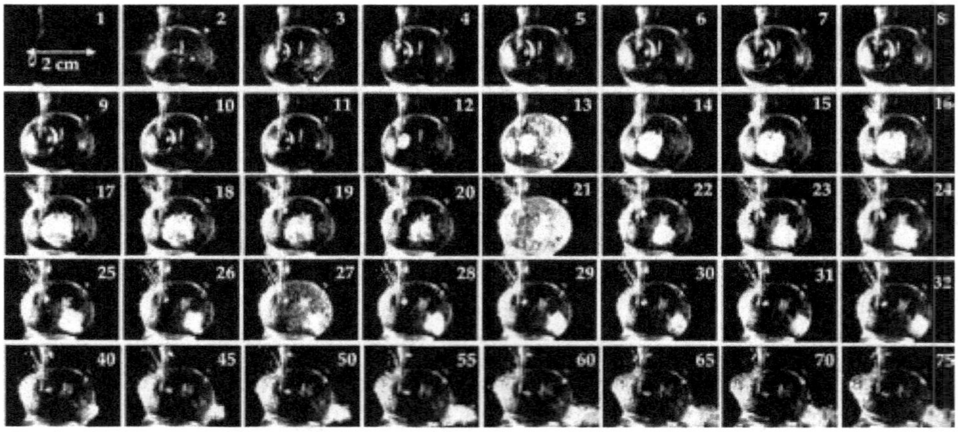

Figure 7.
Visualization of a bubble growth and collapse inside a water drop $R_{d,\min} = 2$ cm, $R_{b,\max} = 1$ cm. The inter-frame is 80 μs (400 μs for the last row).

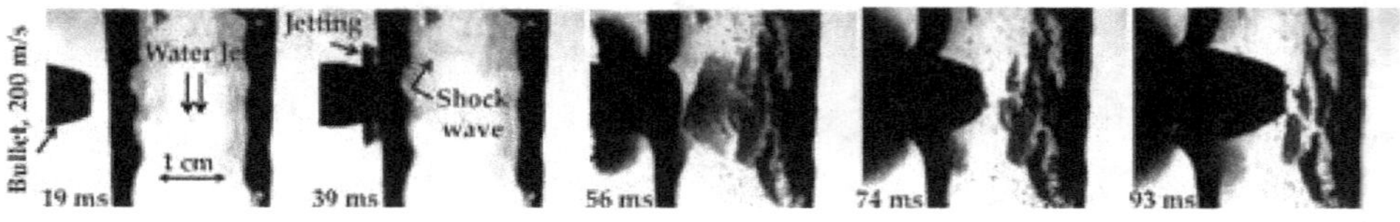

Figure 8.
Visualization of a high-speed impact of a 9 mm bullet on a $\sim$ 2 cm water jet. Evidence of secondary cavitation due to shockwaves confinement.

reveals the sudden formation of a large amount of cavitation bubbles within the jet, shortly after the bullet impact. The occurrence of this cavitation coincides perfectly with the passage of the reflected shockwave, visible on the third frame.

The sequence of **Figure 7** also reveals that as the bubble collapses and rebounds, two opposite jets emerge from the drop. The first one is the so-called counter-jet or splash. The second one results from the water entrained by the bubble as it moves away to the opposite side. Compared to the case of a bubble near a flat free surface (**Figure 4**), the counter-jet is wider and no crown jet is formed. This illustrates the role of the free surface curvature.

The evolution of the bubble radius, obtained by image processing, in the case a bubble in the centre of a 2 cm drop in microgravity, compared with a ground-based experiment in extended water volume is presented in **Figure 9**. We may observe that the collapse time is significantly shorter for the bubble-in-drop. We developed a theoretical model by extending the Rayleigh model and came up with an analytical solution that fits remarkably well with experimental data, as illustrated in the same figure. The details of this theory can be found in [9].

Another fascinating and peculiar outcome of the bubble-in-drop experiments in microgravity is the intriguing similarity found with the type II supernova Cas A, one of the most prominent stellar explosions in our Galaxy in human history [10]. Type II supernovae, also known as "core-collapse supernovae", occur when the iron core of a massive star (>8 solar masses) can no longer support its own weight and collapses within a matter of seconds at a quarter of the speed of light. This event triggers the collapse of the outer shells, causing a violent explosion of the star, temporarily boosting its luminosity a millionfold. Kinematic and chemical analyses of such "supernova remnants" allow us to decipher the physics of the explosion. In the case of Cas A, the pair of jets shown in **Figure 10** remains hard to explain. Despite the limited comparability of these jets to those observed in our water drops, the striking qualitative similarity nonetheless raises the question if Cas A might have been caused by an *eccentric* core collapse, for instance caused by a nearby companion star. Computer simulations are underway to investigate such scenarios.

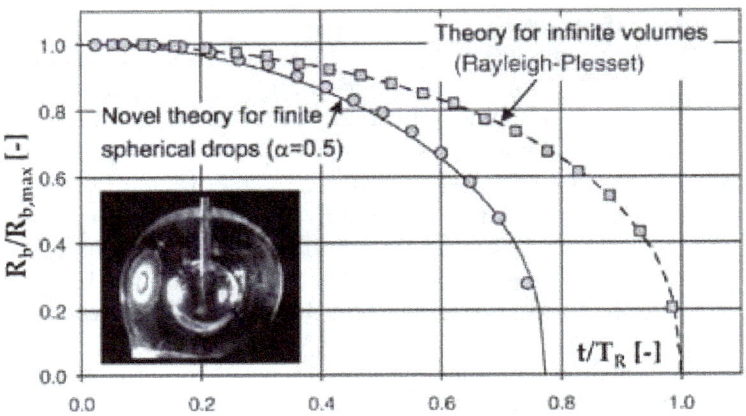

Figure 9.
Evolution of normalized radius of a bubble collapsing in the centre of a ~2 cm water drop. Symbols: experimental data. Dashed line: Rayleigh theory. Solid line: Novel theory for bubbles in drops [9].

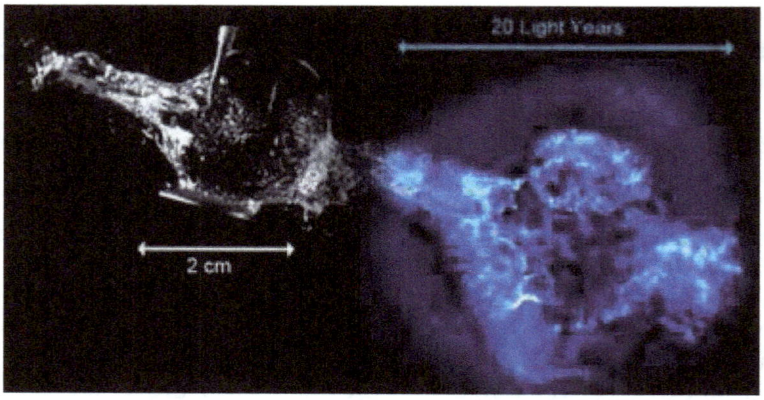

Figure 10.
Analogy of a bubble-in-drop with a massive star explosion (Cas A supernova remnant, viewed by Chandra X-ray Observatory *[10]).*

5. Gravity effects on cavitation bubble dynamics

5.1 Experimental setup: the quest for the most spherical bubble

We provide hereafter an overview of microgravity experiments, covering the periods 2011–2017, during which we took part in seven parabolic flight campaigns (see **Table 1**). We have designed a versatile experimental setup, which produces a single cavitation bubble in still water with a highly spherical shape. The setup is suited for both ground and microgravity flights and fulfills the ESA requirements, related to security and energy consumption as well as volume and payload limitations. We used two conventional racks to host our experiment. The total weight and volume are ~300 kg and ~0.6 m^3, respectively. A special attention was paid to reinforce the mechanical structure of the racks (i) to withstand a maximum acceleration of 9 g, as imposed by the ESA, and (ii) to avoid deformations due to gravity changes, which may alter the delicate optical alignment.

Bubble generation: Our setup generates one bubble at a time in the centre of an 18 × 18 × 18 cm acrylic chamber, filled with distilled water. The bubble is induced by focusing a pulsed laser of 532 nm wavelength, 8 ns duration and a 230 mJ maximum energy per pulse. As illustrated on **Figure 11**, the laser is first enlarged to a diameter of 51 mm using a beam expander. It is then focused with the help of a parabolic mirror immersed in water. The resulting pointlike plasma leads to the growth of a highly spherical bubble. The absolute pressure within the test chamber may be varied from atmospheric pressure down to 0.1 bar, with a vacuum pump.

Obviously, carrying a Class IV laser and 10 L of water in a confined microgravity aircraft does not go unnoticed, and we have to obey strict security rules. To this end, the main rack, which hosts the laser and the water chamber, is sealed to prevent water leakage and generously painted in black to minimize laser reflections. The rack is also equipped with interlocks to cut the electric power whenever the top lid is open.

The entire experiment is controlled by a laptop, which monitors the gravity level and starts the sequence as soon as the predefined level of gravity is reached. Although the duration of the microgravity phase (20 s) is long enough to repeat the test several times, we only do it once. In fact, in the absence of gravity, the residual gas after the bubble collapse remains in the centre of the chamber and does not rise upwards as it does on earth. It is then impossible to focus the laser in a mixture of

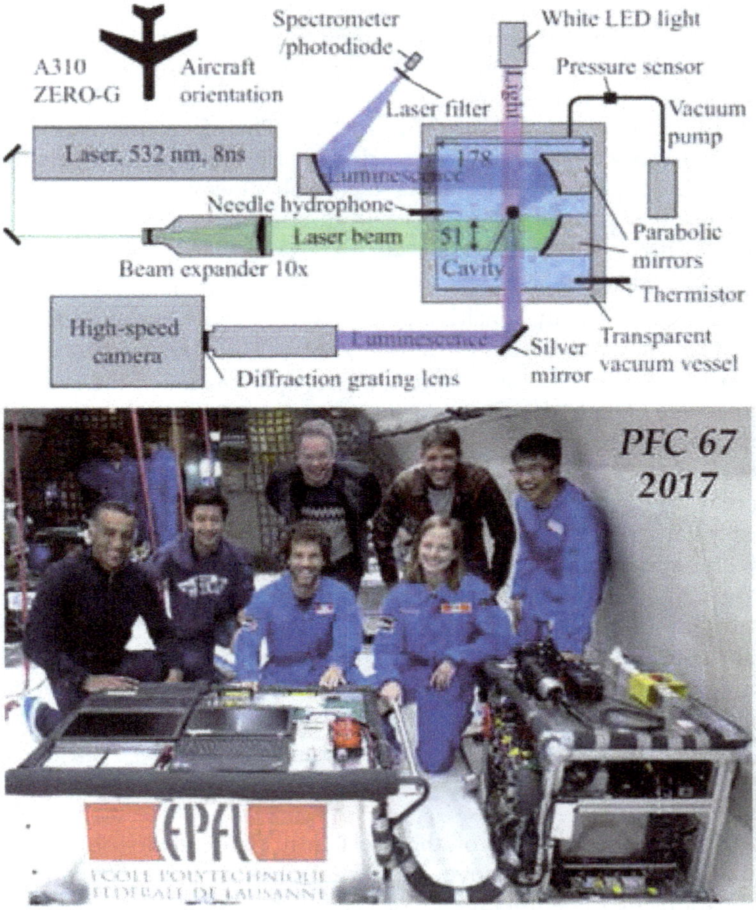

Figure 11.
(Top) schematic of the experimental setup (the dimensions are in mm) and (bottom) view of the two racks attached to the aircraft.

liquid and gas efficiently. Besides the microgravity phase, we also use hyper-gravity phases to widen the parameter space.

We implemented several sensors in the main rack to monitor various parameters during the experiment. A g-sensor is attached to the rack to measure the acceleration with a precision of ~0.2%. The resulting signal is used to start a predefined test sequence as soon as a prescribed level of gravity is reached. This sensor plays a major role in automating the experiment. A typical evolution of the gravity level during parabolic and hyper-gravity sequences is illustrated in **Figure 5**. The water temperature is measured with a thermometer, immersed in the chamber. The temperature is a key parameter for the dynamics of cavitation bubbles since it governs the pressure thresholds and rates of vaporization and condensation as well as the concentration of dissolved gas.

High-speed visualization: The follow-up of the bubble expansion, collapse and rebound as well as the visualization of radiated shockwaves is performed with a high-speed camera, fitted in the main rack as shown in **Figure 11**. A 7 W LED source, opposite to the camera, provides a slightly diverging light beam. Such a configuration requires much less light and makes it possible to view the shockwave passage, because the density gradient at their front deviates the light (shadowgraph). The imaging system can reach up to 10 million frames per second and the minimum exposure time is 50 ns. The image resolution is 400 × 250 pixels for all frame rates. The high-speed performance of Shimadzu camera is due to its complementary metal-oxide-semiconductor (CMOS) burst image sensor, which hosts the pixels and the memory in the same chip. While this technology reduces the transfer time, it offers a limited number of frames (250). Therefore, we have developed a methodology to provide a precise triggering of the camera and explore various phases of its lifetime. To this end, we have used shockwave sensors, light detectors and a GHz oscilloscope to generate accurate trigger signals for the laser pulse ignition and the high-speed camera as well.

Shockwave sensors: The shockwaves generated at bubble initiation and collapse are measured by two highly sensitive piezoelectric needle hydrophones, placed at 34.3 mm and 35.7 mm away from the bubble centre. The measurement bandwidth is ~20 MHz, which makes it possible to follow the rapid change of the pressure during the shock passage. The hydrophones offer a very precise signal to post-trigger the high-speed camera.

Luminescence detection: Another spectacular aspect of cavitation bubble dynamics that we have investigated in microgravity is the luminescence released at the final stage of the collapse. We have tracked luminescence in both time and frequency domains. A major innovation lies in the use of two parabolic mirrors, located inside and outside the water chamber, which improves significantly the signal-to-noise ratio. At the focal point of the second mirror, we have placed a notch filter and an optic fiber, which leads the light to a spectrometer. To allow for time-resolved measurements, the spectrometer is replaced by a fast silicon photodetector, having 1 ns rise time and a range of 200–1100 nm wavelength.

5.2 Major results

We provide here a selection of the most important results obtained with the experimental setup explained above. Interested readers may refer to the more in-depth publications by the *Flash and Splash* team [11–25].

5.2.1 Effects of pressure anisotropy on collapse-induced jets

Thanks to microgravity experiments, we could provide for the first time the evidence of gravity effects on cavitation bubble dynamics. **Figure 12** illustrates the growth and collapse of three bubbles generated with the same laser energy and surrounding pressure at three gravity levels (1 g, 1.6 g and 0 g). We may clearly observe that the bubble in microgravity remains spherical through the collapse and rebound phases while the two other bubbles develop an upward jet during the rebound phase. The latter is more pronounced in hyper-gravity. We may also observe that as the gravity level is increased, the maximum bubble radius decreases, and the centroid of the rebound bubble moves upwards (buoyancy).

The upward jet is due to the difference of hydrostatic pressure between the upper and lower parts of the bubble. The liquid beneath the bubble moves naturally faster towards the bubble centre, leading to interface instability and jet formation. While such explanation holds for large bubbles, it is not clear how far the gravity-induced pressure gradient can influence smaller bubbles. We carried out a large number of experiments involving bubbles of different sizes, subjected to different driving pressures and gravity levels. We have also performed ground-based experiments in the presence of free and solid surfaces. Here again, the bubble size and driving pressure were varied in a wide range. Moreover, we carried out potential flow simulations, using the boundary integral method (BIM). The analysis of a large number of experimental and numerical data, combined with theoretical

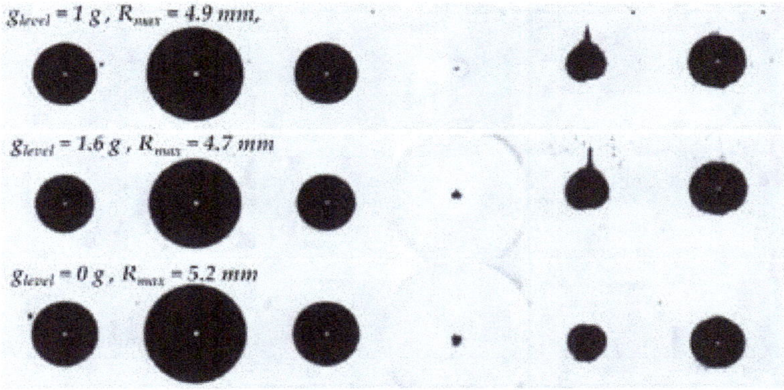

Figure 12.
Three bubbles generated in water with the same laser energy and driving pressure (10 kPa), but at three different gravity levels (adapted from [2]).

considerations, led us to propose a unified framework to predict the bubble deformation and jet properties. Our approach is based on an anisotropy parameter ζ, which is a non-dimensional representation of the so-called Kelvin impulse, i.e. the linear momentum acquired by the liquid during the asymmetric growth and collapse of the bubble. We managed to express this parameter (ζ) in its vector form, for all tested configurations as follows:

$$\zeta = \begin{cases} -\rho \boldsymbol{g} R_0 \Delta p^{-1} & \text{gravitational field} \\ -0.195\gamma^{-2}\boldsymbol{n} & \text{flat rigid surface} \\ +0.195\gamma^{-2}\boldsymbol{n} & \text{flat free surface} \end{cases}$$

where ρ is the liquid density, $\boldsymbol{g}$ is the gravity vector, R_0 is the maximum bubble radius, and $\boldsymbol{n}$ is the unit vector normal to the neighboring surface and directed towards the bubble centre. The so-called stand-off parameter (γ) is defined as h/R_0, where h is the distance between the bubble centre and the boundary.

We have identified three different regimes for the jet induced by a collapsing bubble: weak, intermediate and strong. Weak jets occur for slightly deformed bubbles $(\zeta < 10^{-3})$ and are hardly visible throughout the collapse and rebound. Intermediate jets $(10^{-3} < \zeta < 0.1)$ pierce the opposite bubble wall at the end of the collapse phase and emerge during the rebound. Such jets occur for large bubbles subjected to gravity as illustrated on **Figure 12**. Strong jets $(\zeta > 0.1)$ develop earlier and pierce the bubble way before the end of the collapse phase (see illustrations in the case of a bubble near a free surface in **Figure 4**).

The dynamics of the jet is systematically analyzed through different properties, such as the jet impact time, jet speed, bubble displacement, bubble volume at jet impact and vapor-jet volume. We found that, once properly normalized, these variables reduce to straightforward functions of ζ, in fair agreement with numerical simulations. An illustration is given in **Figure 13** for the normalized jet velocity.

Interestingly, we obtain simple approximations of the jet properties as functions of the parameter ζ, regardless of jet drivers (gravity/neighboring boundaries) and over a wide range of ζ. Here are the approximations found for the jet velocity, the bubble displacement and the jet impact time:

- Normalized jet speed: $U_{jet}/\sqrt{\Delta p/\rho} \approx 0.9\zeta^{-1}$

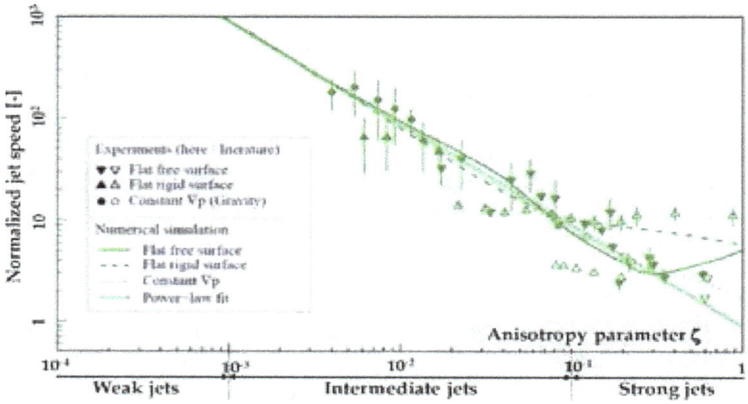

Figure 13.
Normalized jet speed ($U_{jet}/\sqrt{\Delta p/\rho}$) as a function of ζ and γ. Our data (filled symbols) are compared with literature data (empty symbols). The lines are the numerical models. The thick line is the power-law fit (adapted from [3]).

- Normalized bubble displacement: $\Delta z/R_0 = 2.5\zeta^{3/5}$
- Jet impact time scaled by collapse time: $T_{Impact}/T_{Collapse} \approx 0.15\zeta^{5/3}$

5.2.2 Effects of pressure anisotropy on shockwaves

We have found that the formation of shockwaves is also highly sensitive to the bubble deformation induced by pressure anisotropy. In microgravity, the bubble remains spherical throughout the collapse phase with an emission of a single and intense shockwave, as illustrated in **Figure 14**. We have plotted on the same figure the case of a bubble deformed by a neighboring free surface ($\zeta = 0.19$). For such high

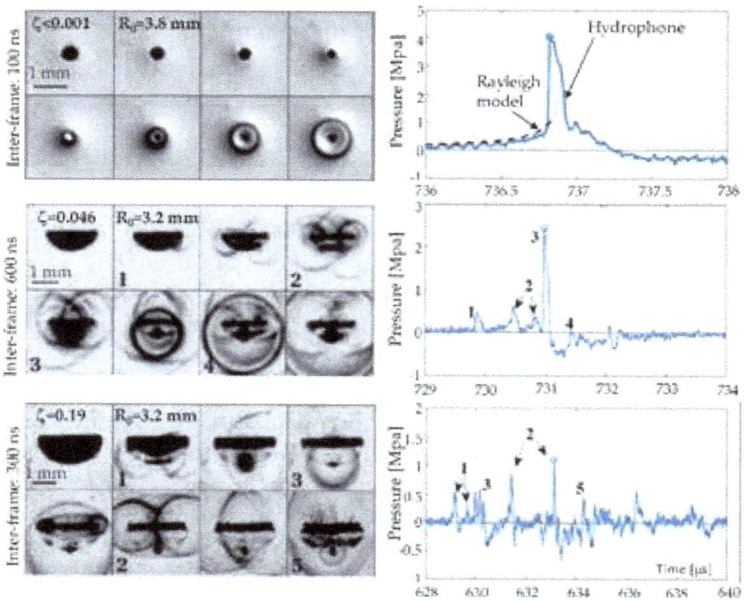

Figure 14.
Visualization of three collapsing bubbles and the corresponding pressure signal recorded by the hydrophone. The shockwaves are denoted by 1, jet impact; 2, torus collapse; 3, tip bubble collapse; 4, second torus collapse; and 5, second tip bubble collapse (adapted from [3]).

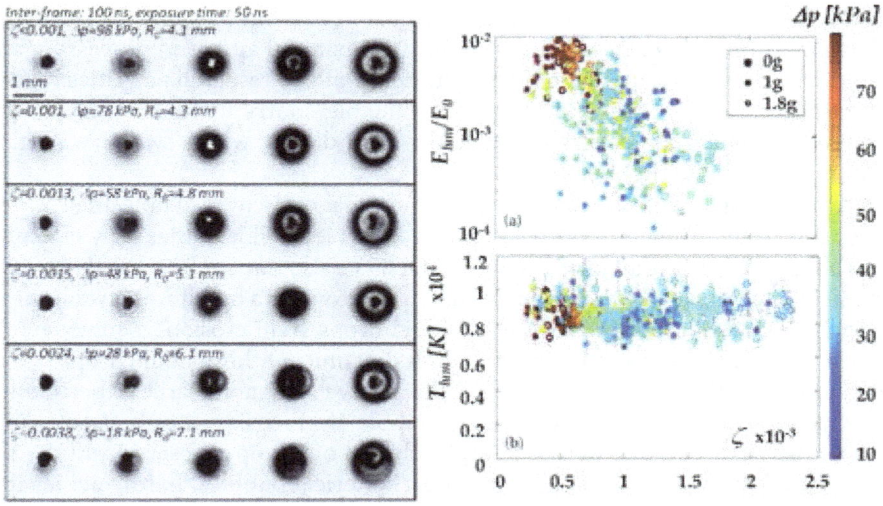

Figure 15.
Left: visualization of luminescence emitted by a collapsing bubble, for different values of ζ. *Right: single bubble luminescence (a) relative energy* E_{lum}/E_o *and (b) blackbody temperature* T_{lum} *vs.* ζ. *Colors indicate the driving pressures, and symbols indicate the different levels of gravity (adapted from [3]).*

anisotropy parameter, a strong jet develops and fractions the cavity into a tip cavity and a toroidal cavity. We observe the emission of several shockwaves related to jet impact and the collapse of the bubble fragments. These events are identified in the corresponding pressure signal. We found that the energy radiated by the shockwaves is maximum for the spherical collapse case and decreases with deformed bubbles.

5.2.3 Effects of pressure anisotropy on luminescence

The direct visualization of luminescence emitted by the collapse of a cavitation bubble in normal gravity is presented in **Figure 15** for different values of the anisotropy parameter ζ. The latter was varied by adjusting the driving pressure. We may observe that luminescence is generated when the bubble reaches its minimum radius, just before the shockwave emission. The figure clearly illustrates how luminescence fades away as the bubble is deformed.

We have processed a large number of broadband spectra of luminescence emitted by individual laser-induced bubbles, collapsing in different gravity-induced pressure gradients. As shown in **Figure 15**, we found that the luminescence energy E_{lum}, normalized by the potential energy of the bubble E_0, varies in a roughly exponential way with ζ. We assumed the blackbody radiation to estimate the temperature reached by the noncondensable gas within the bubble. Our results, presented in the same figure, confirm the high values of gas temperature, which ranged between 7000 and 11,500 K. We may also observe that the luminescence ceases for a threshold value, which coincides with the transition from weak to intermediate jets. Nevertheless, unlike the luminescence energy, we have found no clear variation of the blackbody temperature as a function of ζ. This counter-intuitive result needs further investigations to be clarified.

6. Conclusion

During their brief and violent life, cavitation bubbles may develop powerful phenomena and cause damage in a variety of industrial devices. Nevertheless, by mastering their unique ability to focus energy, cavitation bubbles may be beneficial in many applications, such as cleaning, chemistry, material and food processing and medicine. The present chapter provides a summary of our research activities related to this fascinating topic and underlines the valuable contribution of microgravity experiments. This review covers the past 15 years, during which we have participated in nine parabolic flight campaigns.

We first investigated the case of a bubble-in-drop and discovered how an eccentric collapse leads to the formation of two opposite liquid jets, which amazingly resemble the ones observed during the collapse of giant stars. We also learned how the shockwave confinement generates secondary cavitation, which provides a better understanding of the erosion due to a high-speed impact of liquid drops (rain erosion). Moreover, we developed and validated a new theory for bubble dynamics within a liquid drop.

To explore the final stage of a bubble collapse, we built a versatile experimental setup, which generates a single cavitation bubble by focusing a pulsed laser in a water chamber. The combination of a careful design of the optical setup and the use of microgravity flights let us produce the most spherical bubbles, hardly achievable in ground-based experiments. We also used variable gravity offered by parabolic flights to modulate the hydrostatic pressure gradient and explore its effects on bubble deformation in a broad parameter space. We developed a unified theoretical framework to predict the dynamics of a cavitation bubble, based on a non-dimensional anisotropy parameter (ζ). We have found that the collapse of a spherical bubble ($\zeta \approx 0$) generates a unique and strong shockwave, no jetting and a rather small rebound. As ζ is increased, the bubble becomes more and more deformed with a more complex set of shockwaves, a larger rebound and a microjet that develops earlier and earlier. We also observed a rapid quenching of the collapse-induced luminescence for increasing ζ. Our model predicts how the energy in the initial cavitation bubble is partitioned between the collapse channels, namely, the shockwave, the jet, the rebound and the luminescence. This paves the way to optimize the outcome of a collapsing bubble, depending on the application.

The present chapter was written during the "great lockdown". Outside, the coronavirus disease 2019 (COVID-19) pandemic is raging around the world with dramatic consequences. Who knows? Maybe the solution will come from cavitation bubbles. If carefully injected into the human body and appropriately controlled by an acoustic field, microbubbles can locally deploy their extraordinary power to target and neutralize coronavirus with minimal side effects. Given the state of knowledge on cavitation bubbles, this is not wishful thinking.

Acknowledgements

On behalf of *Flash and Splash* team, I would like to express my gratitude to the European Space Agency for giving us the wonderful opportunity to fly our experiment in microgravity. I am also thankful to the parabolic flight experts for their assistance during the flight campaigns. The project received financial support from the Swiss National Science Foundation (grants no. 116641 and 513234), the European Union (H2020, grant no. 813766), the EPFL Space Centre, the University of Western Australia and Breitling AG.

Notes/thanks/other declarations

The *Flash and Splash* project would not have been possible without the perseverance and commitment of D. Obreschkow, Ph. Kobel, N. Dorsaz and A. De

Bosset, the outstanding contributions of M. Tinguely and O. Supponen and the precious assistance of EPFL technicians (L. Bezençon, M. Raton and R. Fazan). I would like to thank them all warmly.

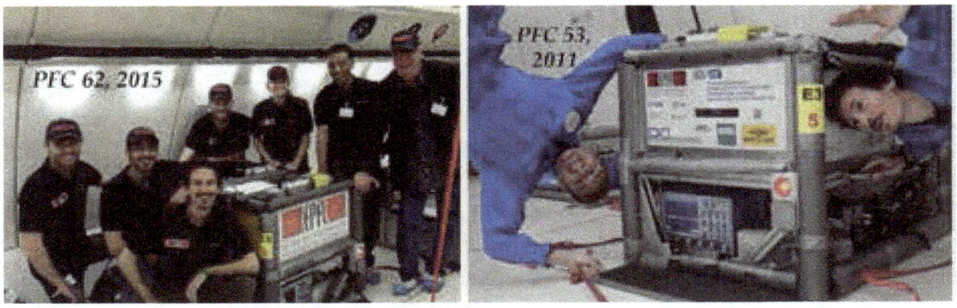

Author details

Mohamed Farhat
Ecole Polytechnique Fédérale de Lausanne, Switzerland

*Address all correspondence to: mohamed.farhat@epfl.ch

References

[1] Robert E, Lettry J, Farhat M, Monkewitz PA, Avellan F. Cavitation bubble behavior inside a liquid jet. Physics of Fluids. 2007;**19**(6):067106

[2] Tinguely M. The effect of pressure gradient on the collapse of cavitation bubbles in normal and reduced gravity. [EPFL PhD thesis N°5674]; Switzerland: Ecole Polytechnique Fédérale de Lausanne; 2013

[3] Supponen O. Collapse phenomena of deformed cavitation bubbles. [PhD Thesis N°8089]; Switzerland: Ecole Polytechnique Fédérale de Lausanne; 2017

[4] Franc P et al. La Cavitation: Mécanismes Physiques et aspects Industriels. France: EDP Sciences; 1995. ISBN: 978-2-86883-451-5

[5] Koukouvinis PH, Bruecker CH, Gavaise M. Unveiling the physical mechanism behind pistol shrimp cavitation. Scientific Reports. 2017;7: 13994. DOI: 10.1038/s41598-017-14312-0

[6] Rayleigh L. On the pressure developed in a liquid during the collapse of a spherical cavity. Philosophical Magazine. 1917;**34**(200):94-98

[7] Field JE, Camus J-J, Tinguely M, Obreschkow D, Farhat M. Cavitation in impacted drops and jets and the effect on erosion damage thresholds. Wear. 2012;**290–291**:154-160

[8] Obreschkow D, Dorsaz N, Kobel P, De Bosset A, Tinguely M, Field J, et al. Confined shocks inside isolated liquid volumes: A new path of erosion? Physics of Fluids. 2011;**23**:101702

[9] Obreschkow D, Kobel P, Dorsaz N, De Bosset A, Nicollier C, Farhat M. Cavitation bubble dynamics inside liquid drops in microgravity. Physical Review Letters. 2006;**97**: 094502

[10] Hwang U et al. A million second Chandra view of Cassiopeia A. The Astrophysical Journal. 2004;**615**: L117-L120

[11] Supponen O, Obreschkow D, Kobel P, Dorsaz N, Farhat M. Detailed experiments on weakly deformed cavitation bubbles. Experiments in Fluids. 2019;**60**:33

[12] Supponen O, Akimura T, Minami T, Nakajima T, Uehara S, Ohtani K, et al. Jetting from cavitation bubbles due to multiple shockwaves. Applied Physics Letters. 2018;**113**:193703

[13] Supponen O, Obreschkow D, Farhat M. Rebounds of deformed cavitation bubbles. Physical Review Fluids. 2018;**3**:103604

[14] Supponen O, Obreschkow D, Kobel P, Farhat M. Luminescence from cavitation bubbles deformed in uniform pressure gradients. Physical Review E. 2017;**96**(3):033114

[15] Supponen O, Obreschkow D, Kobel P, Tinguely M, Dorsaz N, Farhat M. Shock waves from nonspherical cavitation bubbles. Physical Review Fluids. 2017;**2**(9):093601

[16] Supponen O, Obreschkow D, Tinguely M, Kobel P, Dorsaz N, Farhat M. Scaling laws for jets of single cavitation bubbles. Journal of Fluid Mechanics. 2016;**802**:263-293

[17] Koukouvinis P, Gavaises M, Supponen O, Farhat M. Simulation of bubble expansion and collapse in the vicinity of a free surface. Physics of Fluids. 2016;**28**(5):052103

[18] Koukouvinis P, Gavaises M, Supponen O, Farhat M. Numerical

simulation of a collapsing bubble subject to gravity. Physics of Fluids. 2016;**28**(3): 032110

[19] Supponen O, Kobel P, Obreschkow D, Farhat M. The inner world of a collapsing bubble. Physics of Fluids. 2015;**27**(9):091101

[20] Obreschkow D, Tinguely M, Dorsaz N, Kobel P, De Bosset A, Farhat M. The quest for the most spherical bubble: Experimental setup and data overview. Experiments in Fluids. 2013;**54**(4):1503

[21] Sato T, Tinguely M, Oizumi M, Farhat M. Evidence for hydrogen generation in laser- or spark-induced cavitation bubbles. Applied Physics Letters. 2013;**102**(7):074105

[22] Tinguely M, Obreschkow D, Kobel P, Dorsaz N, De Bosset A, Farhat M. Energy partition at the collapse of spherical cavitation bubbles. Physical Review E. 2012;**86**(4):046315

[23] Obreschkow D, Bruderer M, Farhat M. Analytical approximations for the collapse of an empty spherical bubble. Physical Review E. 2012;**85**(6): 066303

[24] Obreschkow D, Tinguely M, Dorsaz N, Kobel P, De Bosset A, Farhat M. Universal scaling law for jets of collapsing bubbles. Physical Review Letters. 2012;**107**(20):204501

[25] Kobel P, Obreschkow D, De Bosset A, Dorsaz N, Farhat M. Techniques for generating centimetric drops in microgravity and application to cavitation studies. Experiments in Fluids. 2009;**47**(1):39-48

3

Tumors in Space: Preparation for Spaceflight

Tricia L. Larose

Abstract

Tumors in Space is a cutting-edge cancer research experiment at the intersection of stem-cell biology and space technology selected by the United Nations Office for Outer Space Affairs and the China Manned Space Agency for a 31-day space mission on board the China Space Station. Anchored in Norway, Tumors in Space includes an international team of exceptional scientists at several European partner organizations including the University of Oslo and the Norwegian University of Science and Technology in Norway, the International Space University in France, the Belgian Nuclear Research Center in Belgium, and Vrije University Amsterdam as well as the Hubrecht Institute in The Netherlands. This chapter first presents our two novel hypotheses including the current state of scientific evidence upon which our hypotheses are based. Following, the seven main steps of our spaceflight preparation are discussed within the context of our 2025 launch date from China. Finally, some thoughts on impact, including support for the United Nation's Sustainable Development Goals and commitment to science communication in the public domain, are given. Tumors is Space is under a programme of, and funded by the European Space Agency with the support of the Norwegian Space Agency.

Keywords: China Space Station, cancer, organoids, microgravity, cosmic radiation

1. Introduction

Spaceflight is a high-risk/high-gain endeavor (**Figure 1**) that is resource intensive. For these reasons, spaceflight experiments are rare, and selection is highly competitive at the international level. Selection criteria for spaceflight experiments include scientific excellence, innovation, feasibility, impact, competency of the scientific team to achieve the proposed research goals, and timely delivery. Moreover, a selected spaceflight experiment must be novel in that it must address a critical issue for which knowledge and understanding are lacking; it must be ground-breaking in nature, both literally and figuratively. In the case of medical research in space, a selected spaceflight experiment must create value for astro-/cosmo-/taikonaut health applied to short- and long-duration space missions and solar system exploration. At the same time, space medicine must also bring value to public health on Earth. In this vein,*Tumors in Space* is a ground-breaking cancer research experiment at the intersection of space technology and stem-cell biology recently selected for a 31-day space mission on board the China Space Station (CSS). Not only will *Tumors in Space* address cosmic radiation as a potential cause of cancer for humans in space, it will also address the side effects of radiation therapy

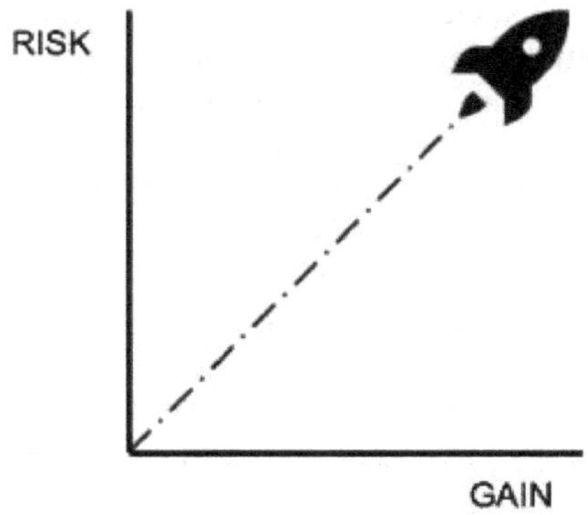

Figure 1.
Spaceflight is high risk and high gain.

For cancer patients on Earth. In addition to radiation biology experiments, *Tumors in Space* will exploit the microgravity environment on the CSS to investigate whether lack of gravity can slow or stop the growth of cancer. Like all selected spaceflight experiments, *Tumors in Space* will undergo several years of preparation before spaceflight launch. This preparation period is broken down into seven main steps over 6 years, culminating in our spaceflight experiment scheduled for a 2025 launch date out of China. This chapter will discuss the different steps of our spaceflight preparation as depicted in **Figure 2**.

2020	2021	2022	2023	2024	2025
Step 1: Ethics, Regulatory Affairs & Information Management					
Step 2: Laboratory, Bioinformatics & Statistical Analysis Pipelines					
Step 3: Spaceflight Hardware Design, Development, Testing & Approval					
Step 4: Ground Experiments			**Step 6:** Parabolic Flights		**Step 7:** Spaceflight
Step 5: Sounding Rocket					

Figure 2.
Tumors in Space *project timeline over a 6-year period from 2020 to 2025 and according to seven steps of spaceflight preparation including launch to China Space Station and safe return to Earth in 2025.*

In addition to these seven steps, other important aspects of *Tumors in Space* will be presented, namely, the scientific background upon which the hypotheses are built, team structure and expertise, the importance of impact including response to the United Nations Sustainable Development Goals, and science communication in the public domain. Let us first begin with a presentation of the problem statement and the supporting evidence upon which the *Tumors in Space* hypotheses are built.

2. The problem statement and supporting evidence

As stated in Section 1, medical experiments in space must address crew health and safety while simultaneously bringing value to medicine on Earth. What follows is a short paragraph on the burden of cancer on Earth, followed by a second paragraph on space radiation risk for crewed missions. To end this section, evidence on the potential use of microgravity for cancer therapeutics will be given.

Interspersed throughout this section are some details on our scientific approach, more of which will be detailed later in this chapter.

The cancer burden on Earth is substantial. Currently, cancer is the second leading cause of death worldwide [1]. As of 2018, an estimated 18.1 million people had been diagnosed with cancer, and nearly 10 million people had died from the disease [2]. Moreover, new cancer cases are expected to increase by 70% over the next two decades [1]. As of 2009, the global cost of treating cancer patients was estimated to be $285.8 billion USD per year, and the indirect costs associated with premature death and loss of productivity was estimated at $1.16 trillion USD per year [2]. Over time, these costs will continue to rise.

The scientific community has made increasing progress toward understanding lifestyle, environmental, and genetic causes of cancer [3], yet approximately 50% of all cancer cases have no known cause. There is no cure for cancer. We must therefore adopt a broad, interdisciplinary perspective and exploit all scientific and technical capabilities to better understand this disease, including the use of orbital platforms [4]. *Tumors in Space* will exploit the natural spaceflight environment on the China Space Station to address cosmic radiation as a potential unknown cause of cancer among crew and to harness microgravity as a potential source for cancer therapy in space and on Earth.

In space, radiation risk is the most dangerous issue for crew health and safety. The spaceflight environment is characterized by high atomic number energy (HZE[1]) ions from three natural sources of radiation that could lead to crew sickness or death [5]. During space exploration missions, crew are exposed to HZE ions from trapped particle belts, solar particle events, and background galactic cosmic rays [5]. When primary HZE ions interact with spacecraft materials, they can generate harmful secondary radiation that can penetrate deeply into the human body. Cancer risk from human spaceflight and exploration is assumed to be high, but the short- and long-term effects of harmful HZE ion radiation during deep space missions are still undetermined and have large uncertainties [5]. As stated by the European Space Agency (ESA), most of this uncertainty is due to poor knowledge on the effects of galactic cosmic rays [6]. *Tumors in Space* will address some of the uncertainty associated with cosmic radiation as a potential cause of cancer among crew. At the same time, advances in space-based cancer research have the potential to improve charged particle therapy in oncology for the benefit of cancer patients on Earth.

All cancers are caused by somatic mutations, meaning an accumulation of genetic alterations that damage our deoxyribonucleic acid (DNA) at the cellular level [7]. Cell damage from exogenous exposures (e.g., HZE ion radiation), as well as endogenous exposures (e.g., circulating hormones), and modifications to DNA can lead to cancer. DNA damage in cancerous cells can be observed as distinct patterns known as mutational signatures. Just like fingerprints are unique to every individual, mutational signatures are unique to specific cancer-causing exposures. Each mutational signature is the outcome of a series of biological processes that include DNA modification or damage, DNA repair or absence of repair, and DNA replication. Using cancer genome sequencing [8], these mutational signatures can be observed, catalogued, and compared to identify unique causes of different cancer types. This was first demonstrated when mutational signatures specific to UV[2]

---

[1] The abbreviation HZE comes from high (H) atomic number (Z) and energy (E).

[2] UV stands for ultraviolet rays from sunlight that induces a unique mutation found specifically in skin cancer.

exposure from the sun were observed in the p53 tumor suppressor gene[3] [9]. Most cancer genomes are now characterized by multiple mutational signatures across multiple mutational processes that generate more than one somatic mutation in each cancer cell [10]. This evidence leads us to our first *Tumors in Space* hypothesis.

Hypothesis 1: We will identify a novel mutational signature of cosmic radiation in healthy and cancer human organoids after 31-days exposure to cosmic radiation in space. The mutational signature of cosmic radiation can then be used as a biological marker for early detection of cancer and cancer susceptibility among spaceflight crew.

Beyond cosmic radiation, the space environment provides unique microgravity conditions that Earth-bound laboratories can simulate but never replicate. Microgravity can be exploited to study biological mechanisms and pathways to improve our understanding of many diseases, including cancer. Previous spaceflight experiments have shown that exposure to the space environment alters immune cell growth and function [11], causes cytoskeleton and cell shape changes [12], and alters cellular gene expression including genes involved in replication and suppression of tumors [13, 14]. This brings us to our second *Tumors in Space* hypothesis.

Hypothesis 2: Exposure to microgravity during a 31-day space mission will slow or stop the growth of cancer.

We will conduct a tumor colony survival analysis which is as simple as counting the number and/or expansion of our organoids after long-term exposure to microgravity and then compare those results to our ground-based controls. It is crucial to block out cosmic radiation that may interfere with this microgravity experimental arm. For that reason, we are building radiation shielding to ensure the cancer organoids exposed to microgravity are not hit by cosmic radiation.

So, the China Manned Space Agency (CMSA) in collaboration with the United Nations Office for Outer Space Affairs has selected *Tumors in Space* for a 31-day space mission on board the China Space Station (**Figure 3**). *Tumors in Space* utilizes space technology and three-dimensional (3D) organoid technology (see further) to study early mutational events in human DNA due to spaceflight exposure. We

Figure 3.
Artist rendition of the China Space Station (credit: China Manned Space Agency).

[3] The p53 gene provides instructions for building a protein that acts as a tumor suppressor. Well-characterized mutations in this gene have been attributed to UV exposure and are found in many types of skin cancer.

expect to find a unique mutational signature of cosmic radiation. We also expect that cancer organoid growth will slow or stop after exposure to microgravity.

3. Going beyond state of the art

As mentioned previously, excellent science in line with the current state of evidence does not fulfill the stringent selection criteria for spaceflight experiments. Space medicine research must be novel, go beyond current scientific paradigms, and have application to crew health in space and public health on Earth. In other words, selected space medicine experiments must go beyond state of the art.

Not only does *Tumors in Space* more than double the mission length of all previous cancer experiments in space, we also use the most advanced and physiologically relevant bio-samples in our experiments—human organoids. Until now the term "organoids" has been introduced without explanation. In fact, organoid technology is so new that many scientists are unfamiliar with the term. So, what exactly is an organoid, and why is an organoid so much more advanced than a cell line that is typically used in medical research?

Human organoids are multicellular, stem-cell-derived, 3D constructs that self-organize to mimic the structure and function of the source tissue [15]. More simply, a human organoid is a 3D clump of human cells and extracellular matrix, taken as a biopsy from living human tissue, but organoids "live" in a lab. What is amazing about organoids is that they can be expanded, cloned and cared for in a way that allows them to replicate and grow while retaining the structure and function of the original source tissue or organ. For example, a colon cancer organoid derived from a tumor biopsy taken (with permission) from a colon cancer patient will not only grow into a 3D structure under laboratory conditions; it will also behave like we would expect colon cancer to behave. Same goes for healthy tissue.

Figure 4 shows an artist rendition of the growth of a colon organoid. A small sample, or biopsy, is taken from the colon tissue of a patient. Both the original colon tissue and the biopsy are 3D in nature. Once the biopsy is removed from the patient, it can be nurtured in the lab so that cell division continues, resulting in larger and

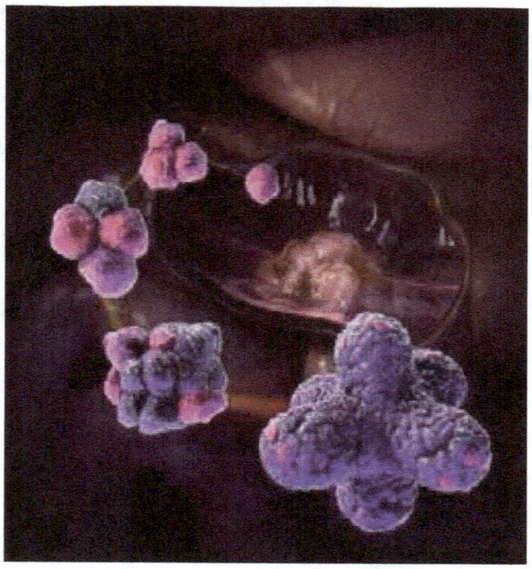

Figure 4.
Artist rendition of colon organoid growth (credit: Hans Clevers, the HUB).

more complex tissue samples that mimic the structure and function of the source tissue (in this case, the human colon). Organoids are much smaller than the organ they mimic but can be up to 5 mm in size. These organoids are the most advanced and physiologically relevant bio-samples available for medical research today. This means that we can conduct cutting-edge cancer research using human organoids rather than cell lines.

Although monolayer cancer cell lines are easily accessible, well understood, and used in all previous cancer experiments in space, they lack spatial cellular organization and physiological relevance [16]. Cell lines cannot mimic conditions found in real humans, whereas organoids can. Using human organoids ensures that our results are applicable to patient diagnosis and treatment. This is crucial since we need to get as close as possible to studying the real human body in space. Organoid technology bridges the gap between cell lines and humans, or in vitro and in vivo. In 2017, organoids were named Method of the Year by *Nature Methods* for their unparalleled potential to advance our understanding of human biology, especially cancer [17]. We are the first team to send healthy and cancer human organoids into space.

4. International collaboration

Before we dig into the seven steps of preparation for our spaceflight experiment, a quick note on international collaboration is warranted. Success of such a complex space medicine experiment requires world experts across many different fields. **Figure 5** is color coded to show collaborating countries, not only according to levels of responsibility with regard to the project but also according to geopolitical boundaries. You can see from **Figure 5** that Norway (in light gray) and China (in dark green) stand apart from the other collaborating countries like the Netherlands, Belgium, and France (in dark blue).

Norway, the center of the figure, is the coordinating country because this is where the principal investigator is located, along with the several members of the *Tumors in Space* team. We also have the support of the Norwegian Space Agency. Furthermore, Norway has identified China as a priority country for technical and scientific collaboration—for example, in satellite and broadband communication, as well as medicine. Norway and China are united by an underlying cooperative effort for the *Tumors in Space* collaboration. China is the host country for the Norwegian *Tumors in Space* experiment that has been selected for a 31-day space mission on board the China Space Station.

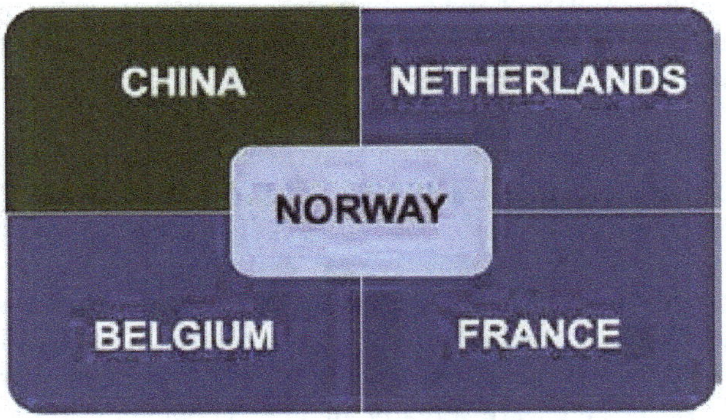

Figure 5.
Tumors in Space *collaborating countries.*

Our additional collaborators are based in Europe and are part of the European Union. The Netherlands is a collaborating country where two critical organizations are located. Firstly, our patient-derived organoids come from Hubrecht Organoid Technology (the HUB) in Utrecht, and some of our ground-based experiments will take place at the European Space Research and Technology Center of ESA in Noordwijk. Our radiation biology experts are located in Belgium at the Belgian Nuclear Research Center where our ground radiation experiments in preparation for spaceflight will be conducted. Not only do they house safe infrastructure for radiation biology experiments, our Belgian colleagues have years of space experience and substantial experience in the safe design of nuclear reactors. For the latter reason, our Belgian colleagues will also participate in the design and development of our spaceflight hardware in close collaboration with Norway and China. Finally, France is the location of the International Space University, to which the principal investigator and other *Tumors in Space* team members are affiliated. Our aircraft parabolic flight campaigns will also be flown from France.

This brief summary shows how important international collaboration is. Without all of these partners, realizing the most advanced cancer experiment ever conducted in space would not be possible. It is because of the unique strengths and expertise of all collaborating partners that the *Tumors in Space* project has become a reality.

5. The *Tumors in Space* experiment in seven steps

The step-by-step preparation of our spaceflight experiment is now described. Each one of these steps could be a chapter in itself. Certainly, some details must be omitted. The success of this entire project relies heavily on the success of each step, both independently and collectively.

5.1 Ethics, regulatory affairs, and information management: step 1

This step ensures proper procedures for the selection, release, and use of intraindividual healthy and cancer organoids according to biobanking protocols and informed consent procedures. Procedures and consents are subject not only to the approval of the HUB biobank in the Netherlands but also from the host institute in Norway, as well as our collaborating institutes where our organoid research will be conducted (the European Space Agency in the Netherlands, the Belgian Nuclear Research Center in Belgium, parabolic flight laboratory in France, and the China Manned Space Agency in China for our spaceflight). Each institute has its own procedure, and each country or set of countries have their own legislations. Shipping bio-samples involves material transfer agreements and customs paperwork, and since human organoids contain human DNA, special ethical requirements apply.

First and foremost, all patients provide written, informed consent to the HUB for future use of their bio-samples in medical research according to very specific guidelines. Research ethics approval for each experiment is required from the HUB and from the host institute in Norway. No experimentation can begin without these ethics approvals that assess our intended use of the samples across all of our collaborating sites in Europe and China. Since the HUB biobank is located in Europe, many of the intricacies with regard to ethics and legislation across our European collaborating sites are already in line. However, human DNA can be imported, but not exported from China. This means that we can ship our organoid samples into China in preparation for our spaceflight launch, but after our spaceflight experiment has been conducted and safely returned to Earth, those same samples cannot be exported to our genomics laboratory in Norway. Consequently, laboratory analysis of our spaceflight samples must be conducted in China, after which time the organoid samples will need to be destroyed and the data will need to be transferred to Norway for bioinformatic and statistical analysis.

5.2 Laboratory, bioinformatics, and statistical analysis pipelines: step 2

In this step, we go from human organoids in a lab that have been exposed to radiation (ground and space) and/or (micro)gravity (ground and space) to laboratory methods at the Genomics Core Facility (GCF) housed in Norway. We then treat the "big data" with bioinformatics tools to break down the noise and scale down the data for statistical analysis. It is important to note that the details in this step are given at the time of writing in March 2020. Genomics as a field is constantly advancing, sometimes exponentially, and so are the tools used to work with the samples and analyze the data. That being said, our *Tumors in Space* experimental methods are still being finalized, but this at least gives you a glimpse into our preparation for spaceflight.

Organoid sample selection: To account for tissue heterogeneity and differences in mutational signatures, we will use paired samples from three different patients. Assuming 7–10 subclones from each starting culture, we would have 21–30 paired samples which will give sufficient statistical power to detect consistent differences in mutational signatures. Subcloning is an essential step for this experimental strategy, as subcloning ensures that mutations occurring prior to subcloning will be fixed in subsequent culture, allowing these mutations to be detected by whole genome sequencing and somatic genotyping.

Organoid cell culture and subcloning: Laboratory procedures will be handled by the Genomics Core Facility at the Norwegian University of Science and Technology (NTNU) in accordance with previously published methods [18]. The CGF provides state-of-the-art high-throughput genomics technology including sequencing and genotyping microarray analysis. The major advantages of using the CGF include highly experienced staff and the use of robotics. Tumor organoids must be cultured in specifically designed medium [18]. After a fixed exposure time, subclones from each source organoid will be cultured for an additional fixed number of cell divisions to allow for the occurrence of random mutations before whole genome sequencing and somatic genotyping [19].

Whole genome sequencing and somatic genotyping: Classic sequencing-based whole genome analyses of mutational signatures (healthy vs tumor tissue) rely on mutations being clonally expanded due to tumor growth, i.e., the mutation occurred in an ancestor cell so that all daughter cells will have this mutation, and the mutation is detectable if a sufficiently large percentage of the tumor cells share this mutational ancestry. Clonal expansion ensures that within an organoid sample, there will be several DNA fragments and sequencing reads that share the same mutation. In this manner, mutations can be distinguished from sequencing errors. This also means that the mutations occurring within the tumor organoid source cells will not be detectable, as these mutations have not yet been amplified by cell division (unless the same mutation occurs at the exact same nucleotide within a sufficiently large subset of the current cells). The sequencing depth determines the percentage of cells that need to share mutations in order to be detected. For example, 100x sequencing depth can detect mutations if the mutations are present in 5% of cells, whereas 1000× sequencing depth can detect mutations present in only 0.5% of cells.

Although polymerase chain reaction (PCR[4]) is a widely used approach, we cannot use this approach for the current study. Even if we increase the sequencing depth to 5x the number of cells, we would get PCR-amplified copies of the DNA in a single organoid sample and would not be able to distinguish PCR errors from mutations. However, with rolling circle amplification, we would always use the same original (circularized) DNA fragment as a template for amplification. In this way, we will have a long DNA fragment with multiple copies of the original DNA fragment, and by sequencing these copies in a single read, we will detect mutations as consistent changes within each copy.

Bioinformatics and statistical analysis pipeline: All somatic changes in whole genome data will be analyzed with mutation calling pipelines developed in Norway. Healthy tissue more than 5 cm from the tumor and consisting of epithelial and connective tissue [18] will be used as the "germline" reference. Mutations considered to be germline will be removed. Mutational signature extraction using non-negative matrix factorization will be performed [20]. Signature attribution in reference to the Catalogue of Somatic Mutations in Cancer [10] and the Pan-Cancer Analysis of Whole Genome [21] will be conducted.

5.3 Spaceflight hardware readiness: step 3

In this step, we are designing, developing, testing, and validating an automated cell culture experimental unit and six-paneled tantalum cover for cosmic radiation protection. Several characteristics make tantalum a good choice for radiation protection, all of which will be presented below. We will use our ground-based platforms, a sounding rocket, and aircraft parabolic flights to test the impact of environmental stressors and operational constraints (e.g., vibration, changes in gravity, changes in temperature, experimental lag time) on the fidelity of our organoid culture technique and the readiness of our spaceflight hardware. Our spaceflight hardware must fit according to requirements and constraints of the biological research module on the China Space Station. Other experimental equipment that we need for our experiment will be hosted by the China Space Station including a temperature-controlled incubator/cooler and a 1 g centrifuge or variable gravity rack. We also need dosimeters to measure radiation levels for the duration of our experiment.

Experimental unit: We are designing and developing a biocompatible-automated organoid culture experimental unit. An essential first step is to complete biocompatibility testing for each material required for the construction of the unit. Direct contact between organoid culture medium and experimental unit materials must not attenuate organoid growth. More specifically, for our experimental unit to be deemed biocompatible, 75% of all organoid samples must survive. Our unit is required to be operational for the full duration of the 31-day experiment, including automated nutrient solution exchange on days 6, 12, 18, and 24. Our unit must adhere to a standard temperature profile (18–38°C for organoid growth, 4°C for fixed samples) and CO_2 saturation level (5%). Finally, our unit must be capable of enduring physical stressors (vibration, g-forces) while maintaining organoid culture and survival.

We cannot procure market-ready experimental units because all previous cancer experiments in space have used 2D monolayer cancer cell lines and *Tumors in Space* uses 3D human organoids. However, units used for previous cancer experiments in

---

[4] PCR is a widely used molecular biology method that makes millions to billions of copies from a DNA sample, allowing amplification of a very small amount of DNA to a larger amount of DNA for a detailed study.

space do provide us with a good foundation for our own design and development. For example, Grimm's lab first tested their own cancer cell line experimental unit during the Shenzhou-8 mission [22], which was then further improved for the SpaceX CRS-8 ISS mission [23]. Grimm's unit sets the standard for safe cell culture and cell nourishment followed by fixation according to a pre-programmed timeframe. From this foundation, we are designing our experimental unit with built-in pre-programmed electronics to control temperature gradients for organoid cell culture medium, as well as automated oxygen, carbon dioxide, and nutrition cycles to keep our organoids alive for the duration of the 31-day space mission.

A small temperature-controlled incubator/cooler with a rack insert: Our experimental units will be fully automated removable inserts designed to interface with a small temperature-controlled incubator/cooler that serves as a miniature laboratory for self-contained, automatic microgravity experiments on cells/organoids—similar to ESA's Kubik[5] on the International Space Station (**Figure 6**) [24]. This equipment will be hosted by the China Space Station. The rack insert of the incubator/cooler will provide passive structure to house our experimental unit. The incubator/cooler will operate between 4°C and 38°C and will permit a 1 g centrifuge insert to allow for simultaneous experiments with 1 g control samples and microgravity samples.

Onboard 1 g centrifuge or variable gravity rack: To thoroughly test our hypotheses and delineate compound spaceflight exposures (microgravity and cosmic radiation), we also require an onboard 1 g centrifuge or variable gravity rack that will allow us to expose our organoids to only cosmic radiation while maintaining the standard gravitational force felt on Earth (1 g). Our organoid medium can stay at 37°C, while in the centrifuge, our experimental unit will fit the standard payload boxes hosted by the China Space Station. A variable gravity rack which contains a centrifuge with rotating containers for biological and fluid experiments [25] may be larger than the standard 1 g centrifuge available within the incubator/cooler insert rack. Either the smaller 1 g centrifuge or the larger variable gravity rack will be used for our experiment, and this decision will be taken in close collaboration with the China Manned Space Agency during our hardware development phase.

Tantalum cover for experimental unit: In order to test whether exposure to microgravity slows or stops the growth of cancer, we need to protect our organoids from cosmic radiation exposure. To do so, we are designing a six-sided radiation shield to cover all sides of our organoid cultivation chamber in order to shield the organoids from cosmic radiation. The size, shape, and mass of the tantalum cover

Figure 6.
Kubik on the ISS (credit: ESA).

---

[5] Kubik (from Russian for cube) is a miniaturized laboratory in a 40 cm cube, installed in the ESA orbital laboratory Columbus module on board the ISS, and is used for several kinds of biology experiments in space.

are being designed according to the size and shape of our experimental unit and the mass constraints of the payload. Several characteristics make tantalum a good choice for radiation shielding of biological experiments in space including very good biocompatibility, excellent corrosion resistance, and very high melting point. Tantalum can also be easily welded into the required shape and size.

Dosimeters: We also need electronic dosimeters to provide passive reading of radiation dosage for the duration of the 31-day experiment. The dosimeters must be sensitive enough to measure high-energy cosmic radiation, lightweight, and compact yet durable enough to be operational in the spaceflight environment [26]. The dosimeters must provide tissue equivalent readings. This means that the electronic data must provide the same information about cosmic radiation dose as we would expect if human tissue, rather than the dosimeter, was directly exposed.

We will work in close collaboration with the China Manned Space Agency to develop, assemble, and test spaceflight hardware according to CMSA standards and quality controls with particular attention to interface compatibility, mass, size, power requirements, and duration of experiment. We will also test our spaceflight hardware during our ground, sounding rocket, and parabolic flight experimental arms.

In preparation for spaceflight, we are running several experiments on research platforms including ground-based space analogues that will produce effects on our organoid models similar to those experienced in space. This includes radiation facilities for ionizing and heavy ion irradiation and the random positioning machine (RPM) to simulate microgravity. It is important to remember that the spaceflight environment can be simulated but never replicated on the ground. We are also preparing for experiments on a sounding rocket and during aircraft parabolic flights. These preparatory experiments are essential to ensure a successful experiment in space. Not only will our organoid cultures and hardware development be tested, but we will also collect important baseline data for later comparison to our spaceflight data. In addition, we will have the opportunity to standardize our shipping, handling, and storage procedures for the organoids, as well as define our laboratory, bioinformatics, and statistical analysis pipelines. Finally, our ground radiation work does not only provide us with a mutational signature of different types of radiation for later comparison with cosmic radiation; it is designed to better understand the side effects of radiation therapy for cancer patients on Earth.

5.4 Ground experiments: step 4

We are using ground-based facilities at the Belgian Nuclear Research Center and ESA's European Space Research and Technology Center to test our hypotheses and theoretical approach and to refine our methodology and operational procedures in preparation for spaceflight. We will collect baseline data on the mutational signature of ionizing and heavy ion radiation and use the random position machine (RPM) to test whether simulated microgravity stops or slows the growth of cancer.

Ionizing radiation: To identify a unique mutational signature of cosmic radiation, we must verify the mutational signature of ionizing and heavy ion radiation for comparison. Using the gamma beam irradiator at the Belgian Nuclear Research Center, we will irradiate the following:

i. Our 1 g control samples

ii. Our samples exposed to simulated microgravity in the RPM

iii. Our experimental unit and tantalum cover

Our 1 g control samples can be placed in direct line of the gamma ray to capture the mutational signature of ionizing radiation. By varying the dose and time of the beam, we will simulate radiation therapy given to cancer patients on Earth and examine whether healthy organoids from cancer patients turn cancerous after dose- and time-dependent exposure to radiation. Next, our organoids will be contained in the RPM, and the RPM will then be placed in front of the irradiator. This will allow us to simultaneously expose our organoids to radiation and simulated microgravity. Finally, we will test the fidelity of our spaceflight hardware by exposing our organoids housed within the experimental unit and protected by the tantalum shielding to a predetermined dose rate of ionizing radiation.

The random position machine: We are using the RPM at ESA-ESTEC to collect baseline data on the effect of simulated microgravity on the growth of cancer organoids. The RPM (**Figure 7**) uses two rotating axes, each with independent motor drives running at random speeds and generating random three-dimensional movements [27, 28] that change the direction of the gravitational vector felt by the organoids such that cumulative gravitational effects are cancelled over time. The maximum angular speed of the inner and outer frame of the RPM generally ranges from 20°/s to 120°/s. Thus, the geometry and size of the container within which the biological system is placed must be carefully considered [29]. The sample container will then be mounted to the inner axis. The RPM can run for the 31-day duration of the experiment, and cultural medium can be replaced when needed.

5.5 Sounding rocket: step 5

We will use a sounding rocket to test the impact of environmental and operational factors (e.g., vibration, changes in gravity, changes in temperature, experimental lag time) on the fidelity of our models and readiness of our hardware. Sounding rockets carry experiments up to 750 km above the Earth's surface and offer up to 13 min of microgravity. After launch, the sounding rocket motor is shut down, and our experiment will be in free fall. Parachutes are deployed on the downward arc to return the rocket and experiments safely to the ground. The rocket and experimental unit will reach a peak gravitational force of 12 g which could impact our organoids, as can launch vibrations and other environmental factors. This will be our first microgravity experiment. Not only will we collect valuable data on the underlying gene expression of our organoids after exposure to microgravity, we will also have the opportunity to test our hardware as part of a rocket launch. At the end of this step, we will use the data to improve our logistics and operations, our organoid models, and the fidelity of our spaceflight hardware.

Figure 7.
The RPM at ESA-ESTEC (credit: ESA).

5.6 Parabolic flights: step 6

Similar to the sounding rocket, we will use aircraft parabolic flights to test the impact of environmental and operational factors on our experiment. The parabolic flight takes place on a specially designed commercial aircraft [30] that flies through a series of parabolas and gives 20–22 s of microgravity on each parabola. The aircraft flies up and down at 45° angles, and at the top of the curve, 20–22 s of microgravity is experienced. About 2 g is experienced during ascent and descent. A typical parabolic flight campaign involves 30 parabolas per flight and 3 flights over a 1-week period. This will be our final step before spaceflight and our last opportunity to make any and all necessary improvements to our experiment.

5.7 Spaceflight experiment: step 7

According to standard operating procedures and in close collaboration with the China Manned Space Agency, we will finalize our preparation for spaceflight in the following manner. Our experimental units will be assembled prior to launch, including several backup units. All hardware will be sterilized and approved for launch. Our samples will be transported to the launch site on dry ice from our biobank in the Netherlands. Our organoids will arrive a minimum 7 days before the launch. After the initial phase of growth, our organoids will be placed into the validated and approved units, and the units will be inserted into the incubator/cooler. The incubator/cooler housing our units will be loaded into the rocket prior to launch (-1 hour to -2 days). Once the payload is ready for flight, no additional handling of the experiment will be required before launch.

The experiment will start and end under constant environmental conditions (e.g., temperature, pressure, humidity) after our experimental units have been installed in the incubator/cooler on the China Space Station. The duration of our experiment is 31 days for organoids exposed to the natural spaceflight environment (microgravity and cosmic radiation), 31 days for organoids exposed to microgravity under the tantalum cover designed to protect our samples from cosmic radiation, and minimum 10 days for organoids exposed to cosmic radiation and Earth gravity with the use of the 1 g centrifuge. Dosimeters will be used to read the radiation level for the duration of the experiment.

The cell nourishment and fixation will be automated and pre-programmed according to experimental needs. Subsequent cold stowage of the fixed samples will be required at the end of the experiment. After completion of the mission and successful environmentally controlled return to the ground, our units will be transported to our laboratories for organoid subcloning, sequencing, and genotyping as described above.

6. Impact

Tumors in Space is a cancer research project at the intersection of stem-cell biology and space technology that will acquire new knowledge on cancer etiology due to the influence of the spaceflight environment. By conducting cutting-edge laboratory-based research in orbit and testing our two novel hypotheses, we are challenging the current scientific paradigm. We aim to spark curiosity and inspire the public by conducting excellent science at the forefront of international research in collaboration with the United Nations Office for Outer Space Affairs (UNOOSA) and the China Manned Space Agency.

Figure 8.
The 17 Sustainable Development Goals (credit: United Nations).

The Sustainable Development Goals: The UNOOSA works to promote the peaceful use of outer space for the benefit of humankind. This includes a collaboration with the CMSA to fly international experiments on board the China Space Station. The 17 Sustainable Development Goals (**Figure 8**) were adopted by the United Nations in 2015 [31] as a global call for action to improve environmental conditions on Earth and socioeconomic as well as health conditions for all humanity. *Tumors in Space* has a role to play in the attainment of the Sustainable Development Goals by 2030 through the peaceful and collaborative use of outer space to ensure healthy lives and promote well-being for all people of all ages.

Science communication in the public domain: Science funded by the people is meant to be delivered to the people, and this is one of the goals of *Tumors in Space*. We have designed a comprehensive science communication and research dissemination plan. The main feature of our plan is our project website that will include open access to our research results and scientific publications, research highlights, a color photo library, team bio-sketches, blog posts, and social media feeds (e.g., Twitter), as well as an up-to-date list of all interviews and feature articles for the *Tumors in Space* project. Our efforts also include educational outreach focused on girls and marginalized youth.

7. Conclusions

This chapter provides an overview of the *Tumors in Space* experiment beginning with the scientific evidence upon which the hypotheses are based. The seven main steps in our spaceflight preparation have been presented including ethics and regulatory affairs, laboratory methods, spaceflight hardware design, ground-based experiments (radiation and simulated microgravity), as well as sounding rocket and aircraft parabolic flight experiments. This takes us to the final preparations for our spaceflight experiment set to be launched to the China Space Station in 2025. Finally, some brief notes on the importance of the Sustainable Development Goals and communication of science in the public domain have been provided. To end this chapter, we provide a note of encouragement. Do share this chapter widely in the spirit of open access to scientific knowledge. We welcome you to follow our journey into orbit as we reach for our ultimate goal: to make novel contributions to cancer risk prediction, diagnostics, and therapeutics for human health during short- and long-duration space missions and public health on Earth.

Acknowledgements

Tumors in Space is represented by an exceptional research team with all necessary expertise and competence to ensure timely progress and ultimate success of the project: Norway, Tricia L. Larose (principal investigator and project coordinator), Carina Helle Berg, Ann-Iren Kittang Jost, Arve Jørgensen, Berge Solberg, Pål Sætrom; China, Mengyun Chen, Yang Yang; Belgium, Sarah Baatout, Bjorn Baselet, Vladimir Pletser, Roel Quintens; France, Ana Diaz Artiles, Ghislaine Scelo, Sergey Senkin, Chris Welch; and the Netherlands, Annelien Bredenoord, Hans Clevers, Jack van Loon.

Some text in this chapter subsection on laboratory methods was previously written by *Tumors in Space* collaborator, Pål Sætrom, from the Norwegian University of Science and Technology, and is reused here with his permission.

Tumors in Space is under a programme of and funded by the European Space Agency with the support of the Norwegian Space Agency. The view expressed herein can in no way be taken to reflect the official opinion of the European Space Agency or the Norwegian Space Agency.

Thank you to my editor, Vladimir Pletser, and to the publisher, IntechOpen, for making this an open access publication.

Author note

"As a final outcome of the application and selection process in response to the first cycle of Announcement of Opportunity under the United Nations/China Cooperation on the Utilization of the China Space Station (CSS) initiative, being implemented by the Office for Outer Space Affairs (OOSA) and the China Manned Space Agency (CMSA) respectively, your proposal entitled "Tumors in Space: Signatures of early mutational events due to spaceflight conditions on 3D organoid cultures derived from intra-individual healthy and tumor tissue", has been fully accepted for implementation on board the CSS."

As announced on 12 June 2019 in Vienna, Austria, during the 62† Session of the Committee on the Peaceful Uses Of Outer Space and signed on 9 July 2019 by Director of United Nations Office for Outer Space Affairs (UNOOSA), Ms. Simonetta Di Pippo.

Author details

Tricia L. Larose[1,2]

1 Faculty of Medicine, Department of Community Medicine and Global Health, Institute of Health and Society, Oslo, Norway

2 Human Performance and Space Department, International Space University, Strasbourg, France

*Address all correspondence to: tricia.larose@flymed.uio.no

References

[1] Stewart BW, Wild CP, editors. World Cancer Report: Cancer Research for Cancer Prevention. Lyon, France: International Agency for Research on Cancer; 2014

[2] Wild CP, Weiderpass E, Stewart BW, editors. World Cancer Report: Cancer Research for Cancer Prevention. Lyon, France: International Agency for Research on Cancer; 2020

[3] Blackadar CB. Historical review of the causes of cancer. World Journal of Clinical Oncology. 2016;**7**(1):54-86. DOI: 10.5306/wjco.v7.i1.54

[4] Becker JL, Souza GR. Using space-based investigations to inform cancer research on Earth. Nature Reviews Cancer. 2013;**13**(5):315-327. DOI: 10.1038/nrc3507

[5] Kanki B, Clervoy JF, Sandal G, editors. Space Safety and Human Performance. 1st ed. Oxford, United Kingdom: Butterworth-Heinemann; 2018

[6] European Space Agency. The Radiation Showstopper for Mars Exploration [Internet]. The Netherlands. Available from: https://www.esa.int/Science_Exploration/Human_and_Robotic_Exploration/The_radiation_showstopper_for_Mars_exploration

[7] Greenman C, Stephens P, Smith R, Dalgliesh GL, Hunter C, Bignell G, et al. Patterns of somatic mutation in human cancer genomes. Nature. 2007; **446**(7132):153-158. DOI: 10.1038/nature05610

[8] Behjati S, Gundem G, Wedge DC, Roberts ND, Tarpey PS, Cooke SL, et al. Mutational signatures of ionizing radiation in second malignancies. Nature Communications. 2016;**7**:12605. DOI: 10.1038/ncomms12605

[9] Olivier M, Hollstein M, Hainaut P. TP53 mutations in human cancers: origins, consequences, and clinical use. Cold Spring Harbor Perspectives in Biology. 2010;**2**(1):a001008. DOI: 10.1101/cshperspect.a001008

[10] Wellcome Sanger Institute. COSMIC: Catalogue of Somatic Mutations in Cancer [Internet]. United Kingdom; 2020. Available from: https://cancer.sanger.ac.uk/cosmic

[11] Gridley DS, Slater JM, Luo-Owen X, Rizvi A, Chapes SK, Stodieck LS, et al. Spaceflight effects on T lymphocyte distribution, function and gene expression. Journal of Applied Physiology. 2009;**106**(1):194-202. DOI: 10/1152/japplphysiol.91126.2008

[12] Lewis ML. The cytoskeleton in spaceflown cells: an overview. Gravitational and Space Biology Bulletin. 2004;**17**:1-11

[13] Zhang ZJ, Tong YQ, Wang JJ, Yang C, Zhou GH, Li GH, et al. Spaceflight alters the gene expression profile of cervical cancer cells. Chinese Journal of Cancer. 2011;**30**(12):842-852. DOI: 10.5732/cjc.011.10174

[14] Lewis ML, Cubano LA, Zhao B, Dinh HK, Pabalan JG, Piepmeier EH, et al. cDNA microarray reveals altered cytoskeletal gene expression in space-flown leukemic T lymphocytes (Jurkat). FASEB Journal. 2001;**15**(10):1783-1785. DOI: 10.1096/fj.00-0820fje

[15] Fatehullah A, Tan SH, Barker N. Organoids as an in vitro model of human development and disease. Nature Cell Biology. 2016;**18**(3): 246-254. DOI: 10.1038/ncb3312

[16] Katt ME, Placone AL, Wong AD, Xu ZS, Searson PC. In vitro tumor models: Advantages, disadvantages, variables, and selecting the right

platform. Frontiers in Bioengineering and Biotechnology. 2016;**4**:12. DOI: 10.3389/fbioe.2016.00012

[17] Method of the Year 2017: Organoids. Nature Methods. 2018;**15**(1). DOI: 10.1038/nmeth.4575

[18] Roerink SF, Sasaki N, Lee-Six H, Young MD, Alexandrov LB, Behjati S, et al. Intra-tumour diversification in colorectal cancer at the single-cell level. Nature. 2018;**556**(7702):457-462. DOI: 10.1038/s41586-018-0024-3

[19] Cibulskis K, Lawrence MS, Carter SL, Sivachenko A, Jaffe D, Sougnez C, et al. Sensitive detection of somatic point mutations in impure and heterogeneous cancer samples. Nature Biotechnology. 2013;**32**(3):213-219. DOI: 10.1038/nbt.2514

[20] Qi Q, Zhao Y, Li M, Simon R. Non-negative matrix factorization of gene expression profiles: A plug-in for BRB-Array tools. Bioinformatics. 2009;**25**(4): 545-547. DOI: 10.1093/bioinformatics/btp009

[21] International Cancer Genome Consortium. The PanCancer Analysis of Whole Genome. [Internet]. 2019. Available from: https://omictools.com/pcawg-tool

[22] Pietsch J, Ma X, Wehland M, Aleshcheva G, Schwarzwalder A, Segerer J, et al. Spheroid formation of human thyroid cancer cells in an automated culturing system during the Shenzhou-8 Space mission. Biomaterials. 2013;**34**(31):7694-7705. DOI: 10.1016/j.biomaterials.2013.06.054

[23] Pietsch J, Gass S, Nebuloni S, Echegoyen D, Riwaldt S, Baake C, et al. Three-dimensional growth of human endothelial cells in an automated cell culture experiment container during the SpaceX CRS-8 ISS space mission—The SPHEROIDS project. Biomaterials. 2017; **124**:126-156. DOI: 10.1016/j.biomaterials.2017.02.005

[24] European Space Agency. Kubik on the Space Station [Internet]. The Netherlands. 2020. Available from: https://www.esa.int/ESA_Multimedia/Images/2018/02/Kubik_on_Space_Station

[25] Wang SK, Wang K, Zhou YL, Yan B, Li X, Zhang Y, et al. The development of the varying gravity rack (VGR) for the Chinese Space Station. Microgravity Science and Technology. 2018;**31**: 95-107. DOI: 10.1007/s12217-018-9670-1

[26] Benton E. Space radiation passive dosimetry. In: The Health Risks of Extraterrestrial Environments. United States of America: NASA; 2012

[27] Van Loon JJ. Some history and use of the random positioning machine, RPM, in gravity related research. Advances in Space Research. 2007;**39**(7):1161-1165. DOI: 10.1016/j.asr.2007.02.016

[28] Herranz R, Anken R, Boonstra J, Braun M, Christianen PC, de Geest M, et al. Ground-based facilities for simulation of microgravity: organism-specific recommendations for their use, and recommended terminology. Astrobiology. 2013;**13**(1):1-17. DOI: 10.1089/ast.2012.0876

[29] Leguy CA, Delfos R, Pourquie MJ, Poelma C, Krooneman J, Westerweel J, et al. Fluid motion for microgravity simulations in a random positioning machine. Gravitational and Space Biology Bulletin. 2011;**25**(1):36-39

[30] Pletser V, Rouquette S, Friedrich U, et al. The first European parabolic flight campaigns with the Airbus A310 ZERO-G. Microgravity Science and Technology. 2016;**28**(6):587-601. DOI: 10.1007/s12217-016-9515-8

[31] United Nations. Sustainable Development Goals. 2020. Available from: https://www.un.org/sustainabledevelopment/sustainable-development-goals/

4

Research in Microgravity in Physical and Life Sciences: An Introduction to Means and Methods

Vladimir Pletser and Thais Russomano

Abstract

Microgravity is the state encountered in a vehicle in free fall, whether on Earth, in low Earth orbit or in deep space. Microgravity research has opened up new possibilities for investigations in physical and life sciences that are necessary to enhance our knowledge of how physical systems and the human body, from cells to whole body systems, perform, react and function in microgravity, which will help prepare for the human exploration of outer space. This chapter presents an introduction of the microgravity environment, which can be obtained using different microgravity platforms, including space missions. Simulation of microgravity effects is also used for research and presented here. The chapter further considers the effects of microgravity on several aspects related to the physical sciences and on the adaptation of the human body to this altered gravity environment.

Keywords: microgravity research, microgravity platforms, microgravity simulation, microgravity effects in physical sciences, human physiology

1. Space orbital environment

Space orbital environment is characterized by several factors that affect experiments in physical sciences and influence the good functioning of all living systems, from cells to humans. The main factors are weightlessness, high-energy radiations, vacuum and temperature differences. These last two factors are generally mitigated by the vehicle yielding the necessary life support to the systems under study. The first two factors on the contrary cannot be completely compensated.

The concept of weightlessness will be developed further.

Perfect protection against high-energy radiations cannot be completely achieved, unless thick shielding walls are installed all around the spacecraft, which is presently excluded in view of launch costs per kg. Nevertheless, a vehicle in low Earth orbit (a few hundred kilometers altitude) stays relatively protected by Earth's Van Allen radiation belts (inner energetic proton belt at 1,000–6,000 km altitude and outer energetic electron belt at 13,000–60,000 km altitude).

To these orbital factors, one should add the conditions at launch and during atmospheric reentry and landing of a spacecraft, i.e. important accelerations and vibrations, that can affect the quality of physiological samples or configurations obtained in microgravity (e.g. for crystals).

2. Weightlessness and microgravity

The state of microgravity, or more correctly micro-weightiness, exists in an orbital vehicle in a state of free fall, i.e. without any force acting on it except for gravitational forces [1]. This means that the vehicle must not be propelled or submitted to any other nongravitational force. Perfect weightlessness is an ideal state practically impossible to achieve. However, microgravity of an excellent quality (typically $10^{-5}g$,

Physical sciences	**Life sciences**
Fundamental physics	Human research
Complex plasmas and dust particle physics	Integrated physiology
Aerosol particle motion	Cardiovascular function
Frictional interaction of dust and gas	Respiratory function
Plasma physics	Body fluid shift
Aggregation phenomena	Central venous pressure system
Materials science	Digestive system
Thermophysical properties	Muscle and bone physiology
Thermophysical properties of melts	Skeletal system
New materials, products and processes	Blood lactate studies
Morphological stability and microstructures	Body mass tests
Physical chemistry	Human locomotion
Aggregation phenomena	Posture
Granular matter	Bone models
Fluid and combustion physics	Neuroscience
Structure and dynamics of multiphase systems	Neurobiology
Pool boiling	Vestibular functions
Heat and mass transfer	Spatial orientation
Dynamics of drops and bubbles	Motion sickness
Thermophysical properties	Motor skills
Interfacial phenomena	Biology
Dynamics and stability of fluids	Plant physiology
Evaporation	Statolith movement
Complex dynamic systems	Gravitropism
Diffusion	Gravireceptors
Foams	Cell and developmental biology
Chemo-hydrodynamic pattern formation	Animal physiology
Combustion	Aging processes
Droplet and spray combustion	Electrophysiological and morphological properties of human cells
Soot concentration	Osteoblast cells
Combustion synthesis	Technology
Laminar diffusion flames	ISS experiment validation
Fuel droplet evaporation	Phase separation technologies for biological fluids
Ignition behaviour	Crew foot restraint
Technology	Crew exercise devices
ISS experiment validation	Urine monitoring system
Metal halide lamps	
Micro-acceleration measurement	

Table 1.
Non-exhaustive list of research fields in microgravity.

where g is the acceleration of weightiness, commonly and erroneously mistaken for gravity[1], with an average value of 9.81 m/s^2) can be achieved in orbit.

Gravity (weightiness) disturbs certain experiments and reduces the field of investigation of some scientific domains. Gravity (weightiness) effects hide other effects pertaining to materials or fluids under study, and that depends often on intrinsic properties of matter or of its state. Convection in fluids, so evident that it is called "natural," is caused by gravity (weightiness) acting on local differences of density caused by differences of temperature or concentration. The resulting Archimedes or buoyancy force induces an ascending motion of fluid zones of lesser density and a descending motion of fluid zones of larger density, creating convection cells in gases, liquids and solids in fusion, yielding disruptive phenomena in separation processes.

Although physical and biological processes are often investigated in hypergravity, e.g. in centrifuge, one knows less what happens in reduced gravity. However, in most cases, one cannot extrapolate from results obtained in hypergravity to microgravity, most of the phenomena being nonlinear in function of the gravity level. One observes many more differences while passing from 1 g to 0 g than between 5 g and 4 g, for example.

Many scientific fields profit from the peculiarities of weightlessness to enlarge their field of investigations. Material sciences, fluid physics and life sciences (biology and physiology) were the first to use microgravity, followed later by many other disciplines (combustion physico-chemistry, crystallography, fundamental physics, critical point phenomena, etc.) in view of varying a new experimental parameter: gravity. Microgravity allows to deepen scientific knowledge in domains that are hardly accessible on Earth.

Table 1 shows some of the scientific fields in which experiments were conducted in microgravity.

3. Physical sciences research in microgravity

Microgravity research allows to study the gravity effects on these different phenomena and the effects of other forces normally masked by gravity on Earth. Weightlessness became an experimental research tool that allows to transpose in microgravity the investigation of phenomena known on Earth but sometimes insufficiently understood, in order to investigate the fundamental processes and to understand their functioning without gravity.

Modifications appear when one studies matter behaviour in weightlessness. One observes on the one hand the disappearance of "natural" phenomena caused by gravity and, on the other hand, the preponderance in microgravity of phenomena that can hardly be observed in normal conditions of gravity. These modifications are particularly important for certain physical, chemical and metallurgical processes having at least one fluid phase: crystal growth, alloy solidification, separation of biological substances, etc.

The main differences that are observed for fluid phases in weightlessness are as follows.

---

[1] The term "gravity" usually designates the force associated with the phenomenon of gravitation. One should therefore in all correctness use here the term "weight", which is the force associated with the phenomenon of weightiness, and the resulting force of gravity and inertia forces existing in the reference frame in which the measure of the weight is done. However, the common (although incorrect) usage of the term "gravity" and its derivatives (microgravity, hypergravity, etc.) instead of "weight" and its derivatives will prevail throughout this introductory chapter.

3.1 Disappearance of separation phenomena

Separation phenomena observed on Earth in multiphase systems that include a fluid phase disappear in microgravity. Sedimentation (precipitation of dissolved or suspended matter) and Archimedean buoyant force (or buoyancy, i.e. the force due to a liquid pressure on a body-immersed volume) disappear. The advantage of the absence of separation in weightlessness is the possibility of obtaining mixtures that are unstable on Earth and material alloys impossible to obtain on Earth or with great difficulty. A disadvantage of the absence of separation in weightlessness is the difficulty of eliminating the gaseous inclusions while, on Earth, degassing is done "naturally" (gaseous zones in liquid matrices go up to the free surface).

3.2 Disappearance of "natural" convection

"Natural" convection disappears in fluids in microgravity. There is no more natural upward displacement of hot zones and downward displacement of cold zones. In fact, there is no up and no down. Other forces become dominant for movements in liquids in microgravity. These forces are linked to superficial or interfacial tension between two liquids. Indeed, such an interface behaves as an elastic "membrane" whose tension is a thermodynamic function of temperature (or concentration for solutions), as shown in **Figure 1**.

For an interface subjected to a temperature difference, superficial tension for most liquids is generally smaller for the hot side than for the cold side. The interface, i.e. the common layer formed by molecules of both fluids, physically moves parallelly to itself from the hot side to the cold side; this membrane deforms itself and slides from the hot side to the cold side. The liquid layers on both sides of the interface are dragged along by viscosity, and a new convection appears,

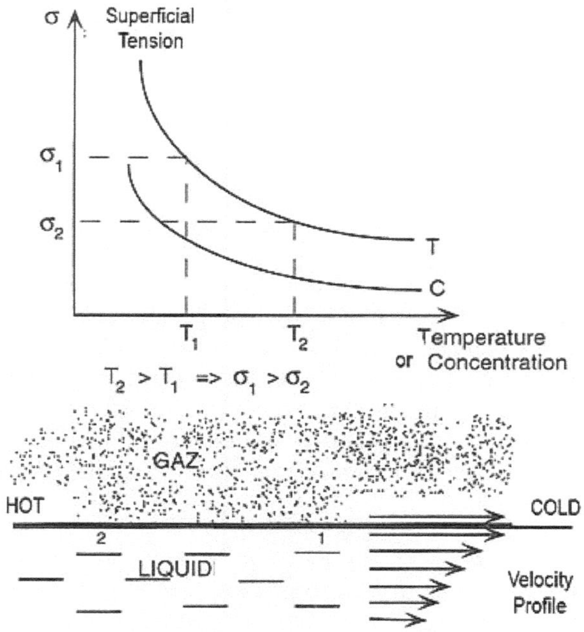

Figure 1.
Liquid/gas interface submitted to a superficial tension gradient, yielding a Marangoni convection cell caused by the physical displacement of the interface membrane from the hot side (point 2) to the cold side (point 1) [1].

called Marangoni convection, after the name of the Italian physicist who studied this phenomenon at the end of the nineteenth century. This phenomenon exists obviously also on Earth, but as its effect is much smaller than those caused by gravity, it is in general negligible and much more difficult to observe. Its study in microgravity allows thus to better understand the fundamental characteristics of liquid behaviour.

It is also because of the absence of "natural" convection that the shape of a combustion flame is different in weightlessness. On Earth, gases produced by the chemical reaction of combustion (e.g. of a candle wick), much hotter, rise, and fresh air oxygen migrate to the combustion centre to feed the combustion process. In microgravity, hot gases have no reason to rise anymore, and the flame is surrounded by a hemispherical ball formed by combustion gases (**Figure 2**), limiting the amount of fresh oxygen transfer.

3.3 Disappearance of hydrostatic pressure

In microgravity, hydrostatic pressure disappears. On Earth, it is responsible for the tendency of fluids to deform under the effect of their own weight, a liquid zone supporting the weight of zones on top. The same phenomenon exists for solids. Structures can be built that would collapse under their own weight on Earth, e.g. crystalline networks (**Figure 3**).

Liquids in weightlessness, abandoned to themselves without any contact with a solid surface, form spherical drops (**Figure 4**), which is the minimal surface enclosing a given volume when subjected to the only forces of superficial tension.

3.4 Possibility of natural levitation

On Earth, crucibles are used to melt alloys, which may contaminate the melt liquid phase. In weightlessness, the liquid phase can be maintained in a contactless levitation, without touching any solid walls, using an electrostatic, magnetic or acoustic confining (**Figure 5**). Many parameters of materials at high temperatures are still unknown and cannot be measured on Earth due to difficulties and limitations caused by crucible contamination and gravity effects.

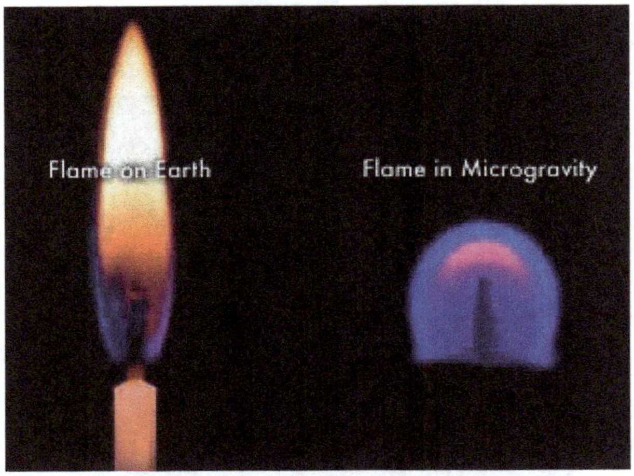

Figure 2.
Flames on ground in 1 g (left) and in microgravity in near 0 g (right). Notice the near-hemispherical shape of the flame in microgravity with the reddish-purple part on top due to some convection caused by small perturbations in the microgravity environment (photo credit: NASA).

Figure 3.
Protein crystals obtained with ESA's Advanced Protein Crystallization Facility during the Life and Microgravity Spacelab mission on NASA Space Shuttle STS-98 in May 1995 (credit: Prof. Martial, University of Liege, Belgium).

Figure 4.
Water drop in free float on ISS (credit: NASA).

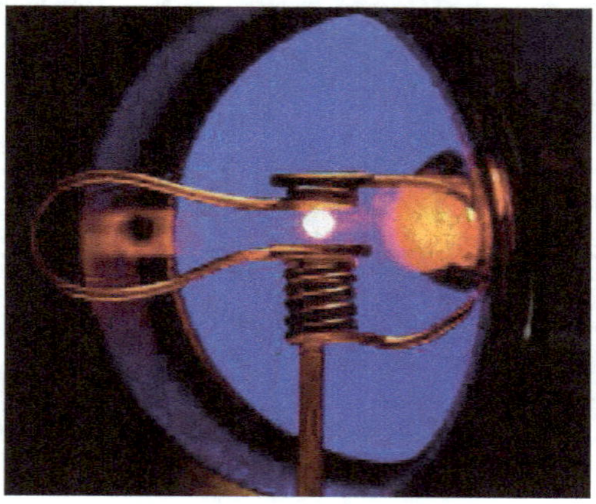

Figure 5.
Core element of an electromagnetic levitator (photo credit: DLR).

The list of the advantages and applications of microgravity to scientific research could be continued at length but is outside of the aim of this publication. The interested reader will find other examples and more details in Refs. [2–5].

4. Physiology research in microgravity

Initially developed in the 1950s and 1960s to support US and USSR space programs, space microgravity medical research quickly evolved. Manned spaceflights very quickly showed physiological changes in astronauts and cosmonauts. The duration of spaceflights has increased throughout the years, from a few hours at the beginning of the 1960s to several months (or even more than a year) today on board the International Space Station (ISS, **Figure 13**). The ISS allows to conduct and to repeat experiments during several years.

New phenomena have been observed on astronauts, some of these effects appearing only after several weeks or months in space. Despite the large number of hours spent in orbit around the Earth by astronauts and cosmonauts from all countries involved in space research and exploration, some problems are still far from being fully understood, and the necessary solutions have not yet been found.

Although physiological systems of human organism function interdependently, one can classify physiological effects of microgravity in four categories:

- Perturbations of sensorial systems related to balance, orientation and the vestibular system
- Modifications of bodily fluid distribution and their impact on the cardiovascular system
- Effects on metabolism and bodily functions
- The adaptive processes of muscular and skeletal systems and their pathological consequences

Relevant knowledge and research on human physiology are presented below, and more details can also be found in Refs. [6–11].

4.1 Balance, orientation and vestibular

On Earth, in a normal gravity environment, the human body has three means to obtain the information of the reference vertical direction and of the top-bottom orientation, characteristic of the gravitational environment on our planet.

The main system is the vestibular system, which is double, located in the inner ear. In one of these organs, small crystals of calcium carbonate called otoliths weigh on a membrane with nervous endings. The semicircular canals form another sensor. Formed by the three canals in planes approximatively orthogonal to each other, a physiological liquid moves by inertia in these canals during a head movement, stimulating nervous endings in the canals. The combination of the information coming from the otoliths and semicircular canals allows the brain to interpret the movement and the position of the head.

The second source of information is the visual system. The visual information allows the brain to recognize the body position with respect to external references (floor, ceiling, walls).

The third information source is the proprioceptive system, constituted of the whole of skin tactile perceptions, articulations and muscle tension. The neck proprioceptive system is the most developed and informs the brain on the position of the head with respect to the rest of the body.

In weightlessness and in absence of accelerated motion, there is no stimulation of the vestibular system. Otoliths are no longer attracted downward by gravity, and the semicircular canals are no longer stimulated. However, the visual and proprioceptive systems continue to function normally. Information sent by these different systems to the brain are incoherent for an organism used to normal gravity and create confusion in the brain zone that normally treats the information on position and orientation. This confusion often yields dizzy spells and nausea and sometime triggers the reflex of emptying the stomach. In short, the subject is sick. This sickness, called space adaptation syndrome, affects most astronauts. On average, one out of two astronauts suffers from nausea during the first few days of spaceflight. After a day or two, the human organism adapts to the new environment, and astronauts can continue to function and work "normally." After the flight, the balance and orientation systems readapt quickly to the Earth's environment.

4.2 Body fluids and cardiovascular system

4.2.1 Loss of body fluids

On Earth, while standing in normal gravity, arterial blood pressure is normally distributed such that, if intracardiac pressure is taken as unity, it is approximately double in feet arteries and two third at head level. While lying down, the distribution of blood pressure is more uniform. Passing from the lying to the standing position yields a blood flow toward the lower part of the body, and blood pressure diminishes in the head. Known as orthostatic postural intolerance, the change of blood pressure is detected by baroreceptors in the vascular system and close to the heart. These receptors send signals that yield, firstly, an increase of cardiac rhythm to compensate the blood volume decrease in head arteries and, secondly, a contraction of arteries in the lower body to diminish the blood flow toward the legs.

In microgravity, gravity does not attract liquids downward anymore, and a redistribution of body fluids takes place. A volume of approximately two liters of body fluids is displaced from the lower extremities to the upper part of the body, increasing the blood volume and pressure in the heart. The volume and blood flow receptors are alerted, and this new situation is interpreted as an overload of the blood system. The reaction of body liquid elimination starts and yields a complex hormonal game, which results in a natural elimination by urine of body liquids. The organism adapts to this new environment, and a new balance is established after 4–5 days.

On the other hand, liquid transfer from lower members toward the upper body has other secondary effects: face swelling due to blood rush in the head, the increase of intraocular pressure, and sinus congestion. These secondary effects disappear up to a certain point after a few days in microgravity. Back on Earth, the organism readapts to a 1 g environment.

4.2.2 Myocardial muscle atrophy and cardiac rhythm

The results of experiments performed with ultrasound echocardiography show a diminution of the left ventricle and auricle volumes during a spaceflight of several weeks. However, after the flight, the cardiac muscle comes back to a normal state.

In microgravity, a decrease of cardiac rhythm and of arterial tension is observed, the heart not needing to pump blood against gravity's downward pull (**Figure 6**).

A high tachycardia (increase of the cardiac rhythm) is observed also at launch, due to psychological stress, but also necessary to compensate the effects of accelerations, in the order of 3–4 *g*, with a maximum of 8 *g*.

4.2.3 Visual impairment and intracranial pressure

Visual impairment and intracranial pressure are another consequence of the upward body fluid shifts, the head filling with blood and other bodily fluids. The various consequences are an increase in intracranial pressure that can cause headache of varying levels of severity and an increase of the intraocular pressure that affects the visual performance and other more minor effects such as congestion of the sinuses. These effects, although observed and investigated for several years, are thought to be temporary as they tend to disappear after return to Earth.

However, intracranial pressure and visual impairment were only recently recognized as more serious as they could impair the performance of astronauts during long-duration 0 g travels in space.

4.3 Physiological functions and metabolism

In microgravity, the main physiological functions are practically unchanged. Astronauts can eat and drink without major constraints. Digestion and intestinal transit are accomplished also nearly normally, except that gravity action is no longer present.

Breathing is also made without too important problems. However, the breathing mechanism is altered: the distribution of inspired and expired gases in the lungs and oxygen exchanges in blood hemoglobin at the level of pulmonary alveoli are modified. The way to breathe is also modified: statistically, in weightlessness, the forced movement of the abdomen contributes more to the breathing mechanism.

Astronauts can also sleep in space. However, daily and sleep rhythms are disturbed. Indeed, on board the ISS in low Earth orbit at 400 km altitude, day and night alternation repeats approximately every 90 min. Astronauts see a sunrise and sunset 16 times per terrestrial 24 h a "day." Psychological and emotional factors and travel excitement intervene also. To remedy it, one imposes a strict and well-established schedule taking into account human natural rhythms. On board the ISS, a three times 8-h schedule

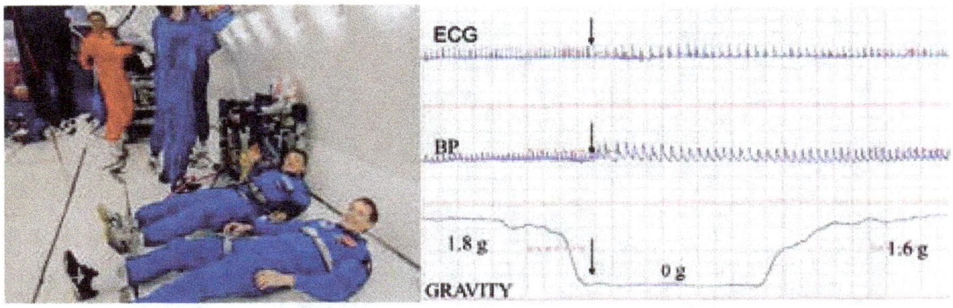

Figure 6.
Experiments during aircraft parabolic flights (left) showed (right) a decrease in heart rate, seen at the beginning of microgravity (arrows), i.e. an increase of duration between successive peaks, corresponding to increased vagal modulation of the heart rate. A sudden increase is also seen in pulse blood pressure (difference between maximum and minimum pressures), indicating an increase in stroke volume (ECG, electrocardiogram; BP, blood pressure) (credit: Left, ESA; right, Prof. A. Aubert, Katholieke Universiteit Leuven, Belgium).

is applied: 8 h for sleep, 8 h for work depending on missions and 8 h for personal time, meals, rests, etc. This schedule is purely theoretical as astronauts on board the ISS spend much more of their time to work, although for long-duration stays on ISS, schedules are loose, and longer rest periods are foreseen some days, generally used by astronauts to watch Earth through windows, mainly the cupola (**Figure 7**).

After long stays in weightlessness, changes are observed in blood composition that can be problematic. Firstly, the number of red blood cells and the hemoglobin level decrease. Secondly, red blood cells of unequal sizes and of abnormal shapes have been also discovered. After 6 months in microgravity in orbit, up to 2% of ovalized red blood cells have been observed in Russian cosmonauts. Thirdly, the immune defense system of astronauts diminishes in microgravity after approximatively 7 days of flight. One observes a reduction of production of lymphocyte T cells (the white blood cells) that intervene in the immune responses and in antibody production. This observation did not find so far a satisfactory fundamental explanation, and this problem could be the one that would impede mankind to adapt to long-duration space travels in microgravity. Astronauts are more prone to infections in space, and they need more time to recover after an infection on ground after their return. The immune system is back to its normal preflight level after a period of 5–10 days after return to Earth.

4.4 Musculoskeletal system

4.4.1 Spine

In microgravity, the first effect that is noticed is the spine extension up to a point that astronauts can gain a few centimeters in height. This is due to the partial decompression of intervertebral discs that do not have to support the weight of the upper body anymore. Back on Earth, after the flight, this effect disappears, and height becomes normal again but with, sometime, the risk of having a nerve blocked between discs and vertebrae. Furthermore, some astronauts complained of back pains during or after a spaceflight, probably caused by this phenomenon of spine extension.

Figure 7.
NASA astronaut Karen Nyberg, Expedition 37 flight engineer in 2013, enjoys the view of earth from the windows in the ESA-built cupola of the International Space Station. A blue and white part of earth is visible through some of the seven windows of the cupola (photo credit: NASA).

4.4.2 Muscular system

The muscular system atrophy is a second consequence, observed after some days in weightlessness. In particular, the most affected muscles are those that control posture and that contribute to support the body weight on Earth. In microgravity, the natural position that astronauts take is a curved position with the legs slightly bent. One floats freely and moves by pushing oneself against a wall, using the action-reaction principle. One notices thus a muscle atrophy, a loss of mass of muscles and the elimination of muscular proteins (**Figure 8**).
By practicing regularly (more than 2 h per day!) and by applying sometime treatments of muscular fiber electrostimulation, astronauts and cosmonauts have no difficulties to readapt upon return to Earth after a more than 6-month mission.

Figure 8.
British ESA astronaut Tim Peake operates the muscle atrophy research and exercise system (MARES) equipment inside the Columbus module. MARES is an ESA facility used for research on musculoskeletal, biomechanical and neuromuscular human physiology to better understand the effects of microgravity on the muscular system (photo credit: NASA/ESA).

4.4.3 Bone demineralization

Bone demineralization, and mainly decalcification, is the most important and serious physiological phenomenon observed in microgravity. Appearing only after 1–2 months in orbit, this could be the second problem that could thwart the hopes of mankind to adapt to space travels in weightlessness.

The loss of calcium is still not completely understood. One knows that decalcification is related to an atrophy of bone fibrous cells containing calcium, corresponding to the part of the bone that allows the marrow to pass. This effect seems to be irreversible once it has started. The rate of calcium loss varies from an astronaut to another and varies also from a type of bone to another. Numerous experiments yield sometime diverging results. On one side, one observes an increase of activity of osteoclastic cells, whose role is to eliminate and resorb elements of bone tissues. On the other side, some results show that bone demineralization would be due to a decrease of activity of osteoblastic cells, responsible for regenerating bone tissues.

This problem of bone decalcification resembles by certain aspects osteoporosis, an illness known on Earth affecting mainly elderly people. This sickness yields a change in the structure (demineralization) of bones, but the composition stays globally the same. The bone loses in thickness, fragilizes and fractures more easily. This shows the importance of conducting research in microgravity on astronauts to better understand this sickness and to contribute in finding a cure for it.

5. The means of microgravity generation

All the means to generate microgravity are based on the principle of free fall; any other method will not result in a real microgravity environment but in a simulated microgravity environment. Microgravity is created in a non-inertial reference frame attached to a vehicle in free fall, in which the resultant of forces other than gravity is null or negligible.

Figure 9 summarizes the different platforms used for microgravity research in an increasing order of microgravity duration.

Figure 9.
Reduced gravity platforms accessible to microgravity researchers (vertical axis, duration of microgravity; horizontal axis, quality of microgravity) (credit: DLR).

Drop tubes and drop towers provide a few seconds (up to 5 s) in the vertical drop mode, where an experimental payload is literally dropped in vacuum or behind a shield to reduce the perturbing effect of air friction.

The level of microgravity obtained in the drop tube of NASA Marshall Centre of 105 m high and 25 cm diameter is in the order of $10^{-6}\,g$ during 4.6 s in a vacuum. In Europe, the ZARM drop tower in Bremen, Germany (**Figure 10**), is 110 m high with a diameter of 3.5 m. Experiment capsules fall during 4.7 s in vacuum, yielding microgravity levels of $10^{-5}\,g$. The microgravity duration can be doubled up to 9.5 s by launching the experiment capsule in a catapult mode from the bottom of the tower upward, falling freely first upward and then downward [12, 13].

Aircraft parabolic flights provide a reduced gravity environment of approximately 20 s, with the major advantage of having human operators and subjects on board. The level of microgravity is typically in 10^{-2} g when attached to the floor structure that can be improved down to 10^{-3} g for a few seconds when left free-floating (**Figure 11**). This important microgravity platform is addressed in the next chapter.

Sounding rocket flights, for which microgravity levels are in the order of 10^{-4}–$10^{-5}\,g$, are used for automated or remotely operated experiments with relatively reduced volumes. Depending on the size of the rocket and the engine used, the duration of microgravity during the ballistic phase of the flights varies between 3 and 14 min [14].

In the near future, suborbital flights will provide microgravity duration in the order of 3–4 min for paying customers but also for microgravity experiments. There are typically two US companies that are working on suborbital vehicles (**Figure 12**): Blue Origin with the New Shephard capsule and a reusable rocket and Virgin Galactic and the SpaceShipTwo spaceplane carried by the WhiteKnightTwo airplane carrier. These two systems would carry passengers and experiments up to an altitude of 100 km or more in a propelled mode and continue in a ballistic mode for approximately 3–4 min after propulsion has stopped.

Manned orbital platforms provide microgravity periods of several years for the International Space Station (ISS, **Figure 13**) [15–17], and the future Chinese Space Station is foreseen to be assembled in orbit in 2022 (**Figure 14**). Residual accelerations are in the order of 10^{-2}–$10^{-4}\,g$, depending on internal perturbations (e.g. crew movements) and external ones.

Figure 10.
The ZARM drop tower in Bremen, Germany. The 146 m high building protects the free fall facility from atmospheric perturbation and wind (photo credit: ZARM).

Figure 11.
During a parabolic flight on board the Airbus A300 ZERO-G during an ESA campaign, several experimental racks are visible to the left and the back, while one of the authors floats freely "upside down." There is no "up" and "down" in weightlessness (photo credit: ESA).

Figure 12.
Two suborbital facilities in development: (left) the New Shephard capsule with a reusable rocket (credit: Blue origin) and (right) the WhiteKnightTwo airplane carrying the SpaceShipTwo spaceplane (photo credit: Virgin galactic).

Figure 13.
The International Space Station (ISS) is the first major international project that includes 14 countries in its realization: The USA, Russia, Canada, Japan and 10 European countries (France, Germany, Italy, Belgium, the Netherlands, Spain, Sweden, Switzerland, Denmark and Norway). With a total mass of 440 tons (but weighing 0 kg ...), the ISS is in low earth orbit between 400 and 450 km altitude at 51.6° inclination. Since November 2000, the station is inhabited by permanent international crews (photo: NASA).

Figure 14.
The Chinese Space Station foreseen to be launched and assembled in the coming years with an assembly completed for 2022. From left, the Tianzhou *(meaning "heavenly vessel" in Chinese Mandarin) cargo freighter docked to the* Tianhe *("harmony of the heavens") core module in the Centre; a piloted* Shenzhou *("divine vessel") vehicle is connected to a node in front of* Tianhe, *which are connected to two scientific modules:* Wentian *("quest for the heavens," at right) and* Mengtian *("dreaming of the heavens," left) (credit: CMSA).*

Space missions of orbital platforms and of sounding rockets require a long preparation, typically of several years, and should be considered for experiments that need a long exposition duration to microgravity. The relatively short preparation time for the use of drop tubes and towers and of aircraft parabolic flights (typically of few days to few months) renders them particularly attractive for short-duration experiments of a few seconds. The utilization of these experimental platforms of earthbound microgravity must be considered as preparatory and complementary to space missions.

Let us insist on the fact that the platforms described in this section do not simulate microgravity but that they really create microgravity, even if it is not always perfect, as all these means are in free fall.

6. The means of microgravity simulation

To the contrary of the means creating microgravity, simulation methods do not allow to really create microgravity. The simulation means allow to obtain experimental configurations in which certain aspects of phenomena can be studied in a way similar to what could be observed in microgravity but without being in weightlessness.

Therefore, these methods have important limitations that reduce their scientific interest to the investigations of some very specific cases. The results obtained by these simulation methods generally complete those obtained in real microgravity. In none of the three following configurations, microgravity is really created as there is no free fall.

The first simulation method was used at the end of the nineteenth century by a Belgian physicist, Joseph Plateau, who gave his name to this method. The principle is simple: it consists in immersing a liquid in another immiscible liquid matrix having the same volumetric mass. By Archimedes principle, the buoyancy exerted by the liquid matrix of volumetric mass ρ_1 on a volume V of a liquid of volumetric mass ρ_2 is directed along the gravity acceleration vector and reads

$$\vec{F}_{fl} = V.(\rho_1 - \rho_2).\vec{g} \tag{1}$$

This force becomes null for $\rho_1 = \rho_2$, yielding results similar to what could be obtained in weightlessness when $g = 0$. In the Plateau configuration, the gravity force is not balanced by inertia forces but by a buoyancy force.

Only static configurations are truly well simulated with this method, e.g. configurations of static equilibrium of liquid zones.

The second simulation method is less known. It consists in balancing locally the force of gravity acting on a body by a magnetic or electrostatic force acting in the other direction. The effects of two fields, the gravitational field and a magnetic or electrostatic field, have to be locally balanced. One sees immediately the limitation of this configuration that would work only for bodies sensitive to magnetic induction or electrically charged. Furthermore, the power needed to maintain these fields is quite important and limits the size of observed configurations. Nevertheless, this method is used sometime to investigate magnetohydrodynamic problems in the absence of gravity effects.

The third simulation method is what is called the dimensionless reduction. This method mainly applies to fluid research for which scientists use a series of dimensionless numbers describing the ratios of different forces acting on fluids. Reducing physical dimensions of an experimental liquid zone greatly diminishes effects caused by gravity in comparison to other forces acting on fluids, e.g. superficial tension force or capillarity forces. One manages to build floating liquid zone of a few millimeters size that allow to study certain phenomena. The main limitations of this method are linked to reduced sizes: firstly, they make it difficult to install precise means of observation and measurement; secondly, they reduce the field of investigation to limited ranges of values of other effects specific to fluids.

Space medical and physiological research does not limit itself to conducting medical experiments in orbit or during parabolic flights but relies also on results obtained by earthbound means. For research on adaptation of the human body to weightlessness, scientists use two simulation techniques that allow within certain limits to recreate the effects of microgravity on the human body. It consists firstly

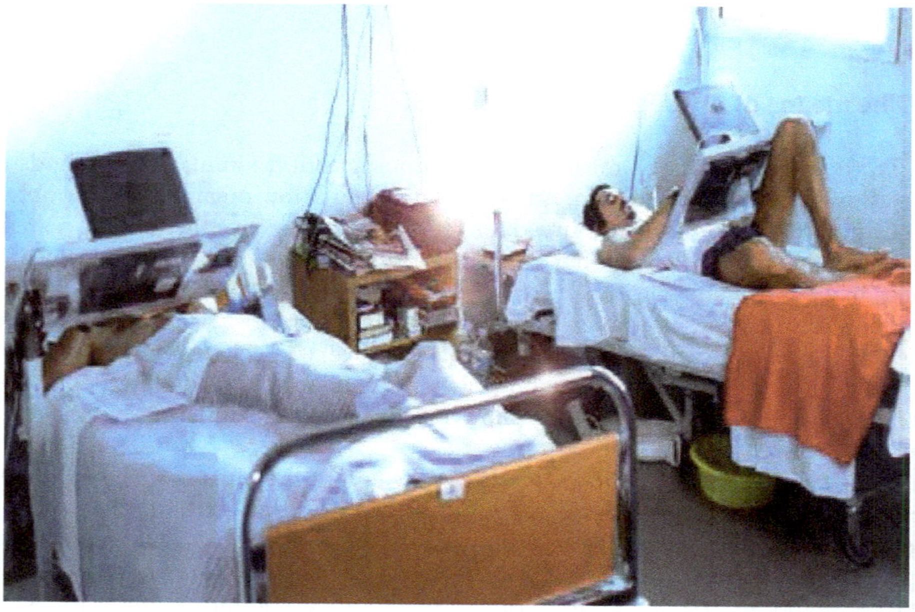

Figure 15.
Head-down bed rest simulates microgravity effects on human physiology. Subjects stay in slightly tilted head-down (typically 6°) beds for weeks or months at a time (photo credit: CNES/ESA).

of immobilization (or hypokinesia, **Figure 15**) in a horizontal position or slightly inclined (head-down) that simulates the shift of body fluids, mainly blood, toward the upper part of the body like in weightlessness.

The second technique is water immersion. As the human body is mainly made of water, buoyancy induces conditions partially similar to microgravity acting on the human body, somewhat akin to Plateau's configuration. A variant of water immersion, called dry immersion, is also used sometime where the subject is placed in an elastic or plastic sheet in a liquid matrix, such that the subject is immersed in the liquid but without direct contact with the liquid.

7. Conclusion

A better understanding of the effects of microgravity on physics and the human body, from cells to body systems, is essential if the human exploration of outer space is to continue. The capacity to conduct research in the microgravity environment provided by spaceflight is fundamental, especially given current plans to expand long-term missions in low Earth orbit and to establish the commercial use of space, together with the ultimate goals of creating a human colony on the Moon and sending a first crewed mission to Mars. Nonetheless, there are many limiting factors that restrict the performance of experiments in space, such as the high costs involved in sending resources and equipment up into space, the safety requirements to which experimental devices must adhere and the small number of astronauts per flight. These constraining factors have motivated the establishment of ground-based research facilities and parabolic flights. The latter presents some limitations in terms of the short period of time of exposure to microgravity given and the hypergravity condition that precedes and succeeds each parabola. However, it is the only provider of microgravity, in which experiments in physics, biology, physiology and medicine can be conducted by human operators and volunteers. Parabolic flights are not a perfect analogue of spaceflight, but they remain a valuable research tool that enables research and testing to take place and a better understanding of the effects of microgravity, assisting academia, the private sector and governments to better design future plans for the human exploration of outer space.

Author details

Vladimir Pletser[1,2,3*] and Thais Russomano[4,5]

1 Blue Abyss, Director of Space Training Operations, Liverpool, United Kingdom

2 Chinese Academy of Sciences, Beijing, China

3 European Space Agency, ESA-ESTEC, Nordwijk, The Netherlands

4 InnovaSpace, United Kingdom

5 Centre for Human and Applied Physiological Sciences, School of Basic and Medical Biosciences Faculty of Life Sciences and Medicine, King's College London, United Kingdom

*Address all correspondence to: pletservladimir@gmail.com

References

[1] Pletser V. Gravity, weight and their absence. In: Springer Briefs in Physics. Singapore: Springer Nature Singapore Pte Ltd; 2018. DOI: 10.1007/978-981-10-8696-0. Available from: https://www.springer.com/gp/book/9789811086953#otherversion=9789811086960. ISBN: 978-981-10-8695-3 [Accessed: 24 April 2020]

[2] Beysens DA, Van Loon J, et al. Generate an extra-terrestrial environment on earth. In: River Publishers Series in Standardisation. Aalborg, Denmark: River Publishers; 2015. ISBN: 978-87-93237- 53-7

[3] Seibert G et al. In: Fitton B, Battrick B, editors. A World without Gravity. ESA-ESTEC; 2001 ESA-SP-1251. Available from: http://www.esa.int/esapub/sp/sp1251/sp1251web.pdf. ISBN: 92-9092-604-X [Accessed: 24 April 2020]

[4] Haerendel G et al. Looking up: Europe's quiet revolution in microgravity research. Scientific American. 2008. Custom publication. Available from: https://www.scientificamerican.com/media/pdf/ESAReader_LowRes.pdf. [Accessed: 24 April 2020]

[5] Myshkis AD, Babskii VG, Kopachevskii ND, Slobozhanin LA, Tyuptsov AD. Low-Gravity Fluid Mechanics; Mathematical Theory of Capillary Phenomena. Berlin Heidelberg: Springer-Verlag; 1987

[6] Clément G. Fundamentals of Space Medicine. New York: Springer; 2005

[7] Clément G, Reschke MF. Neurosciences in Space. New York: Springer; 2008. ISBN: 9780387789507-9780387789491. DOI: 10.1007/978-0-387-78950-7

[8] Aubert A, Beckers F, Verheyden B, Pletser V. What happens to the human heart in space? Parabolic flights provide some answers. Cardiovascular response during gravity changes induced by parabolic flights. ESA Bulletin. 2004;**119**:30-38. Available from: http://www.esa.int/esapub/bulletin/bulletin119/bul119_chap4.pdf. [Accessed: 24 April 2020]

[9] Preidt R. Spaceflight might weaken astronauts' immune systems. HealthDay. 2014. Available from: https://consumer.healthday. com/kids-health-information-23/immunization-news-405/spaceflight-might-weaken-astronauts-immune-systems-690922.html [Accessed: 24 April 2020]

[10] American Society for Biochemistry and Molecular Biology (ASBMB). The human immune system in space. ScienceDaily. 2013. Available from: www.sciencedaily.com/releases/2013/04/130422132504.htm [Accessed: 24 April 2020]

[11] Sutton J et al. Scientific American. 2003. Available from: https://www.scientificamerican.com/article/how-does-spending-prolong/ [Accessed: 24 April 2020]

[12] The ZARM Drop Tower. Available from: https://www.zarm.uni-bremen.de/drop-tower.html [Accessed: 8 April 2020]

[13] Von Kampen P, Kaczmarczik U, Rath HJ. The new drop tower catapult system. Acta Astronautica. 2006;**59**:278-283

[14] Verga A. Sounding rockets. In: Sabbatini M, Sentse N, editors. ESA User Guide to Low Gravity Platforms. 2014. pp. 6.1-6.31 HSO-K/MS/01/14, Iss. 3 Rev. 0. Available from: http://wsn. spaceflight.esa.int/docs/EUG2LGPr3/EUG2LGPr3-6-SoundingRockets.pdf. [Accessed: 24 April 2020]

[15] NASA website. Available from: https://www.nasa.gov/mission_pages/station/main/ [Accessed: 24 April 2020]

[16] ESA websites. Available from: http://www.esa.int/Our_Activities/Human_Spaceflight/International_Space_Station/Highlights/International_Space_Station_panoramic_tour; http://www.esa.int/Our_Activities/Human_Spaceflight/International_ Space_Station Building_the_ International_Space_Station3 [Accessed: 24 April 2020]

[17] International Space Station (ISS). In: Sabbatini M, Sentse N, editors. ESA User Guide to Low Gravity Platforms. 2014. pp. 7.1-7.106 HSO-K/MS/01/14, Iss. 3 Rev. 0. Available from: http://wsn. spaceflight.esa.int/docs/EUG2LGPr3/EUG2LGPr3-7-ISS.pdf. [Accessed: 24 April 2020]

5

A Device for Sampling Earlobe Arterialized Blood in Space and Other Austere Environments

Thais Russomano

Abstract

The coming decades will see a large increase in the numbers of people who will have the opportunity to go into space, whether on traditional Earth-orbiting space stations, tourist spaceflights or proposed space hotels. In addition, humans are likely to be spending longer periods of time in the microgravity of space and the reduced gravity environments found on the moon and Mars, with plans for long-duration spaceflight to reach the red planet and habitation of a moon colony. The anatomy, physiology and psychology of humankind are shaped by the gravity we are subject to on Earth, and it is known that the removal or reduction of this force can have a detrimental effect on our health and wellbeing. Therefore, all steps must be taken to monitor these aspects. Currently, there is no safe and acceptable method to collect arterial blood in space, which can be used to obtain valuable blood gas and blood component variables. This chapter will outline the development of a method for safely collecting arterialized blood in space, the research and steps taken to ensure its suitability and applicability, in preparation for this growing presence of humans in space.

Keywords: space medicine, space physiology, medical emergencies, arterial blood, arterialized blood, blood collector, parabolic flight

1. Introduction

There is an increased need to accurately monitor and medically evaluate human beings in a variety of clinical and research situations in space, with plans for long-duration manned spaceflight, the proposed return to the moon and potential moon and Mars colonies in the future. In addition, greater flexibility in the selection process of astronauts and the advent of space tourism increases the need for adequate health and medical monitoring and evaluation in space, requiring improvements in currently available space medical monitoring systems.

The accurate measurement of arterial blood gas tensions, as opposed to venous, in medical practice and physiological studies on Earth and in space is of particular importance, as these can better reflect alterations in performance of the cardiopulmonary system and related diseases. However, there is currently no suitable method to access arterial blood in microgravity, and consequently, values for blood gas tensions are usually derived from measurements of respiratory gas partial pressures. Nonetheless, the measurements of oxygen saturation by oximetry are not considered comprehensive or accurate enough for detailed research or clinical practice.

The utility of finding a solution to this problem is not in doubt. Physiological findings could be confirmed with greater accuracy and more detailed studies conducted in the future. Clinical emergencies could also be managed with greater facility, resulting in increased safety for all crew involved in space missions. To this end, the arterialized earlobe blood collection technique for evaluating blood gas tensions has been considered for use in space, as analyses of the blood obtained could provide valuable information regarding the diagnosis of a number of medical conditions. This technique was first developed in 1944 and adopted under certain circumstances as an alternative to arterial puncture and arterial cannulation [1]. Nonetheless, the current earlobe arterialized blood collection technique is untested in microgravity, as is the risk of contamination of the environment with blood droplets. Therefore, a series of researches and tests have taken place to validate the suitability of the arterialized blood as an analogue of arterial blood and its suitability for use in microgravity, the creation of an easy-to-use and safe device for collecting arterialized blood from the earlobe, validating its use in ground-based studies on Earth and in microgravity, and determining the space preparedness of the device for surviving the stresses caused by a space rocket launch.

2. Validation of the arterialized blood technique

2.1 Arterialized versus arterial blood

Arterial gas analyses are essential for the clinical evaluation of astronauts, since they provide important physiologic information and can be an important tool for performing disease diagnoses during a space mission. However, currently available devices and methods, such as puncture and cannulation of an artery, are considered unsuitable for use in this scenario.

Arterial cannulation, the positioning of an intra-arterial catheter, is a technique which allows continuous and direct monitorization of blood pressure and frequent sample withdrawal for blood analyses. Arterial blood by means of puncture is usually collected from the wrist or from the inner part of the elbow or other arteries, through the insertion of a needle in a previously cleaned area. The blood then flows into a heparinized syringe, and the needle is removed as soon as enough blood is collected [2].

Both arterial cannulation and puncture are known to be difficult techniques to perform, requiring specialist training, causing pain to the patient and having the possibility of contamination of the environment with blood droplets. Moreover, although low, there is an increased risk of serious complications, such as haematoma, excessive bleeding and infection. Therefore, it is well accepted that the direct sampling of arterial blood is unsuitable for use in many austere environments, such as in space missions [2].

The earlobe arterialized blood technique makes use of the fact that the capillary blood taken from the arterialized earlobe originates from the arterioles and thus has the composition of arterial blood. The technique has been available as a substitute for arterial puncture for more than 60 years in clinical medicine and physiological research. The success of the technique depends upon careful preparation of the earlobe, which is arterialized by rendering it hyperaemic. This can be executed by heating the earlobe or massaging it with a rubefacient cream, thus ensuring free flow of blood from any incision made. The time of preparation varies from study to study, though conventionally it ranges from 3 to 10 minutes, with the standard being around 4 minutes. Ensuring adequate vasodilatation is of primary importance; therefore, if the earlobe is not hyperaemic after 4 minutes of preparation, massage or heating should continue. Conventionally, the skin of the earlobe is cleaned with alcohol, and a puncture, 2–4 mm deep, is made with a sterile blade. The blood is collected in a heparinized capillary tube or cartridge, which is held in such a way that the blood enters anaerobically by capillary action. This blood can then be analyzed using a standard blood analyzer [1, 3–6].

Table 1 summarizes the differences between the two techniques of blood collection via arterial puncture and blood collection from an arterialized earlobe.

2.2 Earlobe arterialized blood technique in microgravity simulation studies

A series of studies were conducted at King's College London, as part of the PhD thesis entitled 'The effect of 3h of 6-degree Head-Down Tilt (HDT) with and without hypoxia and light exercise on lung function' [7], with the aim of evaluating the feasibility of performing this technique in space missions.

There was first the need to establish whether the lower to upper body redistribution of blood that occurs during microgravity exposure, with the subsequent venous congestion of the face and neck of astronauts, could cause contamination of the arterial blood from venous blood, thereby affecting results. The arterialized capillary blood sample technique had not been used previously during ground-based microgravity simulations, parabolic flights or space missions, and therefore, a preliminary study was designed to evaluate the possible effect of the head congestion on the gas tensions of earlobe arterialized blood samples. In order to avoid the cardiopulmonary changes associated with tilting to the 6° head-down position, the ground-based microgravity simulation used, the increase in venous pressure in the earlobe associated with this position was reproduced by inflating a cuff around the neck, with the volunteer in the supine position.

The venous pressure at the earlobe was calculated as the change in the vertical height of the ear relative to the heart on transition from supine to 6° head-down. Assuming a 30 cm distance between the earlobe and the right atrium, the increase

Characteristic	Radial artery	Arterialized earlobe
Level of discomfort	Potentially painful	Virtually pain-free
Potential complications	• Haematoma formation • Hemorrhage • Infection • Potential for reduced wrist mobility • Nerve damage	• Hemorrhage—from the earlobe, and therefore easily controlled with direct pressure • Cutaneous infection at incision site (superficial)
Ease of use	Requires training: currently only physicians and specialist nurses are able to carry out this procedure	Very easy technique to learn and carry out by non-medically qualified personnel
Potential usages	Currently used in hospital setting but only by trained personnel Use in research circumstances is limited by the need for a physician to be available to carry out the technique	Potential for many spheres of use: **Terrestrial:** hospitals, private clinics, rural health centres **Aeronautic:** patient transport **Space:** space station, extraterrestrial bases for research and medical use

Table 1.
Comparison between the characteristics of radial artery puncture and blood collection from the arterialized earlobe.

in hydrostatic pressure at the ear was 2.3 mmHg.[1] The increase in central venous pressure secondary to the headward shift of the blood during head-down tilt was of the order of 3 [8] to 5 mmHg [9], resulting in a total increase in venous pressure on moving from the horizontal to 6° HDT ranging from 5.3 to 8.3 mmHg. Therefore, a neck cuff pressure of 10 mmHg was adopted for the study, which would produce a slightly greater degree of venous congestion of the ear.

The research evaluated seven healthy volunteers, aged 21–36 years. Each volunteer laid supine on a couch and completed three phases of 10 min each, divided into baseline (neck cuff deflated, control), test (neck cuff inflated) and recovery (neck cuff deflated, recovery). During each phase, the respired gases at the lips were sampled continuously, using O_2 and CO_2 rapid response gas analysers, from which their outputs were recorded and used to calculate respiratory frequency, end-tidal PO_2 (partial pressure of O_2) and PCO_2 (partial pressure of CO_2). Two earlobe arterialized blood samples were collected during the last 2 min of each phase, and the PO_2 and PCO_2 were determined using the pH/blood gas analyser. During the performance of the earlobe blood collection, no participant showed apprehension or distress, and there were no reports of complication (skin infection or bleeding) after the completion of the experiment. The healing of the incision was well advanced 72 h following the procedures. These findings are in accordance with those of Spiro and Dowdeswell [10], who found the arterialized earlobe technique to have no morbidity and to be virtually pain-free.

The means (±standard deviation, SD) of the respiratory frequency, end-tidal PO_2 and PCO_2, earlobe arterialized blood PO_2 and PCO_2 and the end-tidal minus earlobe arterialized blood PO_2 and PCO_2 differences before, during (test phase) and after inflation of the neck cuff are presented in **Table 2**.

The findings of this study showed no significant differences in the mean values of respiratory frequency, end-tidal PO_2 and PCO_2 and earlobe arterialized blood PO_2 and PCO_2 between the three phases. During the baseline, test and recovery phases, the end-tidal minus earlobe arterialized blood PO_2 and PCO_2 differences were 7.4 (±2.8) and 1.0 (±0.9), 7.7 (±4.3) and −0.5 (±1.4) and 7.7 (±3.3) and −0.5 (±1.0), respectively. The mean values of the differences found in this study are very similar to those reported in the literature for healthy volunteers breathing air at rest [11, 12].

The findings of this study were very important, as it demonstrated that congestion of the head did not affect the PO_2 and PCO_2 of the arterialized blood taken from the earlobe and the end-tidal arterialized blood differences. Therefore, it is possible to state that raising the venous pressure in the head by 10 mmHg, used to simulate the venous congestion encountered during microgravity exposure, did not cause any deleterious effect on the relationship between the PO_2 and PCO_2 of the arterialized blood sampled from the earlobe and the PO_2 and PCO_2 of the systemic arterial blood [7, 13, 14].

A second experiment was then designed within the scope of the same PhD thesis [7] to further understand the effects of HDT on the earlobe arterialized blood method. Therefore, hypoxia was added to the ground-based microgravity simulation in order to create an extra stressor. The differences between the tensions of oxygen and carbon dioxide in the end-tidal gas and earlobe arterialized blood were examined under two experimental conditions: breathing air (normoxia) and breathing a mixture of 10.7% O_2 in N_2, which is equivalent to breathing air at an altitude of 16,000 feet[2] (hypoxia).

[1] A pressure of 1 mmHg corresponds approximately to 1.33 mbar.

[2] Equivalent to 4876.8 m

	Baseline mean (±SD)	Test mean (±SD)	Recovery mean (±SD)
RF (br/min)	13.1 ± 2.1	13.0 ± 2.2	13.0 ± 2.2
$PETO_2$ (mmHg)	103.6 ± 3.3	107.2 ± 8.7	101.6 ± 5.0
$PabO_2$ (mmHg)	96.2 ± 2.5	99.7 ± 7.1	93.9 ± 3.7
$PETCO_2$ (mmHg)	38.2 ± 4.0	37.8 ± 4.1	38.2 ± 4.1
$PabCO_2$ (mmHg)	38.0 ± 4.0	38.5 ± 4.0	38.8 ± 3.7
PET-ab O_2 (mmHg)	7.4 ± 2.8	7.7 ± 4.3	7.7 ± 3.3
PET-ab CO_2 (mmHg)	1.0 ± 0.9	−0.5 ± 1.4	−0.6 ± 1.0

Table 2.
The effect of inflation of a neck cuff (test) on respiratory frequency (RF, breath/minute), end-tidal PO_2 ($PETO_2$) and PCO_2 ($PETCO_2$), earlobe arterialized blood PO_2 ($PabO_2$) and PCO_2 ($PabCO_2$) and end-tidal minus earlobe arterialized blood PO_2 (PET-ab O_2) and PCO_2 (PET-ab CO_2) differences.

A system was designed for this experiment permitting volunteers to breathe the inspired gas mixture through an oronasal mask. The normoxic gas (air) was supplied to the volunteer from a compressed air cylinder, and the hypoxic gas mixture was produced by mixing appropriate flows of air and nitrogen. The gases were mixed in a 100 L Douglas bag, before being delivered to the participant. The concentration of oxygen in the bag was monitored at 1 min intervals throughout the experiment. The following safety procedures were put in place: a source of 100% O_2 was connected to the gas supply, and the concentration of oxygen in the inspired gas was monitored with an oxygen rapid response gas analyser (alarm set to operate at 10.2% O_2). Arterial oxygen saturation (alarm set to operate at 65%) by means of pulse oximeter and blood pressure and heart rate were continuously monitored with a Finapres device. The ability of the volunteer to respond to simple commands was assessed every 2 min in order to identify any deleterious effect of hypoxia on mental performance and cognition.

Six healthy volunteers, aged 21–26 years, participated in the experiment and were not informed as to whether they were breathing air or the hypoxic mixture until the study was complete. The experiment began with the tilt table placed horizontally, and the individual was asked to lie in the supine position for 30 min (rest period). They were then placed into the required position (either supine or 6° HDT), wearing an oronasal mask and breathing the gas supply (either 20.9% O_2 or 10.7% O_2) for 20 min. For the final 10 min, the oronasal mask was replaced with a mouthpiece, a valve box and a nose clip, the earlobe was arterialized using massage and a vasodilating cream, and two earlobe blood samples were collected. The PO_2 and PCO_2 of the blood samples were immediately determined by means of the pH/blood gas analyser. End-tidal PO_2 and PCO_2 were continuously analyzed via the gas analysers and recorded during the last 10 minutes.

All volunteers completed the study without any untoward effects. The means of the end-tidal PO_2 and PCO_2, the earlobe arterialized blood PO_2 and PCO_2 and the end-tidal minus earlobe arterialized blood PO_2 and PCO_2 differences for each body position during normoxia and hypoxia are presented in **Table 3**.

End-tidal PO_2 and earlobe arterialized blood PO_2 decreased, as expected, from approximately 103 and 94 mmHg during normoxia to 40 and 36 mmHg during hypoxia, respectively, for both positions together ($p < 0.05$). The PET-abO_2, consequently, also decreased from a combined mean of 9.6 mmHg during normoxia to a mean of 3.4 mmHg during hypoxia ($p < 0.05$). The mean end-tidal and earlobe arterialized capillary PCO_2 decreased ($p < 0.05$) during hypoxia in comparison

	$PETO_2$ mean (±SD)	$PabO_2$ mean (±SD)	$PET\text{-}abO_2$ mean (±SD)	$PETCO_2$ mean (±SD)	$PabCO_2$ mean (±SD)	$PET\text{-}abCO_2$ mean (±SD)
Supine, N	101.6 ± 8.8	92.7 ± 8.9	8.9 ± 2.9	43.6 ± 2.9	42.8 ± 3.2	0.77 ± 2.4
6° HDT, N	105.6 ± 4.0	95.4 ± 4.5	10.3 ± 4.0	42.1 ± 2.5	41.9 ± 3.7	0.13 ± 2.4
Supine, H	40.9 ± 4.5*	36.5 ± 3.3*	4.7 ± 1.8*	37.7 ± 3.3*	35.7 ± 3.0*	2.1 ± 3.4*
6° HDT, H	40.8 ± 3.8*	36.9 ± 4.6*	2.1 ± 3.4*	35.8 ± 3.8*	34.3 ± 5.2*	1.4 ± 3.0*

**Different from normoxia; $p < 0.05$. All pressures in mmHg.*

Table 3.
Mean end-tidal PO_2 and PCO_2 (PET), earlobe arterialized blood PO_2 and PCO_2 (Pab) and end-tidal minus earlobe arterialized blood PO_2 and PCO_2 differences (PET-ab) during normoxia (N) and hypoxia (H) for 6° HDT and supine positions.

with normoxia in both positions, due to hyperventilation secondary to the low arterial PO_2. There were no significant differences between the values of end-tidal, arterialized blood and end-tidal minus earlobe arterialized blood differences for PO_2 and PCO_2 when the two positions were compared during either normoxia or hypoxia.

These findings led to the conclusion that the 6° HDT position did not alter the end-tidal minus earlobe arterialized blood PO_2 and PCO_2 differences from those obtained in the supine position during either normoxia or hypoxia, which reinforces the belief that this technique is suitable for use in either ground-based microgravity studies or in space missions.

3. Development and validation of an earlobe arterialized blood collector (EABC) device

The previously presented two studies were pioneering, as they were the first to be conducted during HDT using the earlobe arterialized blood collection technique. It was demonstrated that this technique is feasible for application in space missions or for physiological studies during microgravity simulation on Earth; however, the technique has the possibility of causing contamination of the environment to take place. This could be of major concern, especially in a spacecraft or space station, as blood droplets in microgravity would float with the potential to contaminate fellow astronauts or equipment. Taking this into consideration, a self-contained device was developed that would permit a standardized sampling of earlobe arterialized blood to be safely collected in a microgravity environment by non-medical personnel and without discomfort to the volunteer. The device was developed by the Microgravity Centre in collabo-ration with IDEIA Institute, both from the Pontifical Catholic University of Rio Grande do Sul, Brazil.

3.1 Evolution of the earlobe arterialized blood collector

The vision for the design of the earlobe arterialized blood collector was to develop a device with the following properties:

- Able to produce a suitable incision in the earlobe, such that sufficient flow of blood ensues to allow rapid and easy blood collection.

- The incision should be relatively pain-free and as accurate in depth and position as possible (in as far as these two variables should be predictable and easily adjustable).

- Capillary tubes or cartridges should provide anaerobic blood collection, through being positioned easily, quickly and precisely over the incision made and reducing the potential for contamination of the environment or any other part of the device.

- The device itself should be easy to use in terrestrial, aviation and extraterrestrial environments, with minimal training (user-friendly).

- The device must be easy to apply and remove from the earlobe, allowing quick application of gauze or a similar material to the incision to promote rapid hemostasis.

- The device must be small, lightweight, disposable and low-cost.

The first prototype was constructed in 1999, being 583 g in weight, 102 mm in length and 40 mm in diameter. This first prototype was mainly used to test the concept, and some earlobe arterialized blood collections were performed to evaluate the ability of the EABC to perform the cut and collect blood anaerobically, providing expected arterial gases results from a healthy volunteer (**Figure 1**).

The proof-of-concept success of this first EABC design led to its continued development, with a series of seven devices evolving over a 10-year period, leading to changes and improvements in shape, size, weight and used procedures (**Table 4**).

Figure 2 illustrates the first four generations in the developmental process of the EABC and the final EABC device.

The technique of blood collection is demonstrated in the sequence of six pictures in **Figure 3**, which shows the earlobe arterialization procedure with massage and a vasodilating cream, cleaning of the earlobe skin, placement of the EABC with a cartridge, blood collection and analysis in an i-STAT blood analyser device (Abbott Point of Care Inc., Brazil).

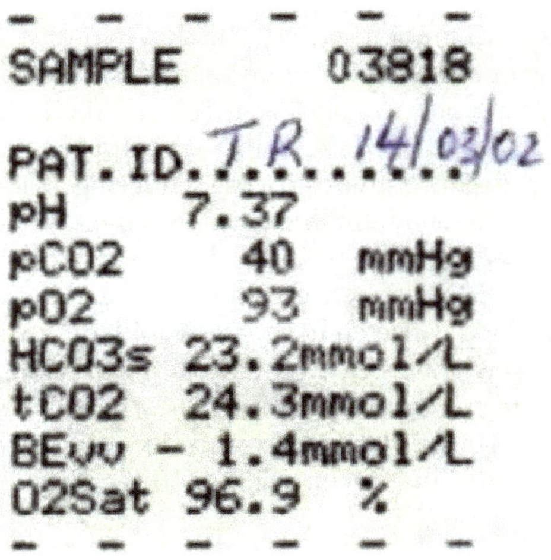

SAMPLE 03818
PAT. ID. T.R. 14/03/02
pH 7.37
pCO2 40 mmHg
pO2 93 mmHg
HCO3s 23.2mmol/L
tCO2 24.3mmol/L
BEvv - 1.4mmol/L
O2Sat 96.9 %

Figure 1.
First earlobe arterialized blood result using the first version EABC.

Version	Dimensions L × Ø (mm)	Weight (g)	Blade model	Blood recipient
1	102 × 94	583	No 11	Capillary tube
2	138 × 40	228	No 11	Capillary tube
3	107 × 27	85	Adapted No 11	Capillary tube
4	104 × 26	42	Adapted No 15°	Capillary tube
5	90 × 23	18	Ophthalmic blade	Without cartridge
6	57 × 26 (55 including cartridge)	29.5	Ophthalmic blade	I-STAT cartridge
7	73 × 26 (55 including cartridge)	28.2	Ophthalmic blade	I-STAT cartridge

Table 4.
Main characteristics of the seven versions of the EABC.

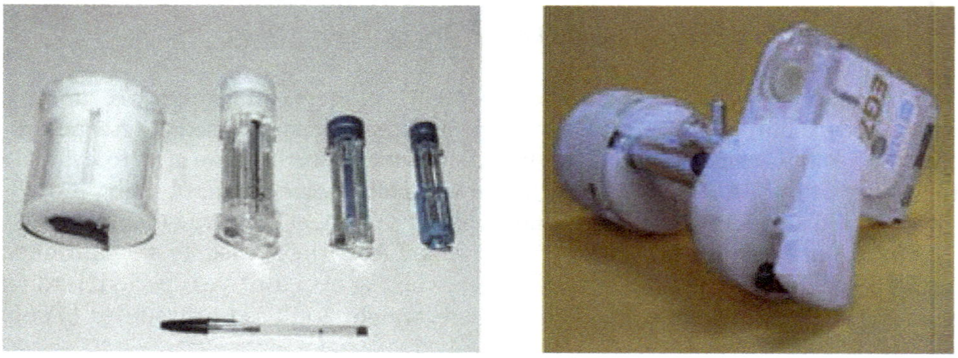

Figure 2.
Evolution of the EABC—First four generations on the left and the seventh EABC device on the right.

3.2 Preliminary EABC validation study

An initial EABC validation research was conducted involving six healthy volunteer students from King's College London, using the second EABC prototype (**Figure 4**) [15, 16].

An 8° HDT was used as a microgravity simulator in combination with hypoxia, equivalent to breathing air at 12,000 ft.[3] Blood samples were collected from the radial artery of volunteers and simultaneously from their arterialized earlobe, after being in the HDT position and breathing a 12.8% O_2 in N_2 mix for 15 min (**Figures 5** and **6**).

The arterialization procedure involved first rendering the earlobe hyperaemic by the application of a rubefacient cream, massaged into the earlobe for a period of 5 minutes. The skin was then cleaned using an alcohol swab and dried with sterile gauze and the second version of the EABC attached to the earlobe. An incision was made in the earlobe and samples of blood collected in the two capillary tubes of the second version of the EABC, simultaneously with the drawing of a 2 mL sample of blood from the radial artery into a syringe lubricated with heparin solution (5000 IU/mL[4]).

The PO_2, PCO_2 and pH of the blood samples were determined immediately using a blood gas analyser (Ciba Corning 238 pH/blood gas analyser, Ciba Corning

[3] Equivalent to 3657.6 m

[4] Heparin is a medication used as an anticoagulant (blood thinner). One unit of heparin is an amount approximately equivalent to 0.002 mg of pure heparin, which is the quantity required to keep 1 ml of cat's blood fluid for 24 hours at 0°C.

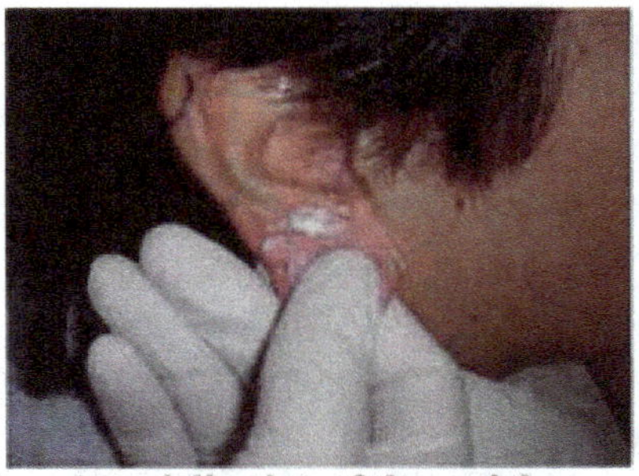

Arterialization of the earlobe:
3 min massage with a vasodilatation cream

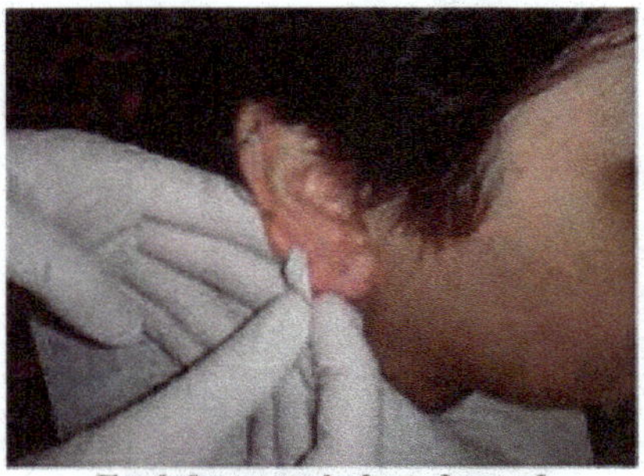

Earlobe area being cleaned
with alcohol swabs

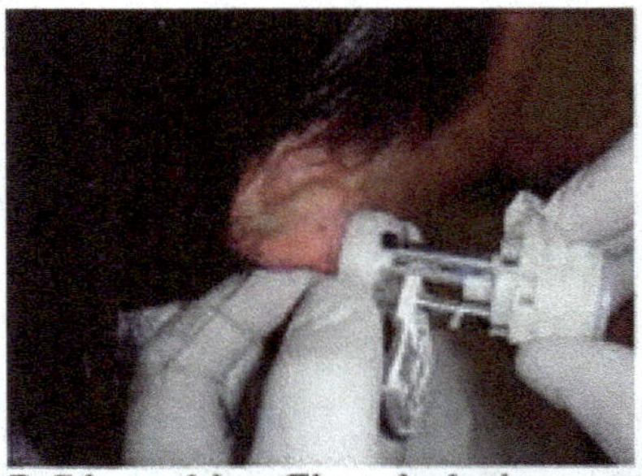

EAB C in position. First clockwise turn for
earlobe incision

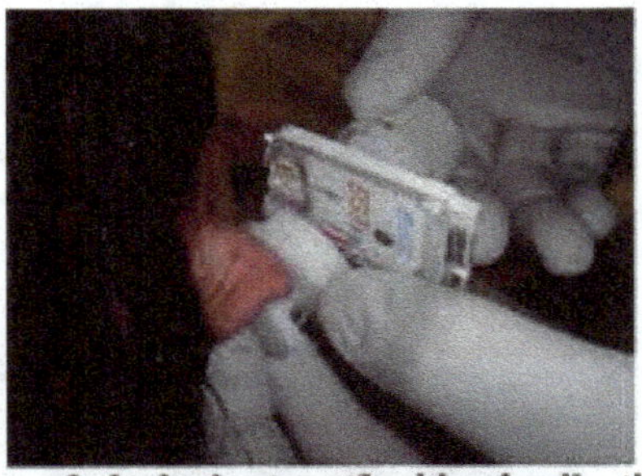

Second clockwise turn for blood collection
(anaerobically) inside the I-Stat Cartridge

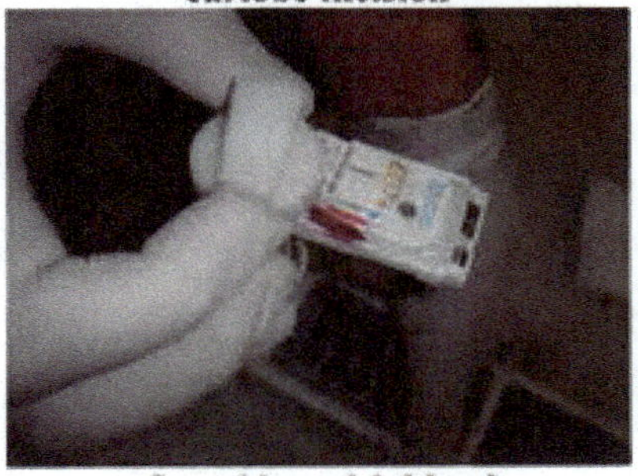

Cartridge with blood

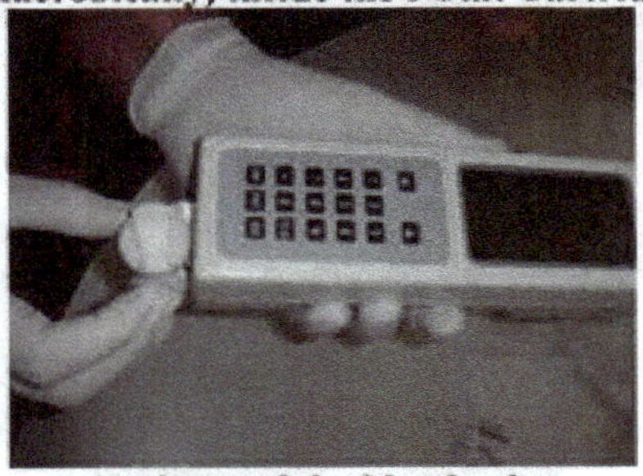

Analyzes of the blood using
a blood gas analyzer (I-Stat)

Figure 3.
Sequence of six pictures showing the earlobe arterialized blood collection and subsequent analysis, placing the EG7 cartridge in the i-STAT device.

Diagnostics Ltd., Halstead, Essex). The mean differences (±SD) in PO_2 between earlobe arterialized and radial artery blood samples were 0.25 ± 1.25 mmHg for PO_2 and 1.0 ± 0.75 mmHg for CO_2; neither difference was significant. There was no difference between the pH values obtained by the two techniques. **Table 5** summarizes the results of the blood analyses.

3.3 Clinical evaluation of the EABC in haemodialysis patients

All EABC clinical studies were funded by the European Space Agency via the Medical Projects and Technology Unit from the Crew Medical Support Office, European Astronaut Centre, Cologne, Germany.

The physiological studies performed during microgravity simulation suggested that the arterialized blood sampled from the earlobe using the EABC may provide sufficiently accurate measurements of the PO_2, PCO_2 and pH of the arterial blood

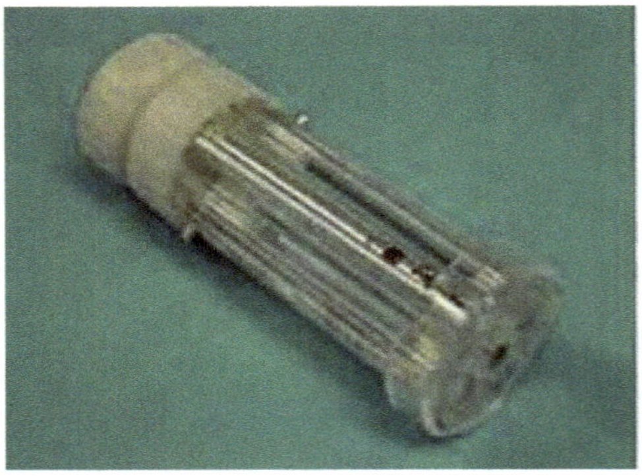

Length = 138 mm
Diameter = 40 mm
Weight = 228 g
Blade number 11
Materials: Acrylic, Polyacetal and Inox
Structure: 4 Stages (Cut, 2 Collect. Gauze)
Cut Size: Length: 3mm – Depth: 1.7mm – Curvature radius 12mm

Figure 4.
Characteristics of the second EABC version.

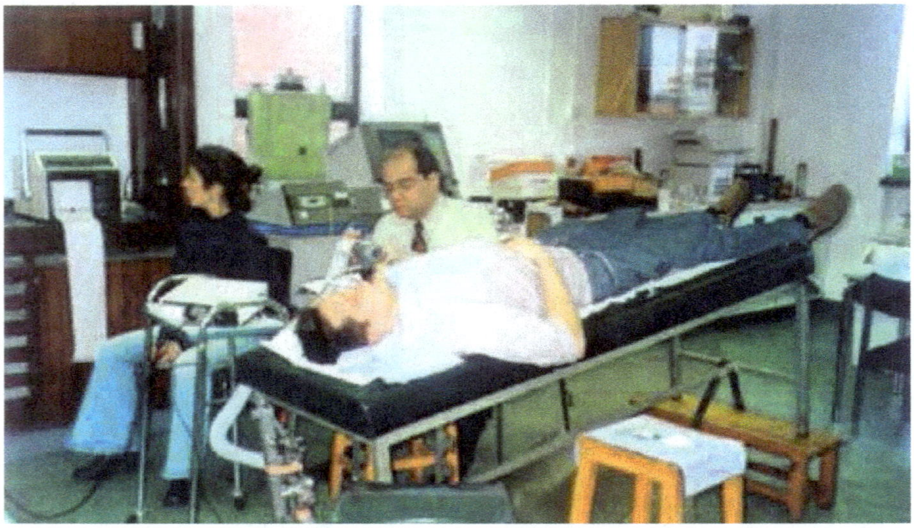

Figure 5.
Volunteer in HDT whilst breathing the hypoxic mixture.

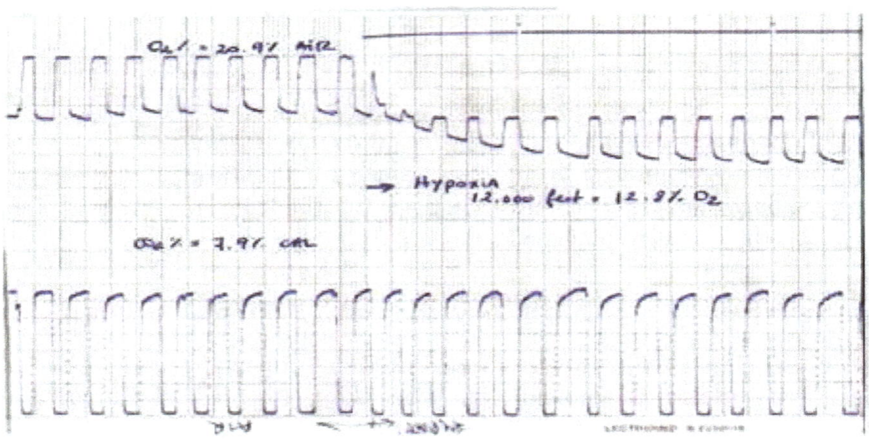

Figure 6.
Example of data being recorded during the beginning of hypoxic exposure (12.8% O_2 in N_2, equivalent to breathing air at 12,000 ft³).

	Radial artery Mean ± SD (range)	Arterialized earlobe Mean ± SD (range)
pH	7.43 ± 0.02 (7.4–7.46)	7.43 ± 0.02 (7.4–7.46)
PO_2 (mmHg)	42.1 ± 3.66 (38–47)	42.9 ± 3.88 (37–50)
PCO_2 (mmHg)	34.1 ± 1.88 (31–37)	33.12 ± 2.38 (29–37)
SaO_2 (%)	79 ± 3.85 (75–84.5)	79.9 ± 3.29 (74–85.6)

Table 5.
Blood gas data for simultaneous radial artery and earlobe arterialized blood samples collected using the EABC.

for clinical or research use in extreme environments, such as space. However, another important step would be to also evaluate the use of the EABC in a clinical setting on Earth, as technology transfer from space to terrestrial application was one of the aims for the use of this pioneering technology.

With this in mind, a first clinical study was conducted involving 12 patients from a hemodialysis clinic, meaning these individuals already had a medically determined need for measurement of arterial blood parameters, including arterial blood gas tensions and acid–base variables, and access to arterial blood was easily provided by an already existing fistula. The main goal was to compare arterial blood variables taken from the arterial side of the arterial–venous fistula with those obtained from the earlobe arterialized blood collected using the seventh version of the EABC. Blood collection was achieved simultaneously from the fistula and the arterialized earlobe in an i-STAT EC8+ cartridge, and the two samples were analyzed using a portable i-STAT blood analyser device (Abbott Point-of-care Inc., Brazil) [17].

In addition to blood parameters, earlobe incision length and subject pain perception were also evaluated. Incision length (mm) was measured with a caliper immediately after blood collection, and the patient pain perception was assessed, using a scale from 0 (no pain) to 10 (maximum perceived pain). **Figure 7** shows a schematic view of the earlobe cut and its measurement during the experiment.

The mean of the differences obtained from the earlobe arterialized and arterial samples ranged from 0.006 (for pH) to 2.8 mg/dL (for glucose). The R^2 was equal or above 0.93 in 10 of the 13 blood variables measured, and the lowest R^2 was for PCO_2 (0.68). Of the 13 blood measurements, 9 presented no significant difference, whilst the 4 that were significantly different (BUN, Cl^-, K^+, anion gap) had their values within normality, presented no clinical implication and did not affect

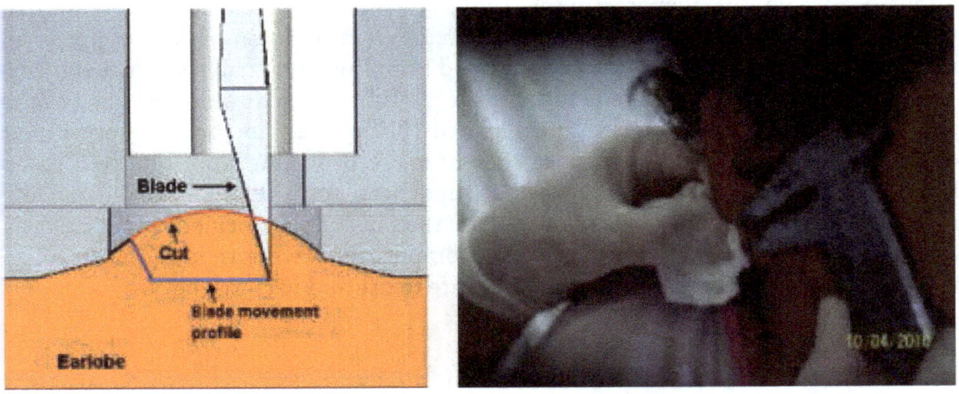

Figure 7.
Schematic view of the difference between cut length and blade movement profile (left) and cut measurement being performed with a caliper.

treatment or diagnosis. The mean (±SD) of the earlobe cut length was 4.4 (±1.3) mm, and the patient perceived pain was classified as minor with a mean of 2.7 points out of 10 points.

These findings were very motivating, as they indicated for the first time that the EABC works in a clinical setting and therefore could be considered a method for safe and easy access to arterialized blood sampling for medical diagnoses, not only in space missions but also on Earth. It led to two further studies, which assessed the use of the EABC in more gravely ill hospitalized patients.

3.4 Clinical evaluation of the EABC in critically ill patients

Two studies were conducted involving critically ill adult patients in intensive care units, aiming to assess the diagnostic and operational capability of the EABC.

A pilot study was first conducted, evaluating the use of the EABC on a cohort of mechanically ventilated adult critically ill patients admitted to an intensive care unit [18]. A comparison was made between the collected arterial blood and earlobe arterialized blood parameters, and the EABC was evaluated for its ability to diagnose acute respiratory distress syndrome (ARDS) in a total of 55 patients.

The results showed a high precision of earlobe arterialized blood samples. The measures of PO_2 demonstrated insufficient agreement levels; however, better agreement was seen for PCO_2 and pH measurements. The findings of this experiment showed a sensitivity of 100% and specificity of 92.3% for diagnosing ARDS using earlobe arterialized blood gasometric measures.

Sampling with the EABC proved to be unsuccessful in 43.6% of cases, due to insufficient blood flow, although this is not a surprising result given the circumstances of the patients and some important factors must be taken into account. The haemodynamic conditions of critically ill individuals and the use of medications that can cause vasoconstriction can negatively impact on the production of adequate peripheral blood flow. Therefore, the earlobe arterialized blood technique, with or without the use of the EABC, would not seem to be the best alternative for the management of patients in an intensive care unit, though it may prove useful in several clinical conditions and other critical care scenarios, such as emergency rooms, advanced medical transportation and pre-hospital care.

A second study was conceived to perform an operational evaluation of the EABC in critically ill patients [19], looking at aspects such as the number of cuts and cartridges required, ratio of sampling failure and success, bleeding complications and storage requirements. Fifty-five ventilated patients hospitalized in an intensive care unit participated in the study. The findings revealed that researchers took 26 min to obtain blood analysis, broken down into 15 min of patient preparation and 11 min for earlobe arterialized blood sampling and analysis. An average of 1.3 cartridges was required to achieve a successful cut of the earlobe. The results also demonstrated that researchers faced difficulties in performing blood collection in 59% of cases, but only 10% of these problems were reported to be linked to the EABC itself, such as superficial cut, blood leak, collector misalignment and vision obstruction. After the cut was performed, homoeostasis appeared to occur quickly, and no major complications were reported. The study results suggest that the EABC is quick and safe to use and user-friendly.

4. Validation of the EABC for space use

It is critically important that any device to potentially be launched into space must be able to withstand the launch process and spaceflight, remaining undamaged. To be considered for use on the International Space Station (ISS) as part of a space mission, the EABC must demonstrate that it can meet the specifications of spaceflight conditions through being submitted to a series of electromechanical tests. The purpose of testing is to expose the EABC to the same circumstances as those encountered during launch onboard a Soyuz rocket and the microgravity environment on the ISS.

The required tests are shock and vibration tests, measurements and mass proprieties, low and high pressure and temperature tests, humidity test and off-gassing evaluation [20]. To confirm its suitability for space use, the following tests were applied:

- Shock and vibration tests were conducted to check the functionality of the EABC after being launched to the ISS onboard the Soyuz. Two EABCs were placed inside a padded container and attached to a shaker and then submitted to different shock and vibration protocols.

- Measurements and mass proprieties must be known to determine precisely the mass and centre of gravity of the EABC.

- Low- and high-pressure and low- and high-temperature tests were performed to verify the physical and chemical stability of the EABC during variations of such conditions.

- Humidity test was applied to check the EABC functionality after the changes in relative humidity.

- Off-gassing levels were determined as different materials can contaminate the spacecraft ambient air and affect air filters, operation of other equipment and even astronaut health.

These tests were conducted at the National Institute for Space Research (INPE), in São José dos Campos, São Paulo, Brazil, with a successful evaluation of the variables tested. The final conclusion of the INPE experts was that the EABC was ready to fly in a space mission, as it is space-proof.

5. Validation of the earlobe arterialized blood collector for use in microgravity

Having validated the EABC through studies performed in simulated microgravity, it was important to further validate the earlobe arterialized blood collection technique and device in an actual microgravity scenario. A study was conceived using the fifth EABC prototype (**Figure 8**, this was the prototype available when the proposal was submitted to ESA) to determine if it could effectively be used in the microgravity environment achieved during the free-fall phase of a parabolic flight (42nd ESA Parabolic Flight Campaign in 2006) [21], without contaminating the aircraft environment with blood products.

A total of eight healthy participants took part in the ESA parabolic flight campaign, acting as both volunteers and researchers. The blood collections took place inside a hood, especially designed by the Microgravity Centre/PUCRS, Brazil, in order to prevent any possible escape of blood to the aircraft environment. The hood had two openings on three sides for the insertion of two gloved hands each side and a larger opening in the front plastic wall for the volunteer to place their face and

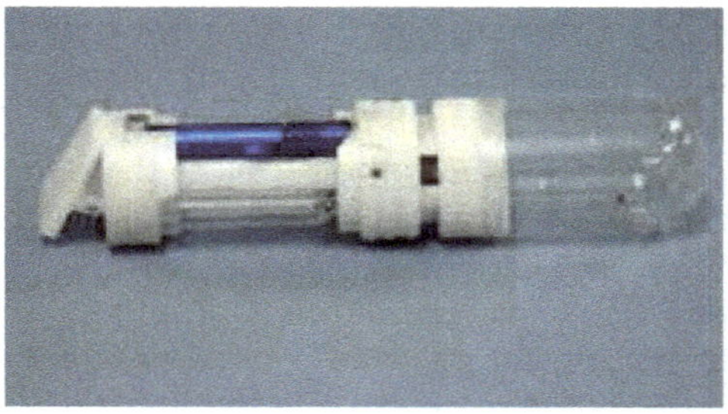

Length = 90 mm
Diameter = 23 mm
Weight = 18 g
Blade 15°
Material: Tecaform AH, Acrylic, Aluminium
Cut Size: Length: 4mm – Depth: 3mm – Curvature radius[1]: 8mm

Figure 8.
Fifth version of the EABC.

be able to breathe, see and talk well. After blood collection, the capillary tube and blood were placed in a hard, human tissue disposal container placed inside the hood at the back (**Figure 9**).

An EABC device was assigned to each of the volunteers, and one or two samples were taken from their earlobes during the 20 s period of microgravity provided by the parabolas. This provided a final study sample of 25 successful earlobe arterialized blood collections in the capillary tubes with a volume of 75 mL (**Figure 10**). Each collection of blood was timed.

The mean (±SD) time for the collection of the arterialized blood from the earlobe during the microgravity phase of the parabolas was 18.9 ± 7.23 s, which was very similar to the time required for the same group of researchers to collect on the ground (mean of 15 s). Researchers reported no difficulties in their ability to handle the EABC under microgravity conditions. It was also observed that no blood products emanated from the EABC, suggesting that the device seals were secure against blood leakage.

The data from this parabolic flight experiment strongly suggests that the arterialized blood from the earlobe can be as effectively sampled using the EABC in microgravity, in much the same way as the blood collections successfully occurred on the ground. Although this first study demonstrated the ability of the EABC to adequately acquire blood in microgravity, the next step required will be to assess the physiological blood variables in the weightlessness phase of a parabolic flight or during the sustained microgravity offered during space missions to ascertain whether this environment will affect such results [22].

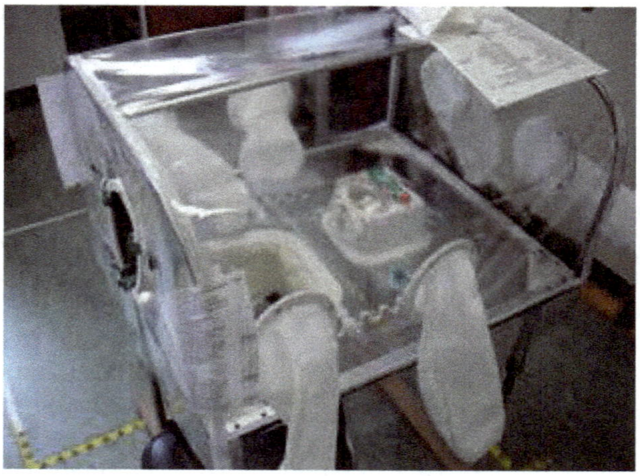

Figure 9.
Hood system designed to avoid any possible blood contamination of the A300 cabin during the experiment.

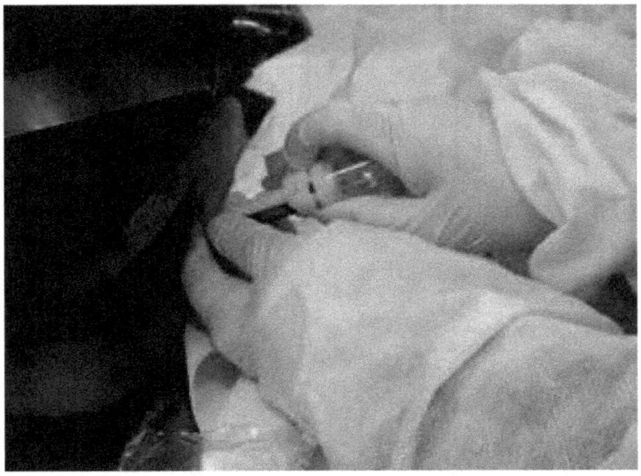

Figure 10.
Arterialized blood being collected during parabolic flight.

6. Conclusion

The earlobe arterialized blood collection was considered for use in space and extreme environments by the author, due to the advantages of the technique, and researches were conducted to evaluate this possibility, with results suggesting it could be applied but at the same time highlighting the chance of blood contamination of the environment. Consequently, a device was developed to prevent this possibility, the earlobe arterialized blood collector, which subsequently underwent a series of tests in simulated microgravity on healthy volunteers and then in clinical practice to also evaluate its potential terrestrial use. Further evaluation was conducted in the microgravity provided by an ESA parabolic flight campaign, and the 'space readiness' of the EABC was assessed through a series of electromechanical tests. In summary, research results suggest the EABC device to be space-proof, easy-to-use and low-cost, enabling the collection of arterialized blood as an alternative possibility to arterial puncture/cannulation in the austere environment of space.

Author details

Thais Russomano[1,2]

1 InnovaSpace, UK

2 Centre for Human and Applied Physiological Sciences, School of Basic and Medical Biosciences, Faculty of Life Sciences and Medicine, King's College London, UK

*Address all correspondence to: thais@innovaspace.org

References

[1] Lilienthal JL, Riley RL. On the determination of arterial oxygen saturations from samples of 'capillary blood. The Journal of Clinical Investigation. 1944;**23**:904-906

[2] Arterial puncture and cannulation. Available from: https://clinicalgate.com/arterial-puncture-and-cannulation/ [Accessed: 02 January 2020]

[3] Godfrey S, Wosniak ER, Courtenay ERJ, Samuels CS. Earlobe samples for blood gas analysis at rest and during exercise. British Journal of Diseases of the Chest. 1971;**65**:68-72

[4] Koch G. The validity of PO_2 measurement in capillary blood as a substitute for arterial PO_2. Scandinavian Journal of Clinical and Laboratory Investigation. 1958;**21**:10- 13

[5] Langlands JHM, Wallace WFM. Small blood samples from ear-lobe punctures. Lancet. 1965;**14**:315-317

[6] McIntyre J, Norman JN, Smith G. Use of capillary blood in measurement of arterial PO_2. The BMJ. 1968;**3**:640-643

[7] Russomano T. The effects of 3 h of 6-degree head-down tilt with and without hypoxia and light exercise on lung function [Ph. D. Thesis]. London, UK: King's College London, University of London; 1998

[8] Lollengen H, Gebhardt U, Beier J, Hardinsky J, Borger H, Sarrasch V, et al. Central haemodynamics during zero gravity simulated by head down bed rest. Aviation, Space, and Environmental Medicine. 1984;**55**: 887-892

[9] Nixon JV, Murray RG, Bryant C, Johnson RL, Mitchell JH, Holland OB, et al. Early cardiovascular adaptation to simulated zero gravity. Journal of Applied Physiology. 1979;**46**:541-548

[10] Raine JM, Bishop JM. A-a difference in O_2 tension and physiological dead space in Normal man. Journal of Applied Physiology. 1963;**18**(2):284-288

[11] Riley RL, Cournand A. Analysis of factors affecting partial pressures of oxygen and carbon dioxide in gas and blood of lungs: Theory. Journal of Applied Physiology. 1951;**4**:77-101

[12] Spiro S, Dowdeswell IRG. Arterialized earlobe blood samples for blood gas tensions. British Journal of Diseases of the Chest. 1976;**70**:263-268

[13] Russomano T, Ernsting J. The arterialized earlobe blood samples for blood gas tensions. A technique for medical emergencies in microgravity. In: 47th International Astronautical Congress, Book of Abstracts IAF/IAA – 96-G.1.12; Beijing, China; 1996. p. 23

[14] Russomano T, Doxey S, Ernsting J. Arterialized earlobe sample for arterial blood gas tensions with and without hypoxia during 6-degree head-down tilt. In: 68th Annual Scientific Meeting Program, Aerospace Medical Association, Book of Abstracts ASMA 192; Chicago, USA; 1997. p. A33

[15] Russomano T, Evetts S, Castro J, dos Santos MA, Gavillon J, de Azevedo DFG, et al. A device for sampling arterialized earlobe blood in austere environments. Aviation, Space, and Environmental Medicine. 2006;**77**:453-455

[16] Sides M, Vernikos J, Convertino V, Stepanek J, Tripp L, Draeger J, et al. The Bellagio report: Cardiovascular risks for space flight: Implications for the future of space travel. Aviation, Space, and Environmental Medicine. 2005;**76**:877-895

[17] Falcao F, Russomano T. Clinical validation of the earlobe arterialized blood collector. Aviation, Space,

and Environmental Medicine. 2010;**81**:1053-1054

[18] Vaquer S, Masip J, Gili G, Gomà G, Oliva JC, Frechette A, et al. Earlobe arterialized capillary blood gas analysis in the intensive care unit: A pilot study. Annals of Intensive Care. 2014;**4**:11

[19] Vaquer S, Masip J, Gili G, Gomà G, Oliva JC, Frechette A, et al. Operational evaluation of the earlobe arterialized blood collector in critically ill patients. Extreme Physiology & Medicine. 2015;**4**:5

[20] Russomano T, Whittle J, Evetts J, Coats E, Vian M, Cardoso R, et al. Assessment of an earlobe arterialized blood collector in microgravity. Aviation, Space, and Environmental Medicine. 2009;**80**:989-990

[21] Project Van Gogh - An Assessment of an Arterialised Blood Collecting Device for Use in Microgravity, Experiment Record N° 8568, ESA website, Erasmus Experiment Archive. Available from: http://eea.spaceflight.esa.int/portal/exp/?id=8568 [Accessed: 11 March 2020]

[22] NASA Systems Engineering Handbook. Available from: www.nasa.gov/sites/default/files/atoms/files/nasa_systems_engineering_handbook.pdf [Accessed: 02 January 2020]

6

G-Jitter, Vibrations, Diffusion: The IVIDIL Experiment

Valentina Shevtsova, Denis Melnikov, Yuri Gaponenko and Aliaksandr Mialdun

Abstract

Experiments aboard the International Space Station (ISS) provide a large but still limited amount of data. A complete set of data is usually returned to Earth on a flash disk a few months later, by which time the experimental facility has already been put into storage or trashed. Thus, scientists have no possibility to repeat the experiment, even if some ambiguities are later found. Therefore, onboard experiments require careful preparation on the ground with multiple tests in the laboratory and in parabolic flights, if possible. Furthermore, during postflight analysis, it is important to clarify all unknown sources of errors. One of the most obvious sources of perturbations on the ISS is g-jitter. Here, we present the preparation and implementation of the Influence of Vibrations on Diffusion in Liquids (IVIDIL) experiment on the ISS, which studied the effect of random g-jitter and given vibrations on diffusion-controlled experiments in liquid mixtures. Since diffusion in liquids is a slow process, only vibrational effects were examined in parabolic flights. A methodology for the analysis of diffusion and thermodiffusion processes was developed in the laboratory.

Keywords: ISS, IVIDIL, parabolic flight, vibration, Soret effect

1. Introduction

The authors of this chapter work at the Microgravity Research Centre of the Free University of Brussels (Université libre de Bruxelles, ULB) and form the core of the research group "Nonlinear phenomena in fluids." Primarily, we are interested in the behavior of fluids in a low-gravity environment and application of the results in Earth-based technologies. We carry out research in the field of physics of fluids, applied mathematics, process/chemical engineering and transport phenomena. More precisely, we are particularly interested in the following topics: vibrational phenomena in fluids, interfacial heat and mass transfer (with application to material science), controlling the flow in thermocapillary liquid bridges embedded in a gas stream, diffusion and Soret effect in free liquids and porous media and transport of impurities in the CO_2 storage. Recently, we have started a new research project related to an environmental issue, such as microplastic uptake by viscoelastic interfaces. Many of these topics trace their roots from the Influence of Vibrations on Diffusion in Liquids (IVIDIL) experiment, detailed below.

By combining theoretical, numerical (commercial codes, "home-made" codes) and experimental tools (essentially implementing methods of optical diagnostics: interferometry, shading, spectrometry), we obtain original results and publish them in high-stand journals. Our other strong points are the development of a fruitful cooperation with European and overseas research groups and the attraction of master and PhD students from all over the world.

2. G-jitter

Access to the microgravity environment provides the opportunity to perform fluid and materials science experiments designed to take advantage of near weightlessness. The term "near" underlines the unavoidable presence of uncontrolled accelerations due to aerodynamic drag, trembling of space vehicles, onboard machinery and crew activity. The fluctuating contribution of g(t) comprises a broad frequency spectrum. These random and periodic gravity fluctuations are referred to as gravity jitters (or g-jitter). The dedicated experimental studies of the effect of g-jitter on fluid systems in weightlessness were rather limited. Until recently, there was an open question: does g-jitter cause a sizeable mean flow, and, if so, does this mean flow promote striations and other defects in materials processing in microgravity? Thus, one of the objectives of one of the first experiments on the International Space Station (ISS) was to present experimental evidence of the role of the microgravity environment and, in particular, g-jitter. For this purpose, the IVIDIL project was proposed to the European Space Agency (ESA) in 2000 by an international team including the Microgravity Research Centre of ULB (Brussels, Belgium), the Institute of Continuous Media Mechanics UB RAS (Perm, Russia) and Ryerson University (Toronto, Canada) and coordinated by the Brussels team. The multiuser facility Selectable Optical Diagnostic Instrument (SODI) was designed and developed by the Belgian company QinetiQ Space (then Verhaert Space). The IVIDIL was the first experiment conducted inside the SODI.

The real need for such an extra experiment may be questioned because one option would be to use scaling analysis. Many analytical studies are restricted by the assumption of linearity, quasi-steadiness or very high/low frequencies. There is always a possibility that some of the aspects of a real experiment are not considered in theoretical hypotheses. Moreover, the failure of some microgravity experiments has been attributed to the effect of g-jitter.

3. Definition of the IVIDIL experiment

When the construction of the ISS was coming to an end, the goal of one of the first onboard experiments was to provide experimental evidence of the influence of the microgravity environment and, in particular, g-jitter, on diffusion-controlled experiments that refer to material science. This experiment was conducted in the SODI belonging to ESA.

The primary idea of the IVIDIL experiment was to measure diffusion coefficients in two binary mixtures in g-jitter environment and under different vibrational frequencies and amplitudes. For this reason, one should create controlled and reproducible concentration non-uniformity in the liquid. If it were generated on Earth, it would diffuse during the waiting time on the launch site that can last several months. It was decided to use the Soret effect for creating a concentration profile in the experimental cell.

The Soret effect (also known as thermodiffusion or thermal diffusion) is a molecular transport of substance associated with a thermal gradient. In response to this gradient, concentration gradients appear in an originally uniform mixture. It extended the objectives of the experiment to the measurements of the Soret coefficient as it has industrially and scientifically driven interests. During each experimental run, a temperature difference is established across the cell, to achieve a separation of components as induced by the Soret effect during 12 h. Then, the temperature difference is switched off, the cell is brought rapidly to isothermal conditions, and the components diffuse back to a homogeneous distribution. While the process takes place, the sample is shaken perpendicularly to the temperature and/or concentration gradient, with frequencies up to 2.8 Hz and amplitudes up to 68 mm.

The application of vibrations to a fluid system with density gradient can cause relative flows inside the fluid. If this gradient results from thermal or compositional variations, such flows are known as thermovibrational or solutovibrational convection, respectively. High-frequency and small-amplitude vibrations are of special interest, as they may create mean flows, an analogue of streaming flows in acoustic phenomena, which contribute to the transport of heat and mass in a time-average sense. Thus, the appearance and progress of vibrational convection in low gravity was another additional target. The test fluids were two different concentrations of water-isopropanol (IPA) with positive and negative Soret effect. The measured values in convection-free environment should be used as standards for ground experiments.

For one of the first experiments on the ISS, it may not sound very exciting, but it is very important for material sciences and may advance many everyday technological processes.

To summarize, the objectives of the IVIDIL experiment were the following:

- To present experimental assessment of the role of g-jitter
- To perform precise measurements of diffusion and thermodiffusion coefficients for two binary mixtures in g-jitter environment
- To disclose the influence of controlled vibrations on the measured values of diffusion and thermodiffusion coefficients
- To investigate vibration-induced convection and, particularly, heat and mass transfer under vibrations

In the frame of preparatory activities, the science team performed laboratory experiments [1–4], parabolic flights [5–8] and numerical calculations [9, 10].

4. Vibrational convection

Vibrational convection provides a mechanism of heat and mass transfer due to the existence of mean flows. This refers to small-amplitude and high-frequency vibrations, i.e. when the period is much smaller than the reference hydrodynamic times. In this case, the flow field can be decomposed into the "quick" part, which oscillates with the frequency of vibration, and the "slow" time-averaged part (mean flow), which describes the nonlinear response of the fluid to a periodic excitation [11].

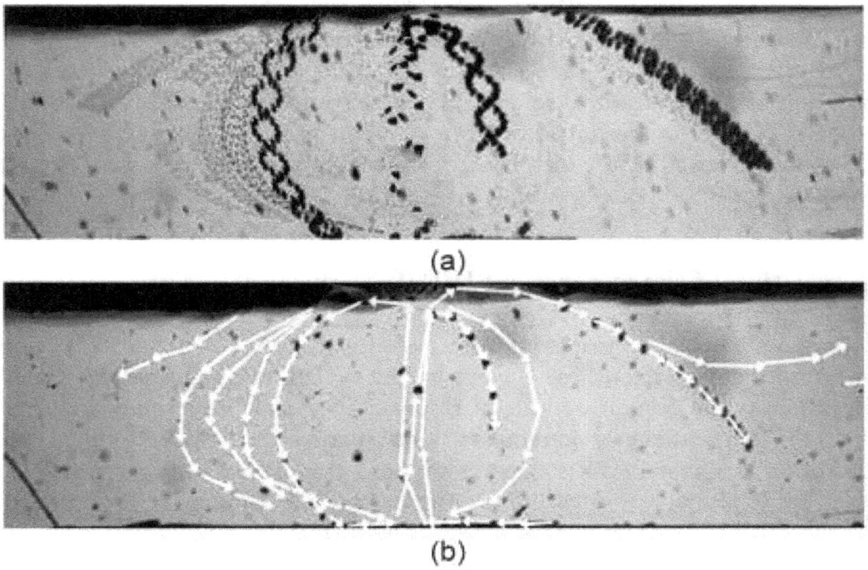

Figure 1.
(a) Image made with 500 superimposed images, keeping minimum pixel intensity for each pixel. The motion of particles illustrates the sum of "quick" and fast part of the flow [12]. (b) Mean flow tracing by selecting images at the same phase of the vibration period [12].

These two types of flow are illustrated by our experiments [12] in **Figure 1**. A few small particles of polytetrafluoroethylene (PTFE) with a characteristic diameter of particles between 20 and 80 μm were present in the cell. The rectangular cavity filled with water-isopropanol mixture was shaken in the horizontal direction with frequency f = 22 Hz and amplitude A = 3.5 mm. Small-amplitude oscillations, as well as average displacement of the particles, can be observed in **Figure 1a**. A fast oscillating contribution seen in the horizontal direction, which has the same frequency as the imposed external vibration, is superimposed on the mean flow with a slow characteristic time scale [6, 11]. The evolution of the mean flow without high-frequency contribution could be observed while selecting images at the same phase of the vibration period. The ensemble of particles in **Figure 1b** represents directly the mean flow without fast oscillations. The arrows in **Figure 1b** highlight the features of the mean flow, clearly demonstrating a flow pattern with two vortices near the cell center.

The effect of the mean flow is most pronounced in the absence of other external forces (in particular, static gravity which has a strong damping effect). In weightlessness, it is an additional way of transporting heat and matter similar to thermo- and solutocapillary convection. Mean flows show some similarity with gravity-induced convection and might serve as a way to control and operate fluids in space [13]. Vibrations can suppress or intensify gravitational convection depending on the mutual orientation of the vibration axis and thermal (compositional) gradient [14, 15].

There have been extensive *theoretical* studies of thermovibrational convection in weightlessness and ground conditions. The fundamental treatise [11] comprises a systematic study of convective flows induced by high- and finite-frequency vibrations in closed and infinite cavities. Thermovibrational convection in square, rectangular and cubic cavities was widely investigated providing a variety of mean flow structures and bifurcation scenarios [16–18]. The influence of vibration on double-diffusive convection with the Soret effect was also analyzed in [19–21].

Experimental studies of vibrational phenomena in weightlessness are rather limited. A series of experiments was carried out onboard the MIR station with the ALICE-2 instrument [22, 23] and DACON [24]. The influence of vibrations on the propagation of a temperature wave from a heat source in a near-critical fluid was investigated. Thermovibrational flows were registered by observing the optical inhomogeneity caused by the distortion of the temperature field. It was not possible to reconstruct this field quantitatively.

5. Preparation of the experiment IVIDIL in parabolic flights

Even though our colleague from MRC had already conducted many experiments in low-gravity environment, for our group, the IVIDIL was the first experiment in weightlessness. We felt an enormous responsibility and proposed to the ESA to conduct the experiment in parabolic flights. Parabolic flights provide repeated periods of approximately 20 seconds of reduced gravity $g \approx 0.01\ g0$ preceded and followed by 20 seconds of hypergravity (up to 1.8 g0), where $g0 = -9.81\ m/s^2$.

Surely, during 20 s one cannot observe the diffusion or Soret process, and the aim of the parabolic flight experiment is to investigate the development of thermovibrational flows in low gravity and verify the existing theoretical studies. During our first parabolic flight, the TEVICON (Thermovibrational Convection [25]) experiments took place in 2007. A large contribution to the success of the experiment was made by Dr. I. Ryzhkov, who worked in our group at that time.

5.1 Theory and numerics

In addition to technical problems and rack design, observation of thermovibrational convection formed in 20 s required serious preparation. The primary task is to find the configurations, where the intensity of thermovibrational convection is significantly larger than that of thermal convection caused by the residual gravity. On the first stage of preparation, we used theoretical estimations and conducted numerical simulation of full Navier-Stokes equations in the closed cubic geometry [8, 9]. Dimensionless variables were introduced by taking the scales of length L, time L^2/ν, velocity ν/L, pressure $\rho_0\nu^2/L^2$ and temperature $\Delta T = T_{hot} - T_{cold}$. The dimensionless equations are written in the form

$$\begin{aligned} &\boldsymbol{u}_t + (\boldsymbol{u}\cdot\nabla)\boldsymbol{u} = -\nabla p + \nabla^2\boldsymbol{u} - (\boldsymbol{Gr} + Gr_v\cos(\Omega t)\boldsymbol{n})T, \\ &T_t + (\boldsymbol{u}\cdot\nabla)T = Pr^{-1}\nabla^2 T, \\ &\nabla\cdot\boldsymbol{u} = 0. \end{aligned} \tag{1}$$

Here, the dimensionless parameters are the Prandtl number, Pr; the Grashof number, Gr; the vibrational Grashof number, Gr_ν; and the dimensionless frequency, Ω:

$$Pr = \frac{\nu}{\chi}, \quad \boldsymbol{Gr} = \frac{\boldsymbol{g}\beta_T\Delta TL^3}{\nu^2}, \quad Gr_v = \frac{A\omega^2\beta_T\Delta TL^3}{\nu^2}, \quad \Omega = \frac{\omega L^2}{\nu}, \tag{2}$$

where A is the amplitude of vibration, $\omega = 2\pi f$ is the angular frequency, $\boldsymbol{g}$ is the gravitational acceleration, β_T is the thermal expansion, υ is the kinematic viscosity, χ is the thermal diffusivity, L is the characteristic size and ΔT is the applied temperature difference.

When a fluid is subjected to high-frequency vibration and density inhomogeneity is caused by the thermal gradient, the vibrational and gravitational convective mechanisms are characterized by alternative dimensionless parameters, such as the Rayleigh number, *Ra,* and the Gershuni number, Gs, its vibrational analogue:

$$Ra = \frac{g\beta_T \Delta T L^3}{\nu\chi}, \qquad Gs = \frac{(A\omega\beta_T \Delta T L)^2}{2\nu\chi}. \tag{3}$$

The ratio Gs/Ra describes the relative importance of thermovibrational and gravitational convective mechanisms. The goal of the numerical study was to understand what kind of mean flow regimes can be observed during the short experimental time of 20 s and assess the influence of residual gravity on the transient process. It should be emphasized that we were interested in mean flows, which can induce heat transfer in a system subjected to external vibration.

It was well known from the previous theoretical studies [11] that in a rectangular cavity under weightlessness, a nonzero mean flow exists at any value of the Gershuni number, *Gs,* when the direction of vibration is perpendicular to the temperature gradient (this configuration was considered in our study). For small values of the Gershuni number, the stationary mean flow is weak and has a four-vortex symmetrical structure (see **Figure 2**, left). When the Gershuni number exceeds some critical value *Gscr*, a flow pattern bifurcates to the pattern with different symmetry: one large diagonal vortex and two small vortices in the corners (see **Figure 2**, right).

The presence of the residual gravity may strongly affect the flow pattern. The numerical simulations have shown that the residual gravity in the Z-direction can destabilize (stabilize) the flow when the gravity vector and the temperature gradient have the same (opposite) directions. Here, the temperature gradient is always codirected with the Z-axis. The *lateral* residual gravity (in the X-direction) is always destabilizing [6]. It holds even if the ratio of the numbers of Gershuni and Rayleigh, which describes the relative importance of vibrational and natural convection due to residual gravity, is very large. For example, the patterns in **Figure 2a** and **b** correspond to Gs/Rax $\sim 2.9{\cdot}10^5$. The sign of lateral residual gravity controls the inclination of the diagonal vortex and the direction of rotation in the steady state as seen from panels (a) and (b).

Next, we examined six different liquid candidates such as water, ethanol, methanol, isopropanol, pentane and transformer oil. The target was to identify the liquid which would provide the largest Gershuni number for the given frequency, amplitude and ΔT. From the definition of the Gershuni number (see Eq. (3)), it follows that this liquid should have large thermal expansion βT but small kinematic viscosity and thermal diffusivity. After detailed consideration, isopropanol has been selected as a working liquid.

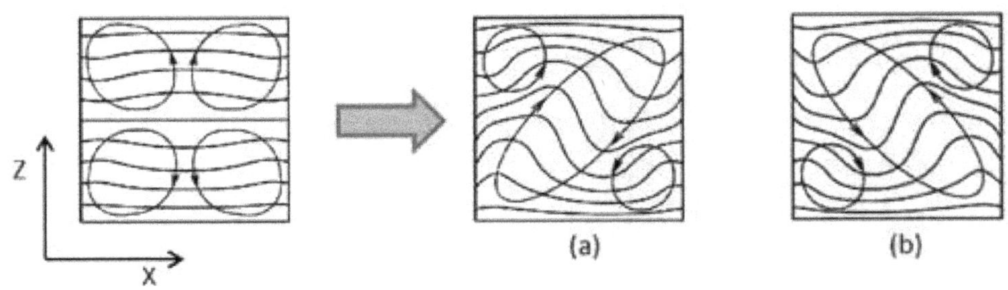

Figure 2.
(Left) Typical flow pattern of vibrational convection predicted theoretically for zero-gravity environment at low Gershuni number. (Right) The flow pattern (a) in presence of lateral gravity; (b) under the action of residual gravity in z-direction.

5.2 Experiment design

To observe thermovibrational convection in microgravity, we have designed a dedicated experimental set-up placed in special rack to be used during parabolic flights (see **Figure 3**, left). Here we provide the basic principles, and the more detailed description can be found elsewhere [5, 6]. The thermovibrational flows were monitored by measuring the temperature field inside the cell by optical digital interferometry. The set-up performance is based on the concept of Mach-Zehnder interferometer shown in **Figure 4**. The light beam of He-Ne laser is expanded by the spatial filter and then splits into two collimated beams of equal intensity by the beam splitter.

We knew about the cell design in the forthcoming IVIDIL experiment, and for the parabolic flight tests, similar (but scaled) cells from Hellma Company were ordered. The cell is made of Suprasil quartz. The internal part, which has a cubic shape with walls of size L = 5 mm, was filled with isopropanol. The external walls of

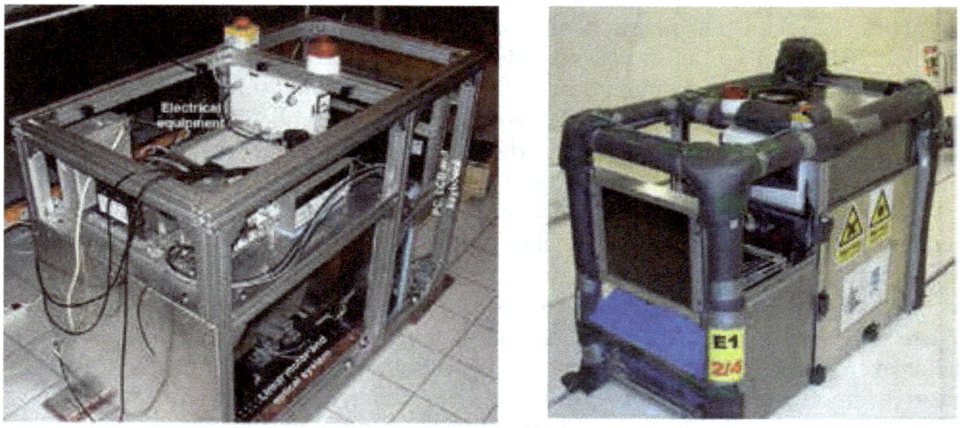

Rack before leaving MRC *Rack is ready to fly*

Figure 3.
Rack designed for the TEVICON-1 parabolic flight experiment in 2007.

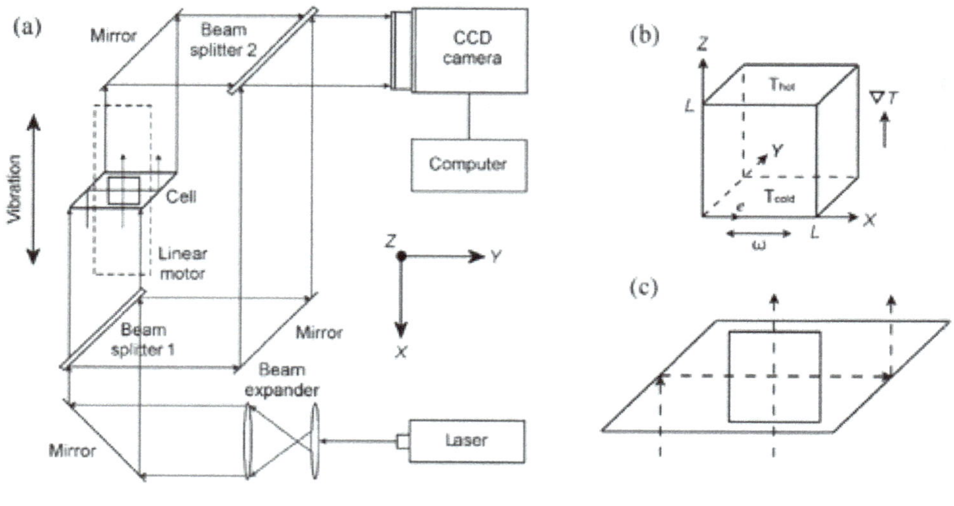

Figure 4.
(a) The scheme of experimental set-up, (b) the cubic cell filled with liquid and coordinate system and (c) the top view of the experimental cell.

the cell are shaped in the form of two transparent prisms (**Figure 4c**) to allow scanning of the front and side views (planes *YZ* and *XZ*, respectively). The beam paths through the cell are shown by the arrows. Due to the smart design of the cell, the objective beam transverses the entire cell in two perpendicular directions.

The top and bottom walls of the cell were kept at constant temperatures *Thot* and *Tcold*, respectively, by Peltier modules. In the experiments, the mean temperature was fixed at 40°C, while the applied temperature difference $\Delta T = Thot - Tcold$ was either 15 K or 20 K. The experimental cell was attached to the linear motor, which performs translational harmonic oscillations in the *X*-direction (perpendicular to the temperature gradient). In the coordinate system associated with the cell, the acceleration applied to the system is the sum of gravitational and vibrational accelerations:

$$\boldsymbol{g}(t) + A\omega^2 \cos(\omega t)\, \boldsymbol{e}, \tag{4}$$

where $\boldsymbol{g}(t) = (gx, gy, gz)$ is the time-dependent gravity vector and $\boldsymbol{e} = (1, 0, 0)$ is the unit vector along the axis of vibrations.

Repetition of the experiments with different vibration parameters and temperature gradients allows to recognize in what extent the thermovibrational convection is sensitive to microgravity and to define some regions of parameters with strong convective flows [6].

5.3 Two busy weeks during the parabolic flight campaign

After long and difficult months of trial and errors in the laboratory, all parts were assembled in a rack and passed numerous tests. Along with real experimental tests, there were a lot of technical reports required by the ESA contractor. Finally, the rack shown in **Figure 3** (left) was placed in a small truck and transported to Bordeaux-Mérignac, along with part of the team.

At our arrival at the Bordeaux-Mérignac airport, we realized that the problems were far from over. The first problem we faced was the thermal controller that was not working. After conducting multiple attempts and tests, we were not able to activate it. The time remaining before flights was rather short. We ordered a new controller on the Internet with an overnight delivery. The next morning, after installing the newly arrived controller, we were ready to cry, as it also did not work. We could not determine the reason. We knew that in our laboratory in Brussels there was another thermal controller, which had been used for several tests, and it worked. We asked Prof. J.C. Legros to bring it by car from Brussels to Bordeaux in the quickest possible way. After installing and testing it, the team cheered with relief: it worked!

Even though the instrument was working, we still needed to go through several control tests, e.g. leak tightness of all the volumes with liquid, safety checks, etc., before being allowed to move the instrument into the plane. One may see our happy faces in **Figure 5** (left), when the rack was allowed into the plane. After a difficult week of problem-solving, we thought it was finally time to fly!

But the obstacles were not over yet. While inside the specialized "Zero-G" aircraft, our rack again needed to undergo a series of control checks by a large team of experts from the *Centre d'Essais en Vol* (CEV, French Test Flight Centre) (see right picture in **Figure 5**). All these controls ended on the eve of the flight experiment day.

The first day of parabolic flights is very exciting. On the one hand, there is the responsibility for the success of the experiment and, on the other hand, a concern

Rack is allowed to be moved to the plane Last safety check

Figure 5.
Rack is transferred to the plane.

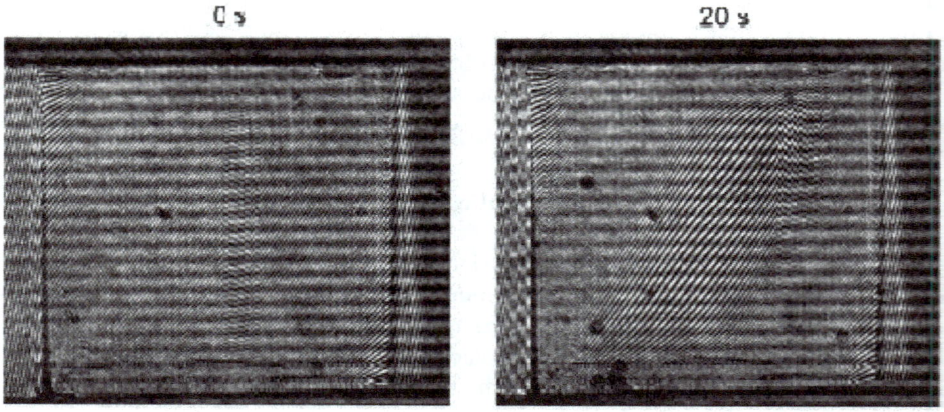

Figure 6.
Interference patterns of the side view of the cell at different times, $f = 4$ Hz, $A = 45$ mm, $\Delta T = 20$ K.

for one's own well-being. After all the effort and the stress of preparation, we were delighted to see that everything went well and without considerable problems. We managed to get excellent scientific results, subsequently published in high-stand journals!

5.4 Scientific and strategic results out of TEVICON experiment

Figure 6 presents the acquired interference patterns of the cell at 0 and 20 s from the start of vibration. During parabolic flight experiments, we watched only black-and-white images and visually drew a conclusion about the power of convection. In the first frame (0 s), the pattern is formed by the interchange of the thin black and white lines (fringes).

These lines are inclined and almost parallel; in addition, they have the same thickness except for the regions near the lateral walls, where temperature slightly deviates from purely conductive state. In the second frame (20 s), we see that the thickness of fringes increases in the diagonal from the bottom left to the top right corner. Around this diagonal, the thickness decreases. In addition, the deformation of fringes becomes larger. All these changes result from the deformation of the temperature field by thermovibrational convection.

To obtain the temperature field, these fringe images were processed by performing 2D fast Fourier transform, filtering the selected band of the spectrum, performing the inverse transform and phase unwrapping. The knowledge of the phase shift gives information about the gradient of refractive index, which is used to reconstruct 2D projections of the temperature field on the front and side view planes. The corresponding thermal fields (the first and last images in color in **Figure 7**) were restored from the interference patterns in **Figure 6**.

The transient development of vibrational mean flows was observed during a large number of microgravity periods with different levels of vibration (each period lasts around 20 s). Measurements of temperature field revealed large deviations from conductive state due to vibrational mean flows. The development of vibrational convection during experimental run with Gs = 71.15 $\times$ 10^3 is shown in **Figure 7**.

The first picture from the left in **Figure 7** (0s) shows the temperature field in the cell (side view) at the beginning of the parabola when vibration is imposed. There are some deviations from a conductive state, which are small in the central region of the cell but become larger near the lateral walls due to heat fluxes through them. One of the reasons for temperature deviations is the weak convection caused by the residual gravity. The residual accelerations in the X and Y directions (perpendicular to the thermal gradient) can slightly destabilize vibrational mean flows. The analysis of different parabolas showed that the deviations from a conductive state were not large. Another reason can be associated with the fact that the walls are formed by two glass prisms, which can absorb heat. Note that thin regions ($\sim$0.2 mm) near the horizontal walls were inaccessible for optical measurements.

The development of thermovibrational flow causes the distortion of the temperature field, which is growing with time. The numerical snapshot taken at 1 s from the start of vibration reveals the mean flow structure with four vortices. The experimental thermal field also indicates the presence of this structure. The flow is not completely symmetrical because of the presence of residual gravity. This result is consistent with the predictions of two-dimensional numerical modeling [6]. The bifurcation from a symmetric pattern (1 s) to an asymmetric one (9 and 20 s) is observed. These images are recorded at different times but correspond to the same phase of vibration, in which the cell was in the focus of the camera.

Here, we would like to emphasize *the strategic* importance of the parabolic flights [25, 26] for the success of the IVIDIL experiment.

The IVIDIL experiment was carried out in the Selectable Optical Diagnostic Instrument, hosted in the Microgravity Science Glovebox (MSG) onboard the International Space Station. The SODI is a modular instrument with experiment-dependent exchangeable cells that are probed by means of optical interferometry (Mach-Zehnder). It is equipped with two Mach-Zehnder interferometers that can

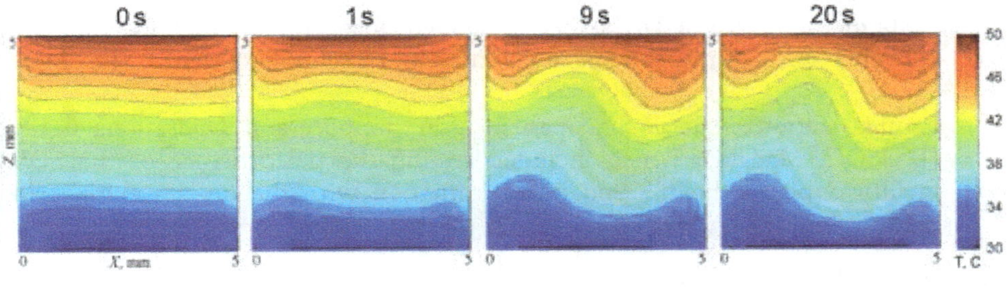

Figure 7.
The evolution of temperature field with time in the side view during the run with f = 4 Hz, A = 45 mm, ΔT = 20 K, Gs = 71,149.

Figure 8.
Science team moves from parabolic flights to the IVIDIL experiment.

be operated at two wavelengths (670 and 935 nm). The SODI allowed selecting a set of optical properties and the fringe spacing, depending on the performance requested by the scientists. In the case of IVIDIL, a total of two cell arrays consisting of two cells each have been analyzed.

This parabolic flight experiment in many aspects has a design similar to IVIDIL; in particular, the camera was mounted to the wall of a box at the end of the vibrational table. The camera had fixed frame rate (24 fps), and the cell vibrated at a given frequency. Thus, when the cell was moving, the camera recorded images both in and out of focus. We discussed with the ESA about a possible need to change the SODI design. However, the design of hardware was impossible to change, but the improvement was made by software on the image treatment. The camera records were synchronized with the frequency of the vibration in each experimental run.

Using all the experience obtained during the parabolic flight experiments, the team focused attention to the scientific part of the IVIDIL preparation [26] (see the team pictures in **Figure 8**).

6. Ground preparation and tests of the IVIDIL experiment

The full program of the IVIDIL experiment tests with participation of the science team consisted of several segments:

a. Numerical simulations

b. Cell filling at the QinetiQ Space company

c. Tests of the entire SODI facility imbedded inside the engineering model of Glovebox at the European Space Research and Technology Centre (ESTEC)

d. Familiarization tests at the Spanish User Support and Operations Centre (E-USOC, Madrid), which included testing the performance of the SODI facility onboard the ISS and its coordination with the on-site engineering model

6.1 Numerical simulations

Numerical programs in support of the IVIDIL experiment have been developed in two ways. The first one was devoted to the calculations of the 3D Navier-Stokes equations using the geometry of the experiment and physical properties of liquid mixtures. For this, two international teams, one from Russia led by Prof. T. Lyubimova and the other from Canada led by Prof. Z. Saghir, have joined the MRC team. The numerical codes were benchmarked between the teams prior to the experiment to be conducted on ISS, and preliminary calculations were performed to select the best experimental parameters [9].

The second way was devoted to the development of codes for image processing and fitting of the experimental results to the analytical solution. Part of it was already done for parabolic flight experiments, but additional work was needed [1–3].

6.2 Cell filling

The cell filling with experimental liquids is a very delicate process. In the experiment with non-uniform temperature, concentration and vibrations, if any bubble would appear, the temperature and concentration fields in the cell would be strongly affected in several aspects. It would lead to Marangoni convection driven by the variation of surface tension on the bubble; in addition, a vibrating bubble would disturb the local flow field and mix the existing concentration gradient in an unwanted and uncontrolled way. Therefore, all experiment liquid solutions are degassed in vacuum chambers to remove all dissolved gases.

The science team prepared the solution mixtures in the MRC laboratories, degassed the solutions and delivered them to QinetiQ Space Company in sealed syringes as shown in **Figure 8**. The cell was filled in the QinetiQ Space by their experienced engineer in the presence of scientists and of an ESA representative. The procedure relies on evacuating the cell volume and then letting the sample be sucked from the syringe into the cell by the vacuum, as there is only one opening for filling. After the preliminary filling, the cell was inspected visually under 20 x magnification, to ascertain that no air is present in the sample. The image of the cell after the first run is shown on the left side in **Figure 9**. The bubbles found were removed manually by slightly tilting the cell array and forcing the bubbles through the filling hole.

The cell was inspected again and again, and after final inspection, the cell was sealed. The developed cell filling procedure brought an excellent result; the cells

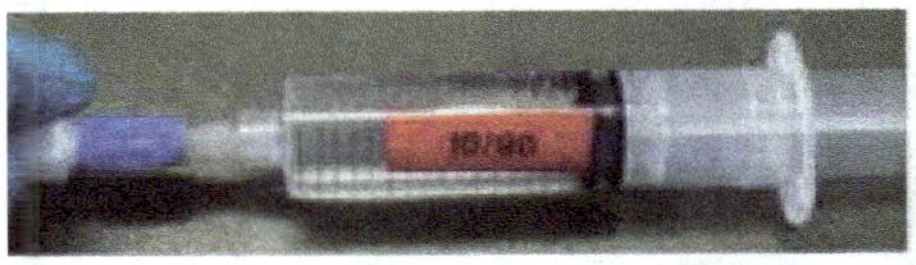

Liquid delivered by the science team

The magnified view of the cell

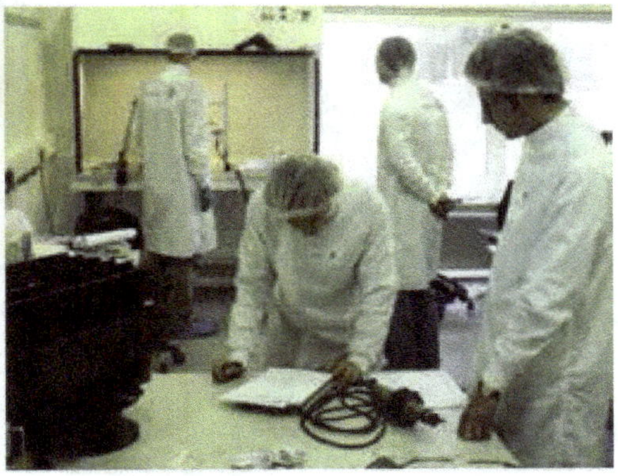

Collaborative work of the science team, ESA and Qinetiq engineer.

Figure 9.
The IVIDIL flight cell filling on May 28, 2009.

were filled on May 28, 2009, and the IVIDIL experiment was completed 8 months later, and all the cells were bubble-free.

6.3 Tests of SODI-IVIDIL facility at the ESTEC

For any experiment on the ISS, at least two identical instruments are usually manufactured: the flight model and the engineering model. Most ground tests are performed using the engineering model. Delays in the manufacture of dedicated highly innovative instruments often occur, and it leads to a limited time for the ground testing. The SODI-IVIDIL facility placed inside the Glovebox model was tested in the ESTEC on February 23, 2009 and February 24, 2009.

The IVIDIL test procedure on the first day included four tests; in each of them, images were recorded during 180 s with sampling rate 10 s:

1. Isothermal cells $\Delta T = 0$, no vibration

2. Temperature gradient ΔT from above and below, no vibration

3. Temperature gradient ΔT from above and below, vibration of $f = 2$ Hz and $A = 25$ mm

4. Temperature gradient ΔT, vibration of highest forcing $f = 2.8$ Hz and $A = 31$ mm

(The tests were repeated two times: with and without personnel near the instrument)

The IVIDIL set-up consists of three principal parts (a cell array, a vibrational table and an optical system) as shown in **Figure 10**, right side. Cell array consists of two cells (as shown in **Figure 4c**) filled with the same binary mixture: the primary cell and the companion one, into which the particles are added. Each cell was monitored by the separate laser diode with wavelength $\lambda = 670$ nm.

After processing of recorded test images at the end of the day, it was found that the variation in contrast over images was quite high and not the same for different laser diodes. The next morning, we warned the developers of optical parts (Lambda-X) about the problems found, and in response, several parameters were changed. For example, the integration time of one of the cameras was increased, the

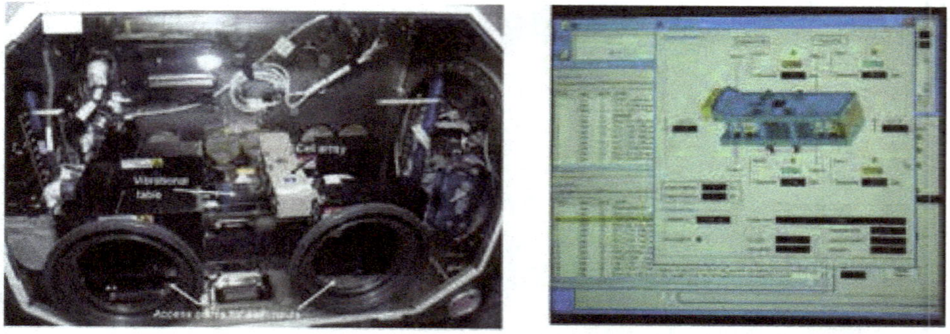

The IVIDIL set-up inside the Glovebox facility

Real-time IVIDIL experiment screenshot at E-USOC

Figure 10.
SODI-IVIDIL on orbit and the experiment monitoring at the E-USOC.

laser temperature was changed, and the electrical current on laser was changed. All these small adjustments have led to a significant improvement in image quality.

The last part of the experiment preparation took place in the Spanish User Support and Operations Centre (E-USOC) in Madrid. To get a better understanding of the experiment hardware, and the possible in-flight changes of experimental parameters, the science team went to the E-USOC in Madrid in July 2009. During these days, the engineering model of SODI-IVIDIL was used to learn about in-flight operations and to simulate experimental runs. Also, telemetry software necessary for observation of the experiment during operations was installed and tested on the computers of the science team. The software for image processing was tested on data acquired with the engineering model. The E-USOC team working with SODI-IVIDIL was very dedicated and was very helpful in conducting all experiment preparations.

To conclude, all preparatory tests made essential contribution to the success of the experiment. However, the IVIDIL experiment overcomes the limits of ground tests which do not reproduce the space environment.

7. The IVIDIL on orbit

Shortly after arrival on orbit, on September 23, 2009, the SODI-IVIDIL was installed inside the Microgravity Science Glovebox in the Columbus laboratory by the ESA astronaut Frank De Winne and his Canadian crewmate Robert Thirsk. The activation of the IVIDIL experiment was scheduled on October 5, and the science teams as well as engineers from QinetiQ (Verhaert) Space were present at the E-USOC. The waiting time of the first image was exciting and nerve-wracking—does it really work? Are the images of good quality? The faces of the team at this breathtaking moment can be seen on the photos in **Figure 11**.

Yes, it worked!

Figure 11.
Waiting for the first response from IVIDIL in the E-USOC control room.

Since this memorable day, 55 experiments were conducted on the International Space Station. The experiment lasted from October 2009 to January 2010. Several interruptions occurred over this time, and they are related to the life onboard. First, it was surprising to learn that if there was a national holiday in one of the countries managing by the experiment, the experiment could run only in regime "unattended" or stopped, like it was during Christmas.

Second surprise was to find out that by default, astronauts do not work on Saturday and Sunday. Once, this nearly led to a critical situation. The first flash disk of the IVIDIL was nearly full and had to be replaced by another one. Among the astronauts onboard, only the ESA astronaut was trained to work with SODI-IVIDIL. His staying on the ISS was coming to an end, and the only chance to replace the disk was to do it on a Saturday. We asked through NASA if the ESA astronaut would accept to replace the disk as part of a volunteer work on a Saturday. He agreed and the disk was replaced.

Each experimental run lasted 18 h, and all the experiments were controlled via telescience provided by the European Spanish User Support and Operations Centre (E-USOC, Madrid), which means that the experiment science run was completely controlled from the ground without intervention of onboard astronauts. Part of the images acquired in each run was transmitted to the ground in real time, and a full set of images arrived on a flash disk a few months later, after completing the experiment. The science team stayed in the E-USOC for a week, until all the tests were completed and scientific runs started. Later, the control images and telemetry data were sent to Brussels on a daily basis on a specialized ftp server. At the very end, after completion of all the planned runs, we had the chance to ask several "nice to have" experiments which were interesting for science after quick analysis of the available results.

8. Scientific results of the IVIDIL

Preliminary results were obtained using a limited amount of data sent via telemetry. Thousands of recorded images returned to Earth on a flash disk a few months later. The experimental results have surpassed our expectations [27–32]. The experiment has been able to trace the variation of concentration of about 0.03%from the initial composition.

8.1 Impact of the onboard g-jitter on diffusion-controlled processes

After the numerous runs of the IVIDIL experiment, it was clearly demonstrated that only the major transient space station vibrations have an impact, such as those due to orbital debris avoidance maneuvers or dockings and undockings of a spacecraft. Furthermore, their impact depended on the duration of events. More common minor movements that are part of daily life aboard a space station did not affect the samples. Let us justify this statement.

The effect of the onboard environment on a diffusion-controlled process can be identified in the following ways: (a) by direct observation of temperature or concentration fields and their smoothness; (b) by comparing fields along two perpendicular views; (c) by reproducibility of experimental results when repeating experiment on different days and thus in a different microgravity environment; (d) by comparing with computations in the absence of perturbations; and (e) by observation of g-jitter-induced convection, in the case that it arises.

In the steady state, when mixture has Soret effect, the concentration field and the temperate field have similar distributions (here ST is the Soret coefficient):

$$C - C0 = -ST\,(T - T0). \tag{5}$$

The steady-state distribution of the temperature field measured onboard is shown in one of the views in **Figure 12**, left side. The temperature field is set within a few minutes, while the concentration field takes more than 10 hours to reach the steady state. Thus, the concentration field is more susceptible to the influence of g-jitter.

The concentration (temperature) field was recorded in two perpendicular directions, as shown in **Figure 4**. Accordingly, in the experiments, when the imposed vibrations are absent or weak, the images should be similar in both directions if they are not perturbed by the microgravity environment. **Figure 12** (right side) provides an appropriate illustration of the absence of perturbations. The central part of the figure presents five isosurfaces of equal concentrations of the experiment at the end of the thermodiffusion step, after 12 h of Soret separation. The midsurface (green) corresponds to the initial concentration $C_0 = 0.90$, and the following surfaces are separated by $\delta C = 6 \times 10^{-4}$.

This 3D concentration field reveals no significant disturbances due to onboard accelerations, except for small ripples of isolines typical of most experiments. The distributions on the sides show averaged concentration fields in perpendicular directions, which are almost identical. Note that these concentration distributions on the sides are similar to the temperature field on the left side of the **Figure 12**.

Several experiments without imposed vibrations were conducted on different days (even months) and for two different binary mixtures. On one of the days, the gravitational environment had stronger perturbations, measured by the Space Acceleration Measurement System (SAMS) instrument. It resulted in a slightly larger scattering of the experimental points [27], but the final results coincided with those from the other days and with the numerical separation for the free-convection case.

Another experiment, with imposing weak vibrational forcing, $g_{os}/g_0 = 7.0 \times 10^{-4}$, was repeated with an interval of 2 weeks. The measured value of the Soret separation, ΔC, which is the concentration difference at the hot and cold walls, was almost identical. Furthermore, the ΔC value for these experiments was similar to

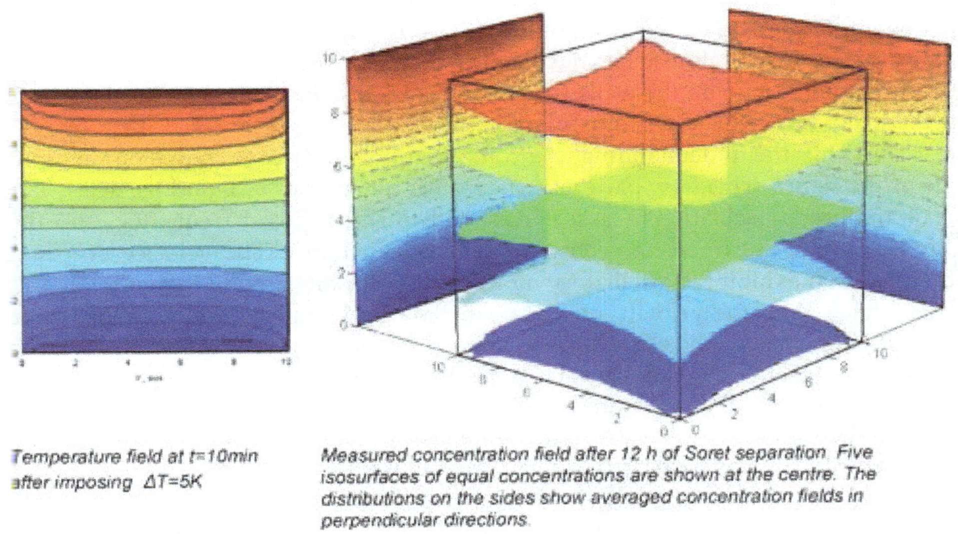

Figure 12.
Measured temperature and concentration field in g-jitter environment of the ISS.

that one obtained in the experiments carried out only in g-jitter environment. These observations emphasize that neither the weak imposed vibrations nor onboard accelerations affect the process of diffusion.

Note that these results may not be applicable in the case of the large peak accelerations that occur on the ISS due to the maintenance of the station or in case of a fluid system with a sharp difference in density (e.g. gas/liquid interface).

8.2 Development of mean flow under the action of periodic forcing

The IVIDIL experiment also demonstrated that, unlike g-jitter, imposed vibrations with constant frequency and amplitude do affect the diffusion process. High-frequency periodic forcing with a zero mean value causes time-averaged (mean or streaming) flows, which substantially affect the regime of mass transfer in a fluid.

The physical mechanisms through which mean flows modify thermal and concentration fields in a binary mixture with a negative Soret effect (water-isopropanol 90–10%) were examined. The evolution of the mean flows is the result of a nonlinear interaction between thermal, solutal and vibrational effects. Because vibrations act on a mixture with non-uniformities in density, the development of the mean flows depends on this gradient. The vibrations were activated 10 min after imposing a temperature gradient ΔT. By that time, the initial density stratification was created by the thermal field in the cell and the concentration gradients very near the solid walls. First, on a short time scale, the vibrations create significant mean flows due to the presence of density gradients primarily induced by thermal non-uniformities. Then, concentration variations caused by a slow Soret effect change the density gradients that drive vibrational convection, which in turn modifies the concentration field. The density gradient caused by a negative Soret effect is opposite to that created by the thermal field and reduces the net density gradient. Thus, the growth of perturbations in a concentration field is decelerated.

Two regimes of mass transfer were identified, depending on vibrational forcing: diffusive for Gs < 1200 and convective for Gs > 1200. In the diffusive regime, the Soret separation decreases linearly with an increase in Gs, and the mean flow begins to affect the distribution of the concentration field. The example of the concentration field and its cross-section in two perpendicular directions at the central part is shown in **Figure 13** when Cs = 910. One may compare this concentration field with that one in **Figure 12**, obtained in g-jitter environment. In the g-jitter environment, the isosurfaces were almost flat, just slightly bended toward the wall due to temperature nonlinearity on the walls. The concentration field in **Figure 13** is deformed by the mean flow which has four-roll structure.

To simplify the understanding, the computed flow pattern is shown on the right side of the figure. This flow pattern is shown via trajectories of liquid particles. The small-amplitude oscillations (fast) correspond to the imposed frequency, and the four large rolls illustrate convective motion (mean flow). The velocity of "fluid particles" that exhibit fast oscillatory motion while moving around the vortex is 100 times higher than the velocity of the mean flow (1–10 mm/s) and is of the same order of magnitude as the characteristic viscous velocity L/ν.

In the convective regime, Gs > 1200, the mean flow is strong enough to trap the concentration near the walls and maintains a constant flow velocity in the bulk. Close to the Cs = 1200, the flow pattern keeps the same four-roll structure which progressively deforms with further increase of vibrational forcing.

In the frame of a postflight study, the vibrational environment (g-jitter) recorded by the SAMS instrument during the IVIDIL experiment was analyzed [33–35] as well as its possible influence of the Soret separation [36]. The study shows that during experimental runs, there were no major perturbations of the microgravity

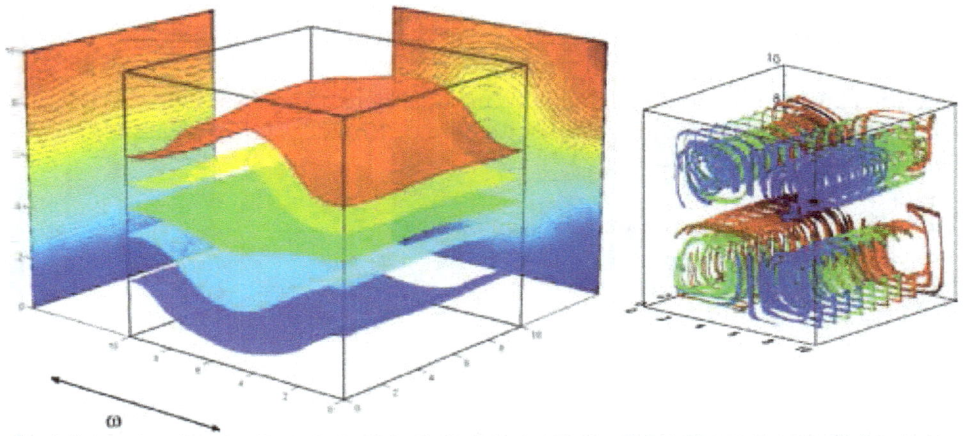

The 3D (tomographic reconstruction) concentration field and its cross-section in two perpendicular directions at the central part in case of imposed vibration

The fluid particles are moving along the mean flow streamlines while performing fast oscillatory motion with small amplitude

Figure 13.
(Left) The concentration field in diffusive regime restored from the IVIDIL data, Gs = 910. (Right) Numerical presentation of the flow pattern for this regime.

environment. However, an unknown frequency equivalent to the third harmonic, but with large amplitude, was recorded during runs at which the strongest vibrational forcing was applied. A more detailed study suggests that, with high vibrational excitations, the linear motor also generates lateral cell motion perpendicular to the main direction. The interactions of these movements lead to the generation of additional frequencies acting on the liquid mixture.

9. Conclusions

In order to take full advantage of the unique opportunity of ISS experiments, the ground-based preparations are indispensable.

Ground experiments are aimed at confirming the operating principle of onboard equipment and motivate the development of codes that will be used to process the microgravity results.

The creation and use of mathematical models is another key element of preparation. Mathematical modeling provides the characteristic scales and the range of parameters that may be beneficial for research. However, any mathematical model contains many assumptions and idealizations. A comparison of numerical predictions and ground-based observations, although perturbed by gravity, can rearrange the set of parameters used in microgravity.

The culmination of ground-based preparation is the tests on parabolic flights. A lot of useful data can be collected from parabolic flight experiments, and they may even call for modification of the onboard installation before heading further into space.

Author details

Valentina Shevtsova*, Denis Melnikov, Yuri Gaponenko and Aliaksandr Mialdun
Microgravity Research Centre, Université libre de Bruxelles (ULB), Brussels, Belgium

*Address all correspondence to: vshev@ulb.ac.be

References

[1] Mialdun A, Shevtsova V. Development of optical digital interferometry technique for measurement of thermodiffusion coefficients. International Journal of Heat and Mass Transfer. 2008;**51**: 3164-3178

[2] Mialdun A, Shevtsova V. Digital interferometry as a powerful tool to study the thermodiffusion effect. Comptes Rendus Mecanique. 2011;**339**: 362-368

[3] Mialdun A, Shevtsova V. Measurement of the Soret and diffusion coefficients for benchmark binary mixtures by means of digital interferometry. The Journal of Chemical Physics. 2011;**134**:044524

[4] Mialdun A, Shevtsova V. Measurement of Soret coefficients: Open questions. Microgravity Science and Technology. 2009;**21**:31-36

[5] Mialdun A, Ryzhkov I, Melnikov D, Shevtsova V. Experimental evidence of thermovibrational convection in low gravity. Physical Review Letters. 2008; **101**:084501

[6] Shevtsova V, Ryzhkov I, Melnikov D, Gaponenko Y, Mialdun A. Experimental and theoretical study of vibration-induced thermal convection in low gravity. Journal of Fluid Mechanics. 2010;**648**:53-82

[7] Shevtsova V, Gaponenko Y, Melnikov D, Ryzhkov I, Mialdun A. Study of thermoconvective flows induced by vibrations in reduced gravity. Acta Astronautica. 2010;**66**:166-173

[8] Melnikov D, Ryzhkov I, Mialdun A, Shevtsova V. Thermovibrational convection in microgravity: Preparation of a parabolic flight experiment. Microgravity Science and Technology. 2008;**20**(1):29-39

[9] Shevtsova V, Melnikov D, Legros JC, Yan Y, Saghir Z, Lyubimova T, et al. Influence of vibrations on thermodiffusion in binary mixture. Benchmark of numerical solutions. Physics of Fluids. 2007;**19**:017111

[10] Melnikov D, Shevtsova V, Legros JC. Impact of conditions at start-up on thermovibrational convective flow. Physical Review E. 2008;**78**(5): 056306

[11] Gershuni GZ, Lyubimov DV. Thermal Vibrational Convection. New York: Wiley & Sons; 1998

[12] Gaponenko YA, Torregrosa M, Yasnou V, Mialdun A, Shevtsova V. Dynamics of the interface between miscible liquids subjected to horizontal vibration. Journal of Fluid Mechanics. 2015;**784**:342-372

[13] Beysens D. Vibrations in space as an artificial gravity? Europhysics News. 2006;**37**:22

[14] Wolf GH. Dynamic stabilization of the interchange instability of a liquid–gas interface. Physical Review Letters. 1970;**24**:444-446

[15] Gaponenko Y, Torregrosa M, Yasnou V, Mialdun A, Shevtsova V. Interfacial pattern selection in miscible liquids under vibration. Soft Matter. 2015;**11**:8221-8224

[16] Savino R, Monti R. Physics of Fluids in Microgravity. London: Taylor & Francis; 2001. p. 178

[17] Hirata K, Sasaki T, Tanigawa H. Vibrational effects on convection in a square cavity at zero gravity. Journal of Fluid Mechanics. 2013;**445**:327-344

[18] Demin VA, Gershuni GZ, Verkholantsev IV. Mechanical quasi-

equilibrium and thermovibrational convective instability in an inclined fluid layer. International Journal of Heat and Mass Transfer. 1996;**39**:1979

[19] Gershuni GZ, Kolesnikov AK, Legros JC, Myznikova BI. On the vibrational convective instability of a horizontal, binary-mixture layer with Soret effect. Journal of Fluid Mechanics. 1997;**330**:251

[20] Zebib A. Low-gravity sideways double-diffusive instabilities. Physics of Fluids. 2001;**13**:1829-1832

[21] Lyubimova TP, Shklyaeva E, Legros JC, Shevtsova V, Roux B. Numerical study of high frequency vibration influence on measurement of Soret and diffusion coefficients in low gravity conditions. Advances in Space Research. 2005;**36**:70-74

[22] Zyuzgin AV, Ivanov AI, Polezhaev VI, Putin GF, Soboleva EB. Convective motions in near-critical fluids under real zero-gravity conditions. Cosmic Research. 2001; **39**(2):175

[23] Garrabos Y, Beysens D, Lecoutre C, Dejoan A, Polezhaev V, Emelianov V. Thermoconvectional phenomena induced by vibrations in supercritical SF6 under weightlessness. Physical Review E. 2007;**75**:056317

[24] Babushkin IA, Bogatyrev GP, Glukhov AF, Putin AF, Avdeev SV, Ivanov I, et al. Investigation of thermal convection and low-frequency microgravity by the DACON sensor aboard the MIR orbital complex. Cosmic Research. 2001;**39**(2):161-169

[25] TEVICON. Transient time of thermal-vibrational convection in reduced gravity, Erasmus Experiment Archive, Experiment Record N° 8990. Available from: http://eea.spaceflight.esa.int/portal/exp/?id=8990 [Last consulted: 19 February 2020]

[26] TEVICON-1. Thermo-vibrational convection in reduced gravity, Erasmus Experiment Archive, Experiment Record N° 9027. Available from: http://eea.spaceflight.esa.int/portal/exp/?id=9027 [Last consulted: 19 February 2020]

[27] Shevtsova V. IVIDIL experiment onboard the ISS. Advances in Space Research. 2010;**46**(5):672-679

[28] Shevtsova V, Lyubimova T, Saghir Z, Melnikov D, Gaponenko Y, Sechenyh V, et al. IVIDIL: On-board g-jitters and diffusion controlled phenomena. Journal of Physics: Conference Series 2011;**327**:012031

[29] Mazzoni S, Shevtsova V, Mialdun A, Melnikov D, Gaponenko Y, Lyubimova T, et al. Vibrating liquids in space. Europhysics News. 2010;**41**:14

[30] Shevtsova V, Gaponenko YA, Sechenyh V, Melnikov DE, Lyubimova T, Mialdun A. Dynamics of a binary mixture subjected to a temperature gradient and oscillatory forcing. Journal of Fluid Mechanics. 2015;**767**:290-322

[31] Shevtsova V, Mialdun A, Melnikov D, Ryzhkov I, Gaponenko Y, Saghir Z, et al. IVIDIL experiment onboard ISS: Thermodiffusion in presence of controlled vibrations. Comptes Rendus Mecanique. 2011;**339**:310-317

[32] Mialdun A, Yasnou V, Shevtsova V, Koeniger A, Koehler W, Alonso de Mezquia D, et al. A comprehensive study of diffusion, thermodiffusion, and Soret coefficients of water-isopropanol mixtures. The Journal of Chemical Physics. 2012;**136**(24):244512-244513

[33] Sáez N, Ruiz X, Gavaldà F, Shevtsova V. Comparative analyses of ESA, NASA and JAXA signals of acceleration during the SODI-IVIDIL experiment. Microgravity Science and Technology. 2014;**26**:57-64

[34] Sáez N, Ruiz X, Jna.Gavaldà, Pallarés J, Shevtsova V. Comparative ISS accelerometric analyses. Acta Astronautica. 2014;**94**:681-689

[35] Sáez N, Jna.Gavaldà, Ruiz X, Shevtsova V. Detecting accelerometric nonlinearities in the international space station. Acta Astronautica. 2014;**103**: 16-25

[36] Gaponenko Y, Mialdun A, Shevtsova V. Experimental and numerical analysis of mass transfer in a binary mixture with Soret effect in the presence of weak convection. European Physical Journal. 2014;**37**:90

7

Aircraft Parabolic Flights: A Gateway to Orbital Microgravity and Extra-Terrestrial Planetary Gravities

Vladimir Pletser

Abstract

Aircraft parabolic flights provide repetitively periods of reduced gravity whose duration depends on the target reduced gravity level and on the type of aircraft used. Typical durations for a large aircraft vary between approximately 20 s for a 0 g environment and 32 s for a Martian g level environment. Parabolic flights are used to perform short duration scientific investigations in weightlessness or at partial-g levels, to train astronauts, and to allow the general public to experience what astronauts' encounters in space during discovery flights. This chapter introduces the aircraft parabolic flights and presents their objectives, the parabolic flight manoeuvre and accelerations, the type of flights organised and a summary of the aircraft used throughout the world.

Keywords: aircraft parabolic flights, weightlessness, microgravity research, moon gravity, Mars gravity

"*Pull-up...* ". The plane pitches up and the horizon tilts through the windows.
"*Twenty...* ". I can't miss it; I'm getting heavier and heavier.
"*Thirty...* ". Around me, most of the researchers are lying down. I have trouble raising my arms. I'm getting heavier and heavier, almost double my normal weight.
"*Injection!*" And suddenly it's the long-awaited and much-repeated miracle. I'm floating freely and flying. I grab one of the straps on the ceiling and with one impulse, I throw myself parallelly to the ceiling to cross the entire cabin of the plane in its length. Below me, the 40 scientists and engineers invited by the European Space Agency to take part in this campaign of microgravity research in parabolic flights are busy. Some are running their instruments and observing the reactions on their computer screens, others do strange things to their subjects who willingly lend themselves to life science experiments.
But already, the pilot announces the aircraft's nose-down angles at the end of his parabola.
"*Twenty*". It's time to think about finding a place to land on my feet.
"*Thirty*". I'm landing between two scientific experiments: one studying manual grip at different gravity levels and the other investigating problems of convection and capillarity in liquids.
"*Pull-out*". It's about time. After landing gracefully on my feet, I look around. Twenty seconds is finally short but at the same time very long, long enough to conduct research in the absence of gravity. The twelve teams that have been working for several months to prepare their experiments are concentrated and are already preparing for the next parabola that the pilot announces, "*One minute*". And so, the cycle starts again.

1. Introduction

Aircraft parabolic flights are used in many countries [1], firstly by space agencies to conduct research and to prepare for space flight missions, and secondly by private operators to provide the general public with the opportunity to experience reduced gravity[1] conditions.

Aircraft parabolic flights are a useful tool for performing short duration scientific and technological experiments in reduced gravity, to train astronauts in this new environment and in associated procedures and protocols, and to test experiments and instrumentation prior to future space missions. The principal value of parabolic flights is in the verification tests that can be conducted prior to space experiments in order to improve their quality and success rate, and after a space mission to confirm or invalidate (sometimes conflicting) results obtained from space experiments.

The main advantages of parabolic flights for reduced gravity investigations are:

- the short turn-around time (typically of a few months between the experiment proposal and its performance);
- the low cost involved (space agencies provide the flight opportunity free of charge to selected investigators);
- the flexibility of experimental approach (laboratory type instrumentation is most commonly used);
- the possibility of direct intervention by investigators on board the aircraft during and between parabolas, and the possibility of modifying the experiment set-up between flights.

Furthermore, aircraft parabolic flights are the only suborbital carrier to provide the opportunity to carry out medical and physiological experiments on human subjects under conditions of microgravity or reduced gravity such as that found on other planetary bodies (Martian 0.38 g, Lunar 0.16 g), to prepare for extra-terrestrial planetary exploration.

This chapter presents first the experimental objectives of parabolic flights with the capability of conducting investigations at Moon and Mars gravity levels in addition to microgravity; second, the parabolic flight manoeuvres, the accelerations and the technical capabilities of a parabolic flight aircraft; third, the type of flights organised, including the safety and medical aspects; and fourth, a summary of the aircraft used throughout the world.

2. Objectives of parabolic flights

An aircraft in parabolic flight provides investigators with a laboratory for scientific experimentation where the gravity levels are changed repetitively, giving

---

[1] The term 'gravity' usually designates the force associated to the phenomenon of gravitation. One should therefore in all correctness use here the term 'weight', which is the force associated to the phenomenon of weightiness, and the resulting force of gravity and inertia forces existing in the reference frame in which the measure of the weight is done. However, the common (although incorrect) usage of the term 'gravity' and its derivatives (microgravity, hypergravity, ...) instead of 'weight' and its derivatives will prevail throughout this chapter.

successive periods of either 0.38 g (Mars gravity) for up to 32 s, or 0.16 g (Moon gravity) for up to 25 s or 0 g for 20 s, preceded and followed by periods of 20 s at approximately 1.8 g (**Figure 1**), where g is the acceleration created by gravity at Earth's surface, on average 9.81 m/s^2.

Parabolic flight objectives pursued by investigators vary. From a scientific point of view, the following objectives can be attained:

- the performance of short experiments for which the reduced gravity is low enough for: qualitative experiments of the 'look and see' type, using laboratory type equipment to observe and record phenomena; and quantitative experiments to measure phenomena in reduced gravity, yielding direct quantitative exploitable results;

- the ability for experimenters to perform their own experiments in reduced gravity with the possibility of direct interventions on the experiment in progress during the low g periods and direct interaction by changing experiment parameters between the reduced gravity periods (**Figure 2**);

Figure 1.
NASA's KC-135 during the pull-up phase (Credit: NASA).

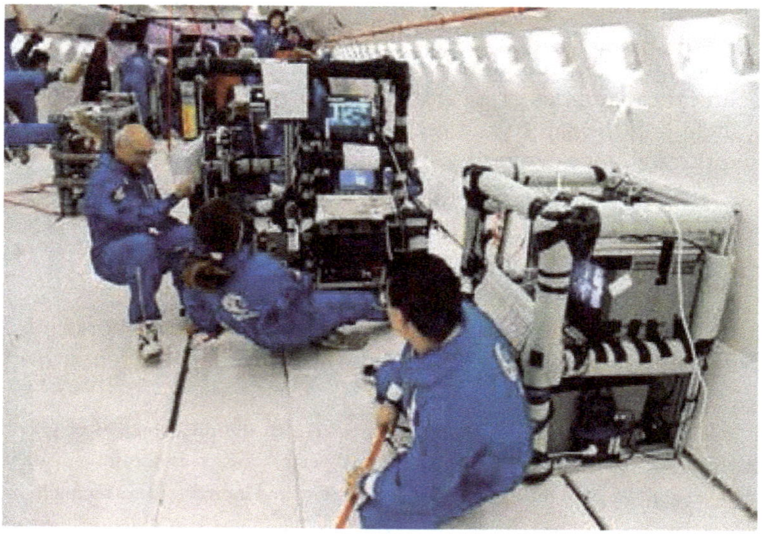

Figure 2.
Experiment on vibrational phenomena during an ESA campaign (Credit: ESA).

- the study of transient phenomena occurring during the transition between high to low and low to high g phases.

Furthermore, for scientific experiments to be performed during space missions, the following goals can be pursued during parabolic flights:

- assessment of preliminary results for newly proposed experiments, which can improve the final design of the experiment hardware;
- test of experiment critical phases on which the experiment success depends;
- for human physiology experiments foreseen to be conducted on astronauts in space, to obtain prior to, or after a space mission, a broader microgravity data baseline by conducting parts of the space experiments on a group of subjects other than astronauts;
- to repeat shortly after a space mission parts of experiments that were not fully satisfactory in space or that yielded conflicting results, giving indications on possible interpretations of experiment results.

Results of all experiments conducted during parabolic flights can be found in ESA's Erasmus Experiment Archive database [2]. Some experiment results of European parabolic flight campaigns are presented in [3–7].

From a technical point of view, in preparing experiment hardware for manned spaceflight or robotic missions, the following objectives can also be achieved:

- test of equipment hardware in reduced gravity;
- assessment of the safety aspects of instrument operation in reduced gravity;
- training of astronauts on experimental procedures and instrument operation (**Figures 3** and **4**).

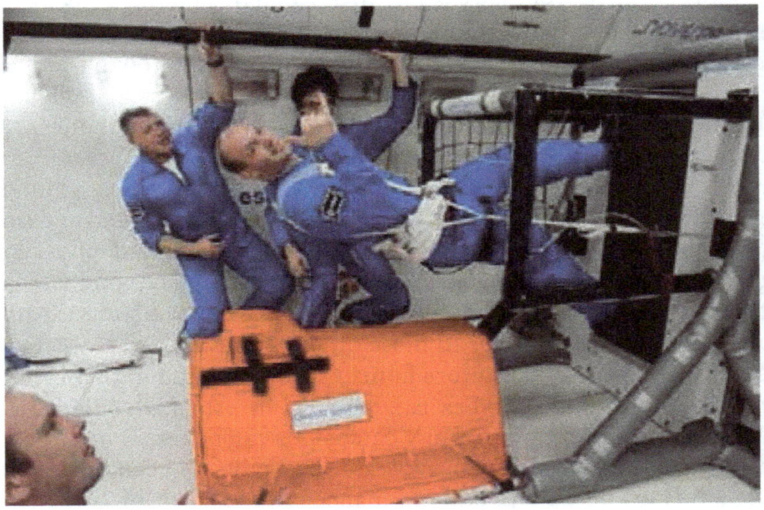

Figure 3.
ESA Astronaut Alexander Gerst training on a treadmill on board the ESA/CNES/DLR Airbus A310 during an ESA campaign before his ISS mission (Credit: ESA).

Figure 4.
Astronaut training for an EVA exit out of a mock-up airlock on board the NASA KC-135 aircraft (Credit: NASA).

Figure 5.
NASA Astronauts Alan Shepard and Edgar Mitchell training on the KC-135 before their Apollo-14 mission (Credit: NASA).

Laboratory aircraft flights allow also to generate partial-g levels between 0 and 1 g. Typically Moon and Mars gravity levels are created during flight manoeuvres that are not exactly parabolic flights but that are still called (albeit erroneously) Moon and Mars parabolas. This approach allowed to answer the growing request of the scientific community to perform complementary research at partial-g levels, and to conduct studies and tests to prepare for future human and robotic exploration missions. The interest in partial-g parabolas is multifold. It allows to complement microgravity experiment results in providing additional data points at intermediate g levels in a variety of scientific fields in physical sciences (e.g., fluid and soft matter physics), life sciences (cell, plant and animal biology, human physiology) and technology. In particular, in Life Sciences, investigations can be conducted on living systems to understand how humans, small animals, cells and plants are affected by a low gravity environment similar to those on the Moon and Mars. Many issues of relevance for the preparation of future human space exploration that includes stays on the surface of planetary bodies can also be investigated (**Figure 5**). Some experiment results of European partial-g campaigns can be found in [8, 9].

Other gravity levels are also achievable during flights and can be used by investigators: e.g., during pull-up and pull-out manoeuvres, periods of 20 s of 1.8–2 g are achieved, and spiral turn manoeuvres provide for longer periods or other levels of high g. These hypergravity periods can be used for certain type of gravity dependent investigations, e.g., in combustion or physiology areas.

Finally, since a few years, parabolic flights are also used in the USA, Europe and Russia to allow the general public to discover the 0 g environment and to enjoy for a few moments the sensations that astronauts experience during their spaceflights.

3. Parabolic flight manoeuvre

For microgravity flights, let us recall first that a body is in free fall if it is subjected to the only force of gravity in an inertial reference frame and this body and all its content are in a state of weightlessness in the non-inertial reference frame attached to the body. Weightlessness thus appears in a non-inertial reference frame, which is in a state of free fall with respect to an inertial reference frame. Weightlessness is a dynamical state that requires a free-fall movement [10]. Note as well that any kind of movement of a vehicle that is subjected to the action of the sole force of gravity is a free-fall movement, and weightlessness is then generated in the vehicle.

Consider now an aircraft flying in Earth's atmosphere, assumed to be quiet. There are usually four forces acting on the aircraft in a straight and level horizontal flight (**Figure 6**):

- its own weight, oriented vertically downwards,
- the aerodynamic lift, induced by the shape of its wings, oriented vertically upward,
- the aerodynamic drag, created by air resistance, oriented horizontally backward,
- the aircraft engine thrust, oriented horizontally forward.

At steady state (dynamic equilibrium), the lift equilibrates the weight and the thrust should be larger than the drag for the plane to move forward.

To come to a free-fall configuration, the pilot should nullify all forces acting on the aircraft other than gravitational forces, i.e., the weight. Therefore, the resulting force along the aircraft velocity vector should be null, i.e., the thrust should exactly balance the drag for the duration of the free-fall manoeuvre. On the other hand, the lift must also be brought to zero. The lift is due to the shape of the wings and to the angle of attack, which is the inclination of the chord line of the wings with respect to the direction of the air flow. To nullify the angle of attack, the pilot must lower the nose of the aircraft to bring the wing chord line parallel to the air flow direction. However,

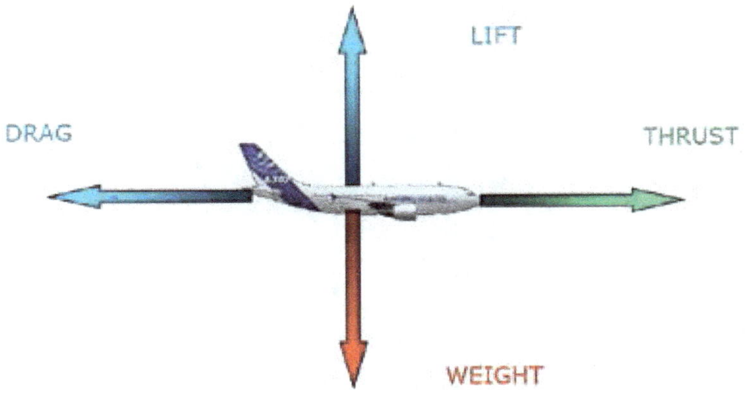

Figure 6.
The four forces acting on an aircraft in a straight and level horizontal flight.

the natural shape of the wing is such that there is always a small remaining lift, even for a zero angle of attack. Therefore, the pilot has to lower the nose a little bit more to bring it to a slightly negative angle of attack, called the zero-lift angle of attack, to completely nullify the aircraft lift. The aircraft is now in a real free-fall state, and as one can imagine, it is not an easy configuration to fly an aircraft.

The microgravity environment is then created in an aircraft flying the following manoeuvres (**Figure** 7):

- from steady horizontal flight, the aircraft climbs at approx. 45–50° (pull-up) for about 20 s with accelerations between 1.8 and 2 g;
- the thrust of each aircraft engine is then significantly reduced for about 20–25 s, compensating the effect of air drag (parabolic free fall);
- the aircraft dives at 45–50° (pull-out), accelerating at about 1.8–2 g for approximately 20 s, to come back to a steady horizontal flight.

One shows easily that the shape of the free-fall trajectory followed by the aircraft is a parabola. Assuming a constant and parallel gravity field and no resistance of the atmosphere, the aircraft at the moment of injection has a certain speed v_0 inclined on the horizontal by an angle α. Gravity being the only acting force on the aircraft, the horizontal component v_x of the velocity is constant throughout the free-fall manoeuvre while the vertical component v_z of the velocity varies with time t

$$v_x = v_0 \cos \alpha \quad v_z(t) = -g\,t + v_0 \sin \alpha \tag{1}$$

Integrating the two velocity components and replacing the time t between them gives the trajectory

$$z(x) = \left(\frac{-g}{2\,v_0^2 \cos^2\alpha}\right)x^2 + (tg\ \alpha)\,x \tag{2}$$

which is a parabola equation, yielding the horizontal distance travelled during the free-fall manoeuvre

$$x_{max} = \frac{\sin v_0^2 \sin\ 2\,\alpha}{g} \tag{3}$$

The duration T of the free-fall phase is found from (1) and (3), yielding

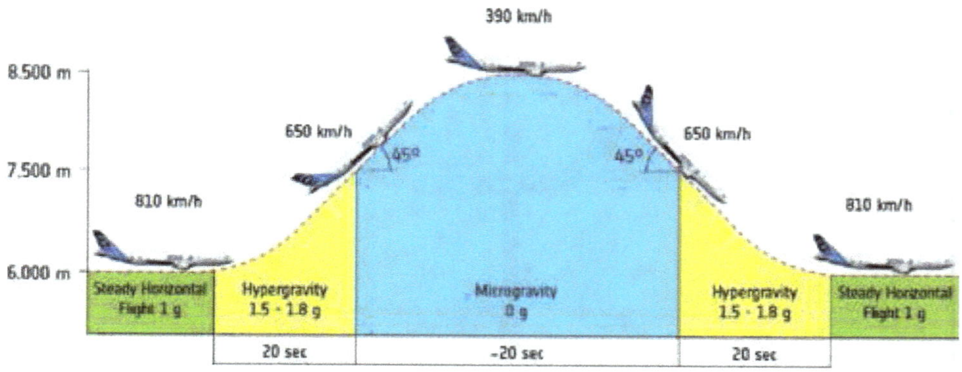

Figure 7.
The parabolic flight manoeuvre of the Airbus A310 (Credit: ESA).

$$T = \frac{2v_0 \sin \alpha}{g} \quad (4)$$

which depends only on the velocity vector at injection, i.e., its norm v_0 and its direction given by the angle α. With typical parameter values for aircraft flying parabolas (velocity at injection v_0 = 500 km/h and α = 45°), one obtains T = 20 s.

In reality, this parabolic arc is an approximation of an elliptic arc which is a part of an elliptic orbit that would intersect the Earth's surface. The relative error committed by this approximation is in the order of 10^{-4}, which justifies that these flights are still called parabolic flights (see [11]).

For partial-g parabolas, the engine thrust is reduced sufficiently to a point where the remaining vertical acceleration in the cabin is approximately 0.16 g for approximately 25 s or 0.38 g for approximately 32 s with angles at injection of approximately 42 and 38°, respectively, for Moon and Mars parabolas.

These manoeuvres are flown separated by intervals of several minutes. The duration of intervals between parabolas can be arranged prior to the flight so as to give enough time to investigators to change an experimental set-up. A typical flight duration is about two to two and half hours, allowing for 20–30 parabolas to be flown per flight, in sets of five, with 2 min intervals between parabolas and with 4–6 min between sets of parabolas. Durations between parabolas and groups of parabolas can be adjusted to the needs of investigators. Parabolas are flown in dedicated air zones controlled by well-trained air traffic controllers. The piloting is usually done manually along the X-axis (aft to front direction) by adjusting the engines thrust, the Y-axis (right to left), and along the Z-axis (floor to ceiling) using visual references utilising coarse (+2 g to –2 g) and fine (+0.1 g to −0.1 g) accelerometers.

Typical acceleration levels are shown in **Figures 8–10** for, respectively, microgravity, lunar and Martian parabolas for the aircraft X, Y and Z axes, measured during a parabola with micro-accelerometers strapped down on the cabin floor structure. During the reduced gravity period of, respectively, microgravity, lunar and Martian parabolas, a transitory phase of a few seconds appears first, with variations of about 10^{-1} g around respectively 0, 0.16 and 0.38 g in the Z direction, followed by a period of respectively approximately 20, 25 and 32 s with acceleration levels of about, respectively, a few 10^{-2}, 0.16 and 0.38 g. Accelerations along the

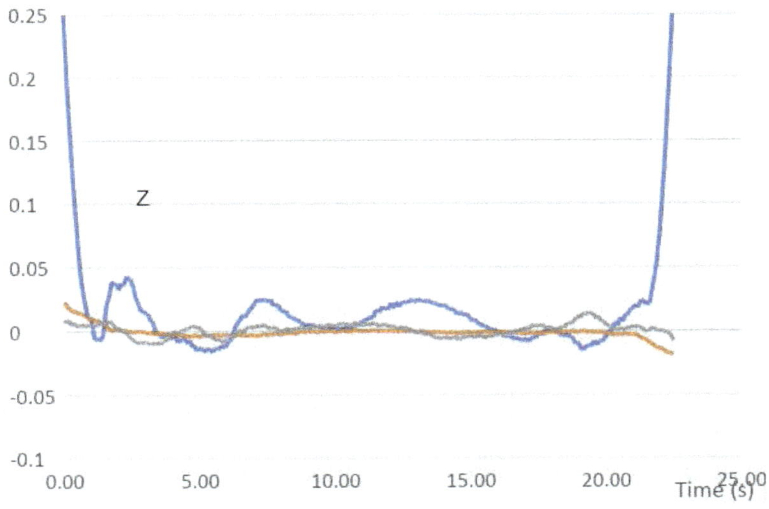

Figure 8.
Acceleration levels for a 0 g parabola.

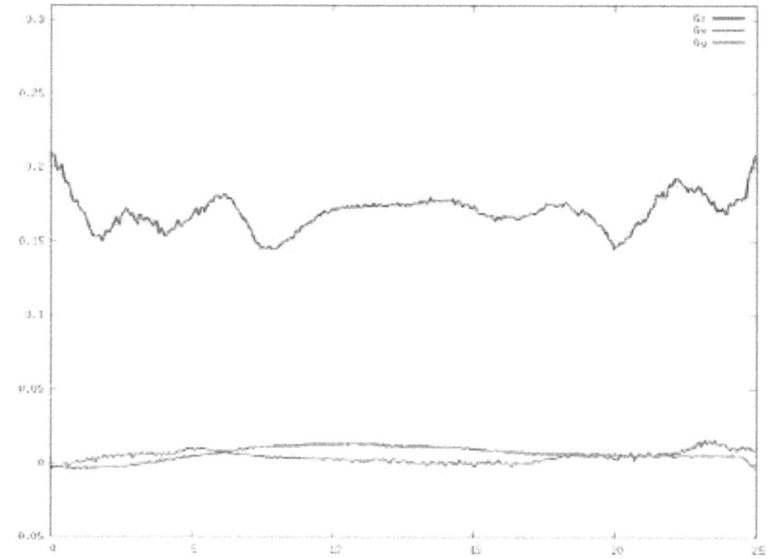

Figure 9.
Acceleration levels for a lunar g parabola for Z axis (brown), X axis (green), Y axis (blue) (Units: vertical axis: 0.05 g; horizontal axis: 5 s).

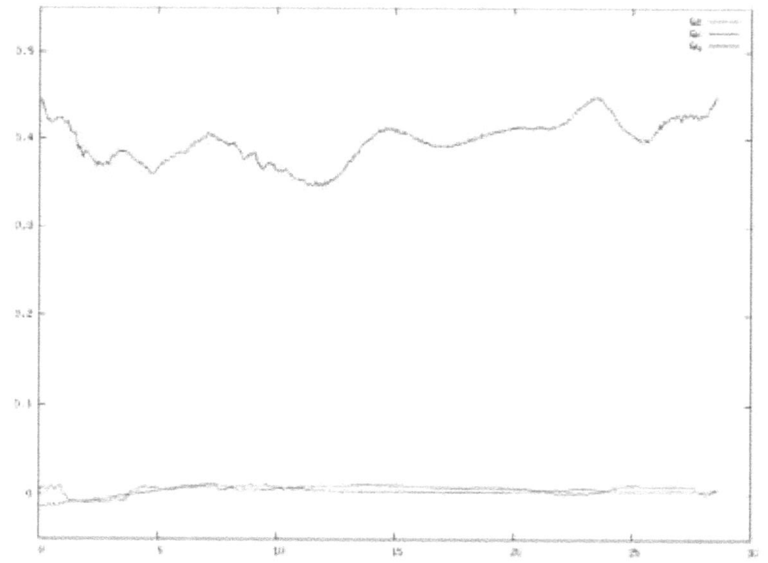

Figure 10.
Acceleration levels for a Martian g parabola for Z axis (brown), X axis (green), Y axis (blue) (Units: vertical axis: 0.1 g; horizontal axis: 5 s).

aircraft longitudinal X-axis (aft to front) and transversal Y-axis (right to left) are less than 10^{-2} g for all parabola types [12].

For microgravity parabolas, the residual accelerations sensed by experimental set-ups attached to the aircraft floor structure are typically in the order of 10^{-2} g, while for an experiment left free floating in the cabin, the levels can be improved to typically 10^{-3} g.

4. Parabolic flight campaigns

As aircraft parabolic flights are considered as test flights, particular precautions are taken to ensure that all operations during flights are made safely and that flying participants are adequately prepared for the repeated high and low gravity environments. All flight participants must pass a medical examination and certifications are verified prior to the first flight. Flight participants attend a mandatory flight briefing before the first parabolic flight.

4.1 Scientific campaigns

Scientists are regularly invited by space agencies to submit experiment proposals to be conducted during parabolic flights, either in microgravity of at lunar or Mars gravity levels. These proposals are evaluated by panels of experts who review them for scientific content and for technical feasibility. Upon recommendation, scientists are formally selected by space agencies and invited to prepare their experiments to be flown on dedicated campaigns.

Prior to a campaign, support is provided for experiment equipment design and related safety aspects. All experiments to be performed and all equipment to be embarked on board the aircraft are reviewed by technical experts several months before the campaign from structural, mechanical, electrical, safety and operational points of view. Technical visits are made to experimenters' institutions to review equipment. A safety review is held approximately 1 month before the campaign. During this review, the integration of all equipment is discussed and the overall safety aspect of the campaign is assessed. Finally, a safety visit is made in the aircraft prior to the first flight to verify that all embarked equipment complies with the safety standards.

During flights, specialised personnel supervise and support the in-flight experiment operations (**Figure 11**). In addition, a Flight Surgeon participates in all parabolic flights to supervise the medical aspect of in-flight operations and to assist participants in case of sickness. Due to the association of flight phases of low and high gravity, motion sickness is quite common among participants of parabolic flights, sometimes hampering them in the conduct of their tasks. Prior to the flights,

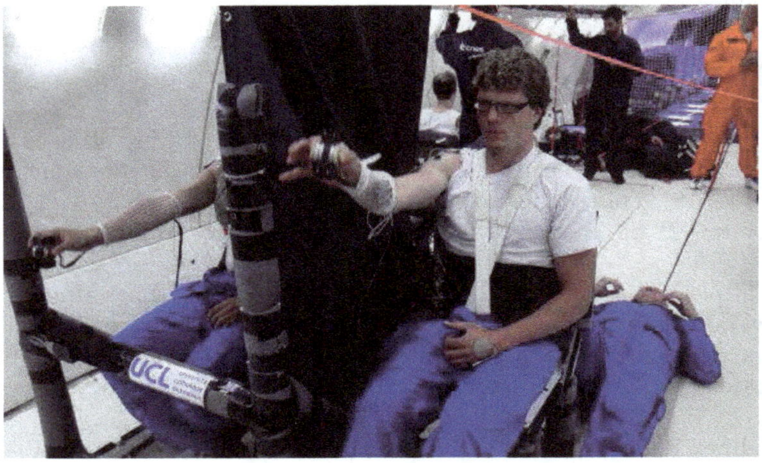

Figure 11.
Seating subjects performing the 'Sensorimotor coordination under partial gravity: movement control and adaptation' experiment (see [13, 14]) (Photo: ESA).

anti-motion sickness medication (usually based on Scopolamine) is made available on the request of flying participants.

The campaign itself takes place over 2 weeks. The first week is devoted to the experiment preparation and loading in the aircraft, culminating with a safety visit at the end of the week to assess that all safety recommendations have been implemented. During the second week, a flight briefing is organised on the Monday afternoon to present the flight manoeuvres, the emergency procedures and medical recommendations, and all experiments are shortly presented by the investigators. The flights take place on the mornings of the following days. A post-flight debriefing is organised after each flight, during which needs and requests of investigators are reviewed and discussed. Due to bad weather or technical problems, a flight can be postponed from the morning to the afternoon or to the next day. Downloading of all experiments takes place on the afternoon after the last flight.

A post campaign workshop is usually organised a few months after the campaign where investigators are invited to present their results. Investigators are further invited to publish the results of their experiments and to communicate their findings online, for example using the Erasmus Experiment Archive database of the European Space Agency, accessible on Internet [2].

4.2 Discovery flights

Since several years, the general public can participate to parabolic flights. Various private companies all over the world commercialise this type of flights.

Flight participants receive a pre-flight briefing similar to the pre-flight briefing of scientific campaigns. A discovery flight usually consists of 12–15 parabolas, including a Martian parabola and one or two Moon parabolas, followed by 10–12 0 g parabolas. A post-flight celebration debriefing is usually held, familiarly called 'regravitation party', where experiences are exchanged and 0 g diplomas are awarded to flying participants.

In the USA, the Zero Gravity Corporation operates a modified Boeing 727-200, named 'G-FORCE ONE' for discovery flights [15]. The Russian Ilyushin IL-76 MDK, located at the Yu. Gagarin Cosmonaut Training Centre at the Star City, near Moscow, is marketed since the 1990s by several private operators for discovery flights open to the public. The European Airbus A310 is also used for flights open to the general public.

5. Type of aircraft used for parabolic flights

Many airplanes are used all over the world to conduct parabolic flights. A review is given in [1, 16]. These can usually be classified in three categories: (1) the large-body aircraft, (2) the medium size aircraft and (3) the small airplanes, jets and gliders.

Large airplanes used for parabolic flights are defined as those aircraft used for flying several (typically 10 or more) experiments and embarking several tens of passengers, either for research purposes or for discovery reduced gravity flights for paying passengers. The large-body aircraft presently in use in the world are:

- a modified Boeing 727-200 of Zero Gravity Corporation in the USA,
- an Ilyushin IL-76 MDK (MDK stands for 'latest modifications' in Russian) in Russia, and

Figure 12.
The Airbus A310 ZERO-G during the pull-up phase (Photo: Eric Magnan/Airborne Films).

Figure 13.
The Dutch NLR Cessna Citation II in flight (Credit: NLR).

- an Airbus A310 (**Figure 12**) used in Europe by space agencies in Europe, the European Space Agency (ESA), the *Centre National d'Etudes Spatiales* (CNES, French Space Agency) and the *Deutsches Zentrum für Luft-und Raumfahrt e.V.* (DLR, German Aerospace Center).

Medium size airplanes are defined as those aircraft used for flying single experiments with several operators and/or subjects. Medium size aircraft in Europe includes:

- the Cessna Citation II (**Figure 13**) based in The Netherlands and operated by the Technology University of Delft and the Dutch National Aerospace Laboratory (NLR) [17];

- a Falcon 20 in Canada operated by the National Research Council's Institute for Aerospace Research (NRC/IAR) [18]; and

- two jets, a MU-300 and a Gulfstream-II jets in Japan, operated by Diamond Air Service [19].

Small airplanes, jets and gliders used for parabolic flights are defined as those aircraft used for flying single passengers and small experiments. These airplanes include presently:

Figure 14.
The Mudry Cap10B aircraft operated by the Aero Club Barcelona Sabadell (Credit: Universitat Politecnica de Catalunya*).*

- a jet fighter F-5E Tiger II aircraft of the Swiss Air Force used in Switzerland for automated science experiments [20];
- a Cessna 206, which was tested in the frame of the MiGrOp project to be used for science experiments in Germany [21];
- a Mudry Cap10B aircraft (**Figure 14**), a two-seat training aerobatic aircraft, used by the *Universitat Politecnica de Catalunya* with the Aero Club Barcelona Sabadell, in Spain for small science and student experiments [22];
- several Schleicher ASK-21 two-seat gliders have been used for students experiments in Belgium [23, 24] and Israel [25–27], with added advantages of a low-cost approach and ease of deployment at close-by geographical locations.

However, as any small or large airplane could basically be used to undergo a parabolic trajectory, it is important to choose carefully which aircraft would be best suited for scientific investigations, in terms of quality and duration of reduced gravity level but also ease of access and technical support from the integration team.

6. Conclusion

Aircraft parabolic flight manoeuvres are a very useful tool to investigate gravity related phenomena, whether in complete weightlessness or at partial-g levels. The quality and duration of microgravity obtained, the flexibility and variety of possibilities for experiments and tests and the easiness in flight preparation make aircraft parabolic flights a unique and versatile tool for scientists and engineers to perform experiments and tests at different gravity levels, from microgravity to hypergravity, including at Moon and Mars g levels. Parabolic flight campaigns for research in reduced gravity throughout the world provided a huge amount of scientific and technical data and knowledge, yielding hundreds of scientific publications, thesis or industrial applications, showing the uniqueness of this versatile tool to conduct gravity related investigations, complementing those on other microgravity carriers and preparing for space missions and for future extra-terrestrial planetary exploration missions.

In addition, opening the access of parabolic flights, traditionally reserved to scientists and astronauts, to the general public increases the perception of the public for space research and exploration, contributing to an enhanced interest for future endeavour of mankind in space.

Author details

Vladimir Pletser[1,2,3]

1 Blue Abyss, Space Training Operations, Liverpool, United Kingdom

2 Chinese Academy of Sciences, Beijing, China

3 European Space Agency, Noordwijk, The Netherlands

*Address all correspondence to: pletservladimir@gmail.com

References

[1] Pletser V, Kumei Y. Chapter 7: Parabolic flights, D.A. Beysens and J. Van Loon Generate an Extra-Terrestrial Environment on Earth, River Publishers Series in Standardisation, Aalborg, Denmark, 61-74, 2015. ISBN: 978-87-93237-53-7. Available from: http://www.riverpublishers.com/pdf/ebook/chapter/RP_9788793237544C7.pdf [Accessed: 08 April 2020]

[2] ESA Website. Erasmus Experiment Archive. Available from: http://spaceflight.esa.int/eea/ [Accessed: 08 April 2020]

[3] Pletser V, Rouquette S, Friedrich U, et al. The first European parabolic flight campaigns with the Airbus A310 ZERO-G. Microgravity Science and Technology. 2016;**28**(6):587-601. Available from: https://link.springer.com/article/10.1007/s12217-016-9515-8 [Accessed: 08 April 2020]

[4] Pletser V, Rouquette S, Friedrich U, et al. European parabolic flight campaigns with Airbus A300 ZERO-G: Looking back at the A300 and looking forward to the A310. Advances in Space Research. 2015;**56-5**:1003-1013. DOI: 10.1016/j.asr.2015.05.022

[5] Pletser V, Pacros A, Minster O. International heat and mass transfer experiments on the 48th ESA Parabolic Flight Campaign of March 2008. Microgravity Science and Technology. 2008;**20**:177-182. Special Issue Two-Phase Systems. Available from: https://link.springer.com/article/10.1007 s12217-008-9039-y [Accessed: 08 April 2020]

[6] Aubert A, Beckers F, Verheyden B, Pletser V. What happens to the human heart in space? Parabolic flights provide some answers. Cardiovascular response during gravity changes induced by parabolic flights. ESA Bulletin. 2004; **119**:30-38. Available from: http://www.esa.int/esapub/bulletin/bulletin119/bul119_chap4.pdf [Accessed: 08 April 2020]

[7] Pletser V. Short duration microgravity experiments in physical and life sciences during parabolic flights: The first 30 ESA campaigns. Acta Astronautica. 2004;**55**:829-854. Available from: http://www.scienced irect.com/science/article/pii S0094576504001353 [Accessed: 08 April 2020]

[8] Pletser V, Winter J, Duclos F, Friedrich U, et al. The first Joint European Partial-g Parabolic Flight Campaign at Moon and Mars gravity levels for science and exploration. Microgravity Science and Technology. 2012;**24**(6):383-395. Special Issue ELGRA 2011. Available from: https://link.springer.com/article/10.1007%2Fs12217-012-9304-y [Accessed: 08 April 2020]

[9] Pletser V, Rouquette S, Friedrich U, et al. Two joint European partial-g parabolic flight campaigns for science and exploration at Moon and Mars gravity levels. In: Paper IAC-13.A2.3.11, 64th IAF Congress; Beijing; 2013

[10] Pletser V. Gravity, weight and their absence. In: Springer Briefs in Physics. Singapore: Springer Nature; 2018. DOI: 10.1007/978-981-10-8696-0. ISBN: 978-981-10-8695-3. Available from: https://www.springer.com/gp/book/9789811086953#otherversion=9789811086960 [Accessed: 08 April 2020]

[11] Pletser V. Are aircraft parabolic flights really parabolic? Acta Astronautica. 2013;**89**:226-228. DOI: 10.1016/j.actaastro.2013.04.019

[12] Pletser V et al. Experimenting at Moon and Mars gravity levels during parabolic flights to prepare for planetary exploration. In: Paper IAC-10-A2.7.4,

61st IAF Congress; Prague; 2010. Available from: www.iafastro.net/iac/paper/id/8411/ext/appendix/IAC-10.A2.7.4.pdf [Accessed: 08 April 2020]

[13] White O, Bleyenheuft Y, Smith A, Thonnard JL, Lefèvre P. Altered gravity highlights central pattern generators mechanisms. Journal of Neurophysiology. 2008;**100**:2819-2824

[14] André T, Lefèvre P, Thonnard JL. Fingertip skin moisture is optimally regulated during object manipulation. Journal of Neurophysiology. 2010;**103**: 402-408

[15] Zero-G Corporation: How It Works. Zero-G Corporation Website. Available from: http://www.gozerog.com/index.cfm?fuseaction=Experience.How_it_Works; http://www.gozerog.com/index.cfm?fuseaction=Research_Programs.welcome [Accessed: 08 April 2020]

[16] Pletser V. European aircraft parabolic flights for microgravity research, applications and exploration: A review. REACH—Reviews in Human Space Exploration. 2016;**1**:11-19. DOI: 10.1016/j.reach.2016.05.002. Available from: http://www.sciencedirect.com/science/article/pii/S2352309316300049 [Accessed: 08 April 2020]

[17] Citation PH-LAB, Partial-Zero-G Facility. TU Delft Aerospace Engineering Website. Available from: http://cs.lr.tudelft.nl/citation/facility/zerog/ [Accessed: 08 April 2020]

[18] Research Aircraft Fleet. NRC Website. Available from: https://nrc.canada.ca/en/research-development/nrc-facilities/research-aircraft-fleet [Accessed: 08 April 2020]

[19] Micro-Gravity Experiment. Diamond Air Service Website. Available from: http://www.das.co.jp/en/service/gravity/index.html [Accessed: 08 April 2020]

[20] Studer M, Bradacs G, Hilliger A, Hurlimann E, Engeli S, Thiel CS, et al. Parabolic maneuvers of the Swiss Air Force fighter jet F-5E as a research platform for cell culture experiments in microgravity. Acta Astronautica. 2011; **68**:1729-1741

[21] Selig H, Gierse A, König G. Parabolic flight with light aircraft. In: Paper IAC-16-A2-5-6, 67th IAF Congress; Guadalajara; 2016

[22] Perez-Poch A, González DV, Lopez D. Hypogravity research and educational parabolic flight activities conducted in Barcelona: A new hub of innovation in Europe. Microgravity Science and Technology. 2016;**28**: 603-609. DOI: 10.1007/s12217-016-9516-7

[23] Caprace DG, Gontier C, Iranmanesh M, Scoubeau M. Glide, without g—A systematic quantification of gliders 0-g flight capabilities. In: Paper IAC-19-A2.5.4, 70th IAF Congress; Washington; 2019

[24] Gontier C, Caprace DG, Scoubeau M, Iranmanesh M, Pletser V. Glide, without g—A systematic quantification of gliders 0-g flight capabilities. Microgravity Science and Technology. 2020. Submitted

[25] Pletser V, Frischauf N, Cohen D, Foster M, Spannagel R, Szeszko A, et al. First Middle East Aircraft Parabolic Flights for ISU participant experiments. Microgravity Science and Technology. 2017;**29-3**:209-219. DOI: 10.1007/s12217-017-9539-8

[26] Pletser V, Frischauf N, Laufer R, Cohen D. First Middle East Aircraft Parabolic Flights for ISU Participant Experiments. In: Paper IAC-17-A2.5.8, 68th IAF Congress; Adelaide; 2017

[27] Pletser V, Frischauf N, Laufer R. Parabolic Flights with gliders as an innovative low-cost platform for microgravity and hypergravity research. In: Paper IAC-17-A2.5.6, 68th IAF Congress; Adelaide; 2017

8

Capillary Driven Flows under Microgravity Conditions: From Parabolic Flights to Space Experiment

Nikolay Smirnov, Valeriy Nikitin and Evgeniya Kolenkina (Skryleva)

Abstract

The chapter investigates imbibition into a porous medium under microgravity condition under the action of capillary forces. The study of capillary effects in terrestrial conditions is often difficult due to the influence of gravity, which makes it necessary to conduct experiments in microgravity. The chapter describes the features of experiments and the processing of experimental data in two types of microgravity experiments: during parabolic flights and at space stations in terrestrial orbit. During parabolic flights, a highly permeable artificial porous medium consisting of glass balls of the same size was considered (such a medium makes it easy to visualize experimental results and simulate media with different permeability using balls of different diameters). In experiments in orbit, the flow of fluids with various physical properties in natural sands was considered. The chapter also describes the mathematical modeling of such processes and presents the results of numerical simulations and their comparison with experimental data. The possibility of determining rheological relationships for capillary forces being functions of governing parameters in porous medium on the basis of data obtained from experiments in microgravity is demonstrated.

Keywords: microgravity, imbibition, parabolic flights, porous medium, capillary effects

1. Introduction

Investigation of capillary driven seepage under microgravity conditions is of great interest for space applications and terrestrial engineering and technology. On space platforms, capillary transport of liquids is observed in various devices/processes such as purification filtration systems, heat pipes, and fuel transport from tanks in weightlessness. Capillary forces turn to be the major mechanism driving the feeding fluid to the plant's roots in hydroponics plant growth systems in space.

On the other hand, microgravity investigation of seepage processes gives a deeper insight in the fundamentals of this nonequilibrium phenomenon, thus providing a better understanding of seepage processes for terrestrial applications, such as oil recovery. Microgravity conditions allow us to study capillary effects in large pores, which are impossible in terrestrial conditions due to the influence of gravity, which distorts the shape of the interface and changes effects.

Under terrestrial conditions, the capillary forces are partly shielded by gravity effects, but, nevertheless, are still present. Accurate investigation of wettability effects in immiscible fluids filtration requires abandoning all other effects as far as possible. Determination of some rheological relationships for accounting the influence of capillary forces in porous media is possible by performing experiments under reduced gravity conditions. Such relationships can then be introduced into already existing models, integrating heat and mass transfer through porous media.

The process of imbibition of viscous fluids into a porous medium depends essentially on capillary effects and instability, which may develop on the displacement front in case of multiphase flow. Accounting for capillary forces is critical for the description of the motion of liquids in porous media. The study of the capillary effects under ordinary conditions is difficult because of the significant effect of gravity on the seepage process in big pores, wherein visual registration is possible. Therefore, in this chapter, we consider the flow of liquids in a porous medium under microgravity conditions during parabolic flights. The problem of multiphase seepage in porous media is very well developed. There is a lot of different models based on the modified Darcy law [1–6]. Experimental and theoretical studies of the flow of fluids in porous media and natural sands under microgravity conditions have been carried out, and mathematical models describing these processes have been developed [7–11]. Another important factor, which influences the displacement, is the instability developing at the front. Initially flat interface of the liquids is broken; some “fingers” of the liquid break through, which causes liquid being entrapped inside the porous medium. The unstable liquid displacement is well studied [12, 13]. Experiments on the flow of liquids through a porous medium under microgravity conditions under the action of capillary forces are described in [9]. The motivation for the present study is investigating the flow of fluid in an artificial porous medium containing heterogeneities under the action of capillary forces. Professor Jean-Claude Legros (**Figure 1**) and his student Eric Istasse were engaged in the determination of capillary characteristics in porous media, and they

Figure 1.
Professor Jean Claude Legros (left) with Professor Nikolay Smirnov (right).

conducted experiments on imbibition of fluids into porous media during parabolic flights. In this chapter, the results of these experiments are reviewed and compared with the results of numerical simulation. In the case of a porous medium formed by relatively large particles, it becomes important to take into account additional inner terms in the Darcy equation. The corresponding mathematical model is described here. Models without inertia effects, described in detail in [2–4], describe fairly accurately slow flows in classical low-permeable porous media, but they are not suitable for processes discussed here.

In this chapter, we focus on seepage flows in porous media with inhomogeneity. The presence of heterogeneity brings to the nonuniformity of the seepage flow and the displacement front, which can lead to nonlinear physical results. Such nonlinear physical results are described in detail in [14], where experiments on the imbibition of liquids into a natural porous medium under microgravity conditions are considered. Results showed that a point located higher along the length of a sample can be reached by a liquid faster than a lower one. In a natural porous medium, we can only make hypothesis about the reasons of such a behavior; while using an artificial porous medium assembled of transparent glass balls, one can visually register the effect of heterogeneity on the capillary driven seepage flow.

2. Experimental investigations imbibition into artificial porous medium

Parabolic flight campaigns were organized (and funded) by the European Space Agency. The parabolic flights are achieved using the French Airbus A300-ZeroG, an aircraft especially transformed to a rather big microgravity laboratory. Detailed description of experimental procedures and techniques can be found in [15]. Below we provide a shortened version of the description.

Microgravity conditions encountered during the parabolic flight exercise are very special ones, and anybody will certainly understand that they are far from usual ground-based laboratory conditions. The design of a dedicated experimental payload must thus integrate the specific aspects of reduced gravity environment. For example, flight procedures should be simplified as much as possible, rendering in-flight operations easy to perform. Mass and overall dimensions of the experimental setup are also limiting factors that have to be carefully considered during the payload design.

A typical parabolic flight campaign is usually scheduled around three successive flights days, preceded by 10 days of experiment integration and eventual "last minute" modifications. The short period of time between the flight opportunities may impact the intrinsic conception of the experimental devices, e.g., experimental cells, payload maintenance tasks, and so on. Furthermore, the aircraft safety rules are very strict, and any boarding experiment must first pass a detailed control procedure, performed by the safety crewmembers.

All these aspects are part of a weightlessness flight program and have to be kept in mind during the preparation of a flight opportunity.

2.1 Instrumental setup description

Several experimental parameters were fixed before setting the design of the experiment:

- We consider porous samples made of glass spheres having various granulometric size distributions. The regular shape of the glass spheres will eliminate potential anisotropy of the porous sample, simplifying thus the investigation of the observed flows. Another advantage is the translucent character of the glass spheres that enables us to visualize in some way internal fluid motions that are usually inaccessible by direct visualization. Finally, choosing spherical constituting solid particles makes it possible to realize a lot of different samples with various well-known intrinsic properties, i.e., porosity, permeability, and mean characteristic pore radius. Overall dimensions of the porous samples ($75 \times 50 \times 200$ mm^3) have been determined taking into account the available duration of microgravity.

- We investigated several fluid systems. The simplified two-phase flow model, presented in the previous section, is experimentally investigated in the case of a water-air system and in the case of a water-alkane system. We choose the iso-octane, also known as 2,2,4-trimethylpentane, for the hydrocarbon, because it is less dense than water, completely immiscible with it. Iso-octane is less wetting the porous matrix as compared with water. This liquid-liquid system investigation enables us to vary the surface tension parameter without modifying the wettability properties of the porous medium. Furthermore, the viscosity of iso-octane is less than the water one. Hence, we limited the appearing of viscous fingering phenomenon that tends to greatly destabilize the moving liquid-liquid interface. The stable shape of the interface is indeed a necessary condition to neglect the dispersion fluxes appearing in our mathematical model.

- The boundary conditions mentioned for the mathematical model, namely no hydrostatic pressure difference between bottom and top of the porous sample, were satisfied. We have also seen that it is necessary to avoid as far as possible the capillary effects at the level of the fluid reservoir. This was achieved by saturating completely the reservoir with the wetting fluid. Note that this reservoir must be able to deform when the capillary flow occurred in microgravity, otherwise we would have observed the formation of an air bubble, which, if growing, would have been able to block the entrance of the porous sample for the fluid. Hence, the reservoir consisted of a deformable plastic bag that remained full of liquid during flight conditions.

- During the parabola, the transition period between the pull-up phase and the microgravity phase induced undesired effects that greatly influenced the flow through the porous sample. Capillary driven filtration started indeed immediately at the beginning of the transition period. At this particular moment, we did not have clear initial conditions for the subsequent creeping under reduced gravity. Hence, it was very important for the filtration modeling to restrict as much as possible the transient flow occurring during this transition phase. It was preferable to wait for acceptable reduced gravity conditions to start up the flow. The experimental setup incorporated simple technical solutions to control these undesired effects. The hydraulic scheme of the experiment indeed included electromagnetic valves that allowed us to close hermetically the porous medium. Consequently, if a capillary flow was starting during the transition phase, it would be rapidly damped by the air pressure increase, which was resulting from the decrease of the volume accessible for the gaseous phase inside the cell. Once in real microgravity, we simply opened the valves to allow the capillary driven flow to invade the solid matrix.

- The effect of temperature on surface tension and viscosity properties of the fluids was not considered in the present study. The aircraft air-conditioning system maintained the cabin temperature around 25°C during flight operations.

2.2 Payload description

The experimental payload has been completely designed and developed by the scientific and technical teams of the Physical Chemistry Department of the Free University of Brussels. It incorporated all the necessary devices to study various capillary driven flows under reduced gravity conditions, and its basic features will be now briefly detailed.

The payload consisted of two racks integrating the experimental setup (main rack) and the porous samples confined in dedicated experimental cells (storage rack). The main rack included the following:

- An electrical panel, mandatory to interface the experiment with the aircraft power supplies.
- A video system, including black and white video camera and high-quality video recorder.
- A hydraulic scheme, i.e., a network of valves and tubes necessary to handle the eventual fluid feeding of the experimental cells.
- A temperature and pressure acquisition system was loaded and allowed to work autonomously during the flight. Data were simply downloaded after landing.
- A residual g-level data acquisition system, coupled with the video recording system, enabling video incrustation of residual g-level directly through the video signal.

The main rack measured $880 \times 650 \times 1440$ mm^3 (length $\times$ width $\times$ height) and integrated in fact two experiments, with a total mass of approximately 300 kg.

The experimental rack was designed to investigate two experimental cells simultaneously. This was a very convenient feature that made it possible to directly correlate two porous samples having different properties under the same micro-gravity conditions. As the flight planning was organized in 6 sets of 5 successive parabolas, we investigated the experimental cells by pair, and we video recorded 5 successive capillary creeping for them. Longer breaks between two successive sets of parabolas were used to perform the cells exchange.

Following this scientific strategy, we were able to study the influence of the preimbibition of the porous sample. Starting from a completely dry porous medium, the capillary creeping occurred during the first parabola, followed directly by a drainage process during the pull-out phase. The amount of liquid being trapped inside the porous matrix was depending on the surface tension properties of the fluid-fluid pair under investigation. We observed experimentally that the seepage «creepability» parameter Ψ [7, 15] was significantly influenced by the preimbibition. But, once the porous matrix has been imbibed, following successive imbibition was not modifying essentially its microscopic surface state. Successive capillary seepage during a set of parabolas would invade the porous matrix higher and higher, simply because the fluid encountered less resistance in the already preimbibitted zone and the latter tended to increase during successive capillary driven imbibition. The number of operations that experimenters had to perform during the parabola itself was intentionally small. The parabolic flight maneuver, with its succession of hypergravity and low gravity periods, revealed to be physio-logically tiring. In these conditions, the experimental procedures should be as simple as possible to allow the experimenters to anticipate these repetitive gravity variations. For the capillary creeping experiment, the most difficult tasks were the cells exchange, and the fluid management to adjust correctly the initial fluid levels inside for the porous samples. These tasks were executed before the pull-up phases. As we continuously video recorded the experimental cells, the experimenters had only to wait for the optimal reduced gravity level to switch on the electromagnetic valves, thus enabling the flows through the samples.

2.3 Experiment details and results

We concentrate in this section on the peculiarities of seepage flow in a media containing permeability inhomogeneity. The zones of different permeability were arranged in the experimental cells (**Figure 2**) using glass balls of different diameters. Porous samples were hermetically confined inside the experimental cell, whose overall dimensions were $74 \times 50 \times 315$ mm^3. Each cell had a separate fluid reservoir made of a deformable plastic bag. The reservoir fluid capacity was sufficient to fill in totally the porous sample. The experimental cell was connected at its bottom side to a cylindrical connector. This connector was 70 mm long and served as a hydraulic junction between the porous sample and the reservoir, filled in with a water-dye mixture. The diameter of the channel zone has been chosen large enough, i.e., 37 mm, to allow reasonable fluid feeding to the porous medium during microgravity periods. During the flight, the capillary driven filtration was video recorded, using a high-quality video equipment. The back-light system allowed to visualize the phase interface separating water and air.

Three experiments are reported here. In all cases, we considered the imbibition of fluid into the porous media that has been prewetted throughout previous parabolas. Zones with different permeability were formed by spheres with diameters of 2 and 6 mm. The artificial porous medium within each zone was composed of glass balls of the same diameter. Thus, the permeability of the porous medium consisting of spheres (d = 6 mm) was $K_1 = 8.5{\cdot}10^{-9}$ m^2, while permeability in the low permeable zone (d = 2 mm) was $K_2 = 2{\cdot}10^{-9}$ m^2.

In the first experiment, the whole cell was filled with a homogeneous porous medium consisting of glass spheres with a diameter of 6 mm (**Figure 3**). The graph (**Figure 4**) shows the position of the interface as a function of time. By reducing the gravity level, the capillary driven imbibition of fluid into porous sample begins. During the observation period, the interface did not exceed the height of 40 mm.

Figure 2.
The experimental cell for artificial porous medium.

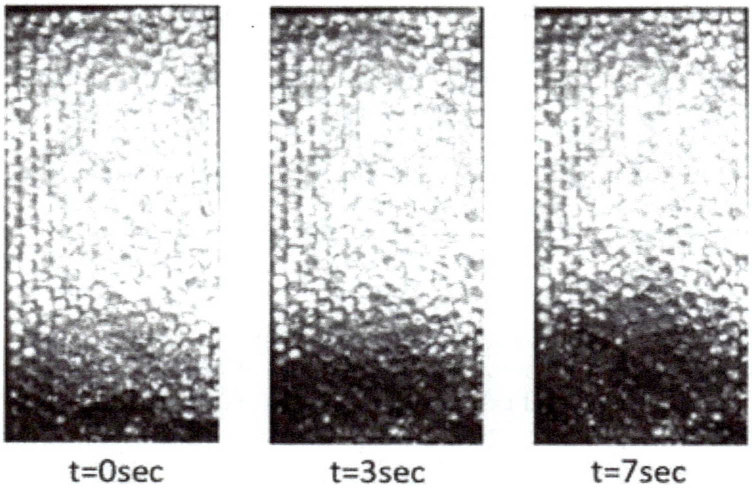

Figure 3.
The flow of liquid in a porous medium (permeability: $8.5 \cdot 10^{-9}$ m²; porosity: 0.458) under microgravity conditions (displacing fluid: viscosity: $1 \cdot 10^{-3}$ Pa·s, density: 10^3 kg/m³; displaced gas: viscosity: $1.5 \cdot 10^{-5}$ Pa·s, density: 1.4 kg/m³; surface tension: 0.0582 N/m).

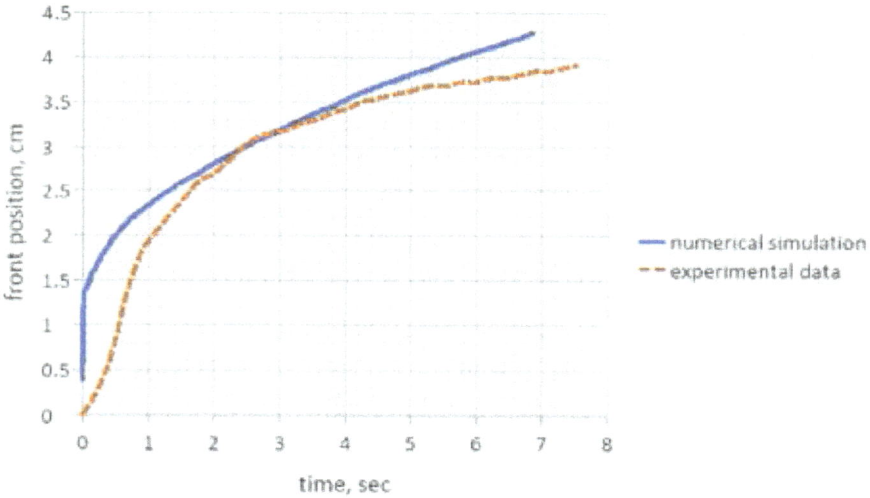

Figure 4.
The position of the interface as a function of time for a flow in a homogeneous medium (permeability: $8.5 \cdot 10^{-9}$ m²; porosity: 0.458; displacing fluid: viscosity: $1 \cdot 10^{-3}$ Pa·s, density: 10^3 kg/m³; displaced gas viscosity: $1.5 \cdot 10^{-5}$ Pa·s, density: 1.4 kg/m³; surface tension: 0.0582 N/m).

In the second experiment, the lower layer of the medium (with a height of 4 cm) consisted of spheres with a diameter of 6 mm, and the upper layer consisted of spheres of smaller diameter (2 mm) (**Figure 5**). The position of the interface is shown in the graph (**Figure 6**). The process of imbibition is analogous to the case of a homogeneous medium until a zone of low permeability was reached. On crossing the border of high permeable and low permeable zones, the velocity of phase interface drastically increased upon entering the zone of lower permeability small sphere assembly. After a while, the interface velocity again decreased. The acceleration of the imbibition front is explained by the fact that the capillary forces in the medium consisting of small spheres were higher. Successive slowing down of the interface was due to the decrease of permeability. The effects of increasing capillary

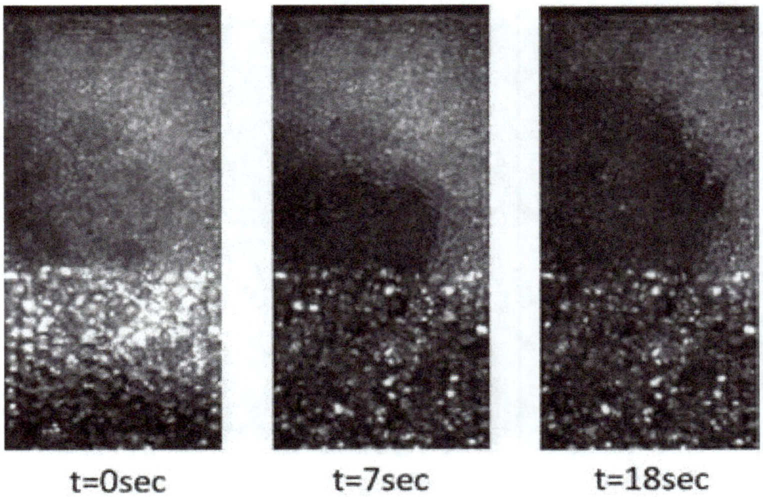

Figure 5.
The flow of liquid in a porous medium consisting of two layers (bottom layer: height: 4 cm, permeability: $8.5 \cdot 10^{-9}$ m², porosity: 0.458; upper layer: permeability: $2 \cdot 10^{-9}$ m²; porosity: 0.466) under microgravity conditions (displacing fluid: viscosity: $1 \cdot 10^{-3}$ Pa·s, density: 10^3 kg/m³; displaced gas: viscosity: $1.5 \cdot 10^{-5}$ Pa·s, density: 1.4 kg/m³; surface tension: 0.0582 N/m).

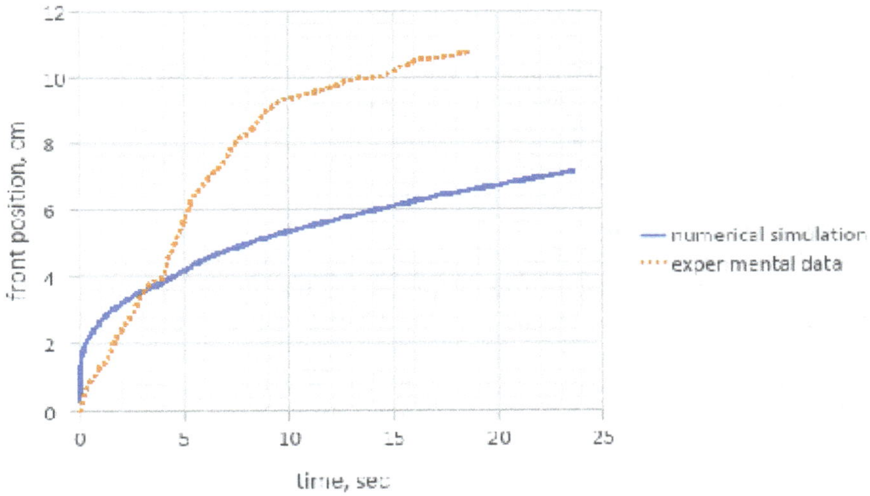

Figure 6.
The position of the interface as a function of time for a flow in an inhomogeneous medium (bottom layer: height: 4 cm, permeability: $8.5 \cdot 10^{-9}$ m², porosity: 0.458; upper layer permeability: $2 \cdot 10^{-9}$ m², porosity: 0,466; displacing fluid: viscosity: $1 \cdot 10^{-3}$ Pa·sec, density: 10^3 kg/m³; displaced gas viscosity: $1.5 \cdot 10^{-5}$ Pa·s, density: 1.4 kg/m³; surface tension: 0.0582 N/m).

forces and drag forces acted on different time scales in unsteady-state flows: capillary forces increased by a jump, and then sustained constant value, while drag forces grew linearly on increasing the fluid penetration depth.

In the third experiment, the cell was filled with spheres (d = 6 mm) and included a zone ($30 \times 20 \times 30$ mm³) with a lower permeability in the central part of the cell near the left wall, filled with glass beads (d = 2 mm) (**Figure 7**). The phase interface remained relatively flat in a homogeneous zone. On approaching the zone of lower permeability, the interface becomes curved, and the capillary creeping was faster in the zone of low permeability.

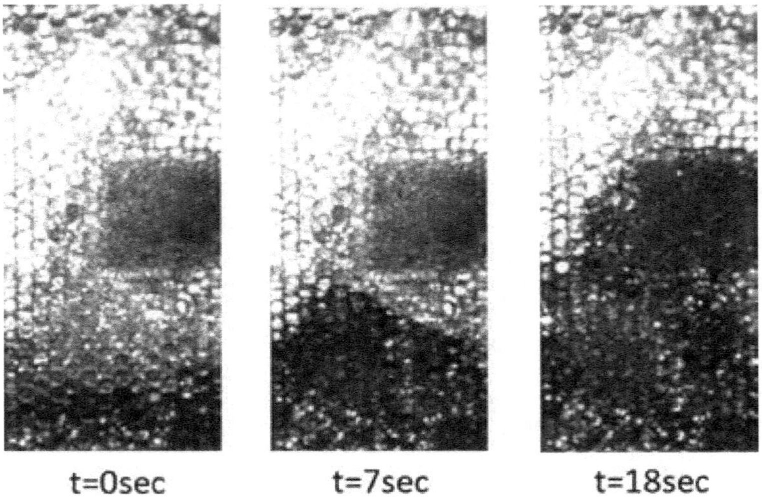

Figure 7.
The flow of liquid in a porous medium (permeability: $8.5 \cdot 10^{-9}$ m²; porosity: 0.458) containing an insert ($3 \times 2 \times 3$ cm³) with a low permeability (permeability: $2 \cdot 10^{-9}$ m²; porosity: 0,466) under microgravity conditions (displacing fluid: viscosity: $1 \cdot 10^{-3}$ Pa·s, density: 10^3 kg/m³; displaced gas: viscosity: $1.5 \cdot 10^{-5}$ Pa·s, density: 1.4 kg/m³; surface tension: 0.0582 N/m).

3. Mathematical model and numerical investigation

This section introduces the mathematical and numerical modeling used for the experiments. A more detailed description of can be found in [16].

3.1 Balance equations

The flow of two incompressible fluids in porous media is considered, without thermal effects taken into account. The imbibition is modeled by the Darcy law, considering the capillary effects at the boundary of the phases. Non-stationarity is taken into account. Permeability and porosity depend on the spatial co-ordinate and on the fluid pressure in the pores. The mathematical formulation of the problem describes the displacement of one fluid by another due to the pressure drop on the sides of the sample or due to a given flow on one side and also due to capillary effects.

The mass balance equations for the phases are as follows:

$$\frac{\partial}{\partial t}(\varphi \rho_k s_k) + \frac{\partial}{\partial x_j}(\rho_k u_{k,j}) = 0 \tag{1}$$

Here, φ is the porosity, ρ_k is the intrinsic density of the kth phase, s_k is the phase saturation, t is the time, $u_{k,j}$ is the jth component of the seepage velocity of the kth phase, and x_j is the component of the radius vector (point coordinates). The equations are summed up over the repeated index j.

Note that the intrinsic phase velocity $v_{k,j}$ is expressed through the seepage velocity as:

$$v_{k,j} = \frac{u_{k,j}}{\varphi s_k} \tag{2}$$

Unlike the seepage velocity of a phase, intrinsic phase velocity is not determined where there is no phase (the seepage velocity at that place is zero).

Both fluids are incompressible. Dividing Eq. (1) by ρ_k yields the reduced form:

$$\frac{\partial}{\partial t}(\varphi s_k) + \frac{\partial u_{k,j}}{\partial x_j} = 0 \tag{3}$$

The average volume seepage velocity u_j is defined as:

$$u_j = u_{1,j} + u_{2,j} \tag{4}$$

Summation of the Eq. (3) using the definitions (4) leads to an equation for the total phase seepage velocity:

$$\frac{\partial \varphi}{\partial t} + \frac{\partial u_j}{\partial x_j} = 0 \tag{5}$$

In the case of porosity not changing with time, the Eq. (5) reduces to the condition of the solenoidality of the fluid seepage velocity field:

$$\frac{\partial u_j}{\partial x_j} = 0 \tag{6}$$

The momentum equations from which the Darcy laws for each phase are derived are as follows:

$$\frac{\partial}{\partial t}\left(\rho_k \varphi s_k v_{k,j}\right) + \frac{\partial}{\partial x_j}\left(\rho_k \varphi s_k v_{k,i} v_{k,j}\right) + \varphi s_k \frac{\partial p_k}{\partial x_j} = -\frac{\mu_k}{K_0} v_{k,i} + F_k \tag{7}$$

where μ_k is the dynamic viscosity, K_0 is the absolute permeability, p_k is the pressure in the phase, and F_k is the interaction force with other phases of the fluid. The right-hand side of (7) corresponds to the sum of the force of interaction with the skeleton or solid phase and the forces of interaction with other phases of the fluid.

To obtain Darcy equations, the inertial terms (i.e., non-stationary term $\frac{\partial}{\partial t}\left(\rho_k \varphi s_k v_{k,j}\right)$ and convective term $\frac{\partial}{\partial x_j}\left(\rho_k \varphi s_k v_{k,i} v_{k,j}\right)$) are usually neglected in (7), and the interaction forces with other phases, except the skeleton or solid phase), are modeled by modifying the basic interaction force with the skeleton by introducing relative permeability. We will neglect only the convective component of inertia, leaving the non-stationary component. After this, the Darcy equations are modified, containing the time derivative of the velocity:

$$u_{k,i} = -\frac{K_0 K_k^R}{\mu_k}\left(\frac{\partial p_k}{\partial x_i} + \frac{\rho_k}{\varphi s_k}\frac{\partial u_{k,i}}{\partial t}\right) \tag{8}$$

We assume the relative permeability of the phase, depending on its saturation s_k and, in general, on the saturation of other phases: $K_k^R = \theta_k \bullet s_k$, where θ_k is a proportionality coefficient. It should be noted that since the relative permeability is zero in the case of zero phase saturation s_k, the singularity in this case in (8) has no place. So the modified Darcy equations can be written as follows:

$$u_{k,i} + \frac{K_0 \rho_k}{\varphi \mu_k} \bullet \theta_k \frac{\partial u_{k,i}}{\partial t} = -\frac{K_0 K_k^R}{\mu_k}\frac{\partial p_k}{\partial x_i} \tag{9}$$

Denote the parameter, which is in front of the time derivative of the velocity and has the dimension of time, as T_k; in general, it depends on its position in space, as does absolute permeability, but not on the current distribution of variable parameters (saturation, velocity, etc.):

$$T_k = \frac{K_0 \rho_k}{\varphi \mu_k} \tag{10}$$

The Darcy equation is then written in the following form:

$$u_{k,i} + T_k \frac{K_k^R}{s_k} \frac{\partial u_{k,i}}{\partial t} = -\frac{K_0 K_k^R}{\mu_k} \frac{\partial p_k}{\partial x_i} \tag{11}$$

The parameter T_k can be called the characteristic inertia time. In the case of permeability K_0 being sufficiently small, as in ordinary porous media, this time is negligible and inertia can be neglected. But if the porous medium consists of stones/balls with a sufficiently large diameter, then such a time for low-viscosity liquids can be several seconds or more; it will most likely be impossible to neglect inertia.

Locally, the pressure difference between the phases is determined by capillary pressure, which depends only on the phase saturation, besides the skeleton data and physical data of the surface separating the phases. Hereinafter, we assume that there are only two fluid phases in the pore space, so that:

$$p_1^c(s_1) = p_2 - p_1 = -p_2^c(s_2), \qquad s_1 + s_2 = 1 \tag{12}$$

3.2 Modeling relative permeability and capillary pressure

The relative permeabilities K_k^R are calculated using the Brooks-Cory model [17]:

$$K_k^R = \begin{cases} k_k^0 S_k^{n_k^0}, & s_k \geq s_k^{res} \\ 0 \quad , & s_k < s_k^{res} \end{cases}, \quad S_k = \frac{s_k - s_k^{res}}{1 - s_1^{res} - s_2^{res}} \tag{13}$$

Here, $k_k^0 > 0$ and $n_k^0 > 0$ are the model parameters, and the effective saturation S_k is determined by the residual saturations of $0 \leq s_k^{res} \leq 1$ $(s_1^{res} + s_2^{res} < 1)$. It should be noted that for $0 \leq S_k \leq 1$, a porous medium is impregnated with both fluids; the phase saturation at $s_k \leq s_k^{res}$ is not described with this model. Further, we always assume that impregnation with both fluids takes place, and under these conditions, the zero alternative of (13) can be disregarded. It can also be noted that $S_2 = 1 - S_1$, and thus index 1 is omitted further and at reduced saturation, the effective saturation of the second phase is expressed through the first. For the effective saturation and relative mobility of the phases not based on the model (13) and definition (14), we have the following expressions:

$$\begin{cases} S_k = \dfrac{s_k - s_k^{res}}{1 - s_1^{res} - s_2^{res}} \\ M_1 = \dfrac{k_1^0}{\mu_1} S^{n_1^0} \\ M_2 = \dfrac{k_2^0}{\mu_2} (1 - S)^{n_2^0} \end{cases} \tag{14}$$

Capillary pressure from the kth phase is expressed through the J-function of Leverett [1] and some other parameters like:

$$p_k^c = \frac{\sigma \cos \alpha_k}{\sqrt{K/\varphi}} J(s_k) \tag{15}$$

where σ is the coefficient of surface tension, and α_k is the angle between the wall from the phase k and the interface (wetting angle). If this angle is less than $\pi/2$, then the phase is wetting; if not, it is nonwetting.

Suppose the definition (15) holds for both phases with the same J-function. In this case, (12) leads to:

$$\cos \alpha_k J(s_k) = -\cos \alpha_{k'} J(s_{k'}) = \cos \alpha_k J(1 - s_k)$$

where the primes refer to nonwetting phase and, because of the connection between the wetting angles, we get

$$J(s) = J(1 - s) \tag{16}$$

None of the known models of the Leverett function imply the right side of (16). Consequently, the expression (15) should be considered only for one of the phases, but from the side of the other phase, the capillary pressure is determined by (12). In most of models, a wetting phase k ($\alpha_k < \pi/2$) is modeled using (15), while the nonwetting phase k' makes an adjustment to the definition of capillary pressure:

$$p_{k'}^c = \frac{\sigma \cos \alpha_{k'}}{\sqrt{K/\varphi}} J(s_k) \tag{17}$$

Using both expressions (15) for the wetting phase and (17) for the nonwetting, we express the capillary pressure using data from phase 1 (omitting the index 1 of capillary pressure, wetting angle, and saturation):

$$p^c = \frac{\sigma \cos \alpha}{\sqrt{K/\varphi}} \bullet \begin{cases} J(s), \alpha \le \pi/2 \\ J(1-s), \alpha > \pi/2 \end{cases} \tag{18}$$

Let the J-function of Leverett be modeled by two parameters $C_J > 0$ and $a_J > 0$ and effective saturation:

$$J(s_k) = C_J S_k^{-a_J}, \quad \alpha_k \le \pi/2 \tag{19}$$

Boundary conditions used:

At the inflow $z = 0$, we set the pressure $p = p_{in}$ and maximal saturation $s = 1 - s_{res}^2$.

At the outflow, we set a zero pressure and zero saturation normal derivative.

At the impermeable walls, we set to zero the normal components of phase velocities $u_{kn} = 0$, and this corresponds to zero normal derivatives of pressure and saturation.

To demonstrate the role of inertial effects in this problem, two calculations are carried out: with and without inertia (**Figure 8**). It is seen that the dynamics of imbibition on the left and right graphs are significantly different. When inertial effects are taken into account (left side), the graph has first a shape of a quadratic function of time (0–0.02 s), then linear (0.02–0.06 s), and then as a square root (0.06–2 s). When approaching zones with different permeability, these shape sections repeat: quadratic (2–3 s), linear (3–3.9 s), and square root. This result is consistent with the results presented in [7]. When inertial effects are not taken into account (right), the graph has the form of a monotonously increasing function close to a square root with a small break when passing through the border of zones with different permeability (4 cm).

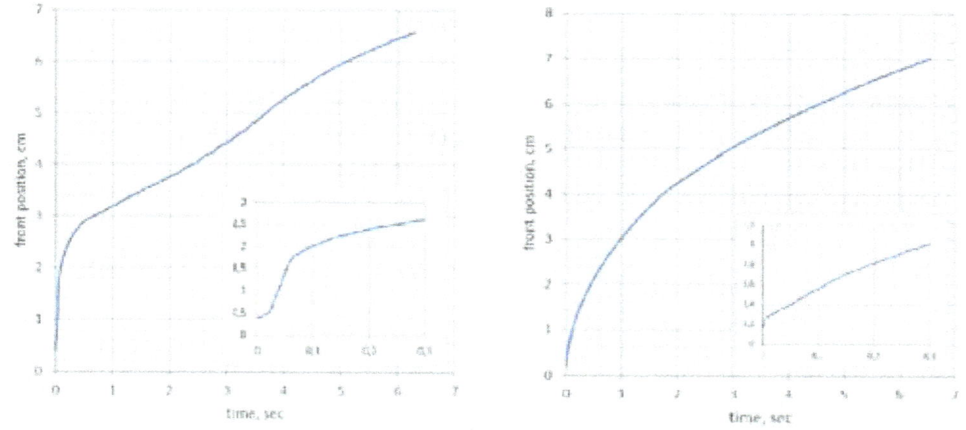

Figure 8.
Positions of the interface as a function of time for a flow in an inhomogeneous medium with inclusion of inertia (left) and excluding inertia (right) (bottom layer: height: 4 cm, permeability: $8.5 \cdot 10^{-8}$ m^2, porosity: 0.45; upper layer: permeability: $2 \cdot 10^{-8}$ m^2, porosity is 0.45; displacing fluid: viscosity: $1 \cdot 10^{-3}$ Pa·s, density: 10^3 kg/m^3; displaced gas: viscosity: $1.5 \cdot 10^{-5}$ Pa·s; density: 1.4 kg/m^3; surface tension: 0.0582 N/m).

4. Comparison of numerical calculations with experimental data

The microgravity quality was poor for the present experiments that does not permit us to perform quantitative comparison of results with the outcomes of numerical modeling, which were carried out for ideal conditions. Nevertheless, the qualitative results of the calculations are similar to the experiment.

In the case of flow in a homogeneous medium (**Figure 4**), the displacement is uniform gradually slowing down. In the experiment, the speed falls down more than in calculations. This is because the mathematical model does not take into account the acceleration of gravity g, which is close to zero in the experiment, but not equal to zero.

In the case of an inhomogeneous medium (**Figure 6**), when crossing the border of zones with different permeability, the graph has a complex space-time

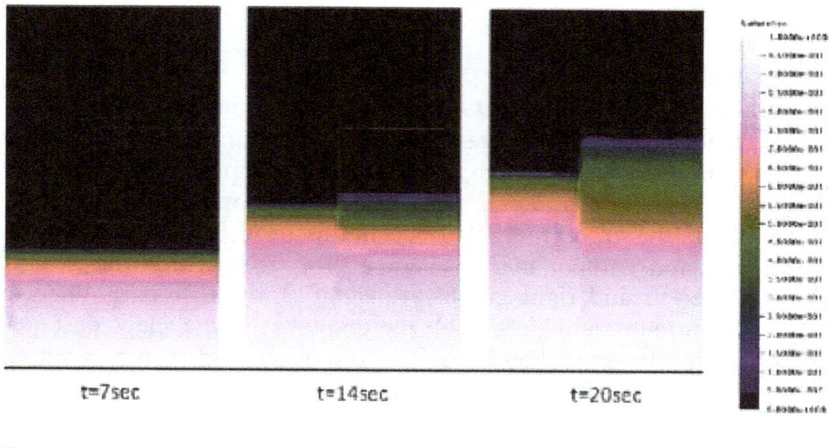

Figure 9.
The result of numerical modeling of fluid flow in the region containing an insert with a low permeability (media permeability: $8.5 \cdot 10^{-9}$ m^2, porosity: 0.458; insert: dimensions: $3 \times 2 \times 3$ cm^3, permeability: $2 \cdot 10^{-9}$ m^2, porosity: 0.466; displacing fluid: viscosity: $1 \cdot 10^{-3}$ Pa·s, density: 10^3 kg/m^3; displaced gas: viscosity: $1.5 \cdot 10^{-5}$ Pa·s, density: 1.4 kg/m^3; surface tension: 0.0582 N/m).

dependence form: quadratic (3–4.5 s), linear (4.5–5.5 s), then as square root. The main characteristic feature of the imbibition front passing through the boundary of media is its rapid acceleration (, it moves faster than if the medium were uniform, as in **Figure 4**). This corresponds to the results of the experiment.

The distribution of the saturation of the displacing liquid at different instants for the calculation, when the region contains an insert with a low permeability, is shown in **Figure 9**. This calculation corresponds to the experiment shown in **Figure 7**. In calculation, as in the experiment, having reached the insertion, the liquid begins to flow through the insert faster than in the zone with a higher permeability. The calculation also shows that the dispersion in a less permeable medium is higher (**Figure 9**).

5. Space experiments on capillary driven imbibition

Carrying out experiments on capillary seepage in space allows observing imbibition under conditions of higher quality microgravity and for longer periods. Such experiments were described in detail in [9, 14]. Here, we give a brief description of these experiments and show the possibility of determining empirical constants in mathematical models based on experimental data.

5.1 Experiment payload

Experiments on capillary driven seepage of oil in natural porous medium (sand) were performed within the frames of the MIRROR GAS programme. The experiments were prepared and supervised by Dr. D'Arcy Hart, C-CORE (Memorial University of Newfoundland), Drs. Laurier Schramm and Fred Wassmuth, Petroleum Recovery Institute (PRI, Calgary). The MIRROR payload is shown in **Figure 10**.

Liquid seepage was observed in three sample cylinders with the same porous medium: mixture of 20% (by weight) kaolinite (a sort of clay) and 80% silica sand. The fluids moving in each of the three cells were crude oil, lubricating oil, and polymer in distilled water.

The media and fluids were transported separately into orbit. An experiment cell consisted of a reservoir and a soil cell, with a sliding gate dividing the two (see **Figure 11**). At the beginning of the experiment, the stepper motor retracted the sliding gate, exposing the soil to the reservoir fluid, whereupon the liquid began to leak under the action of capillary forces under microgravity conditions. The level of fluid in the soil was detected using 48 fiber optic sensor probes located along the length of the sample. The results of the experiment are shown in **Figure 12** (black asterisk markers).

5.2 Mathematical model for multiphase seepage

We consider the seepage of a liquid through the area of length L. The capillary forces are taken into account. The nonlinear convection-diffusion equation for a porous medium is [12, 13].

$$\varphi \frac{\partial s}{\partial t} + \frac{\partial}{\partial x}\left(F - D\frac{\partial s}{\partial x}\right) = 0, \tag{20}$$

where s is the saturation of the displacing fluid, φ is the porosity, D is the diffusion coefficient, and F is the convection flux.

Figure 10.
MIRROR payload ready for GAS container integration at Kennedy Space Center. Capillary flow experiment cells are in the right-hand part of the upper section.

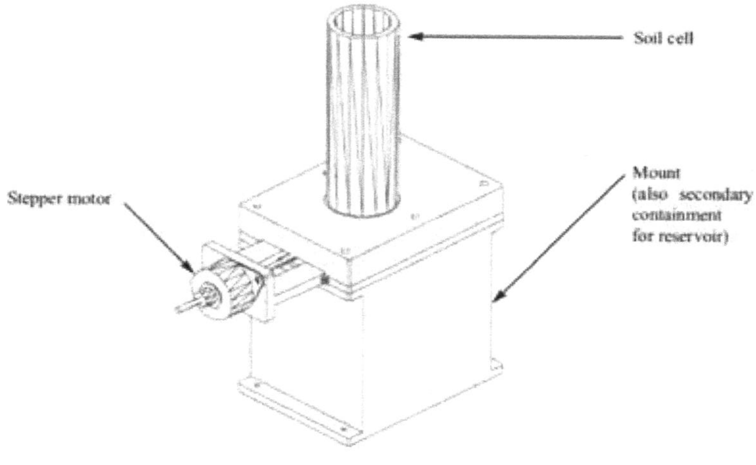

Figure 11.
Experiment cell for space experiments on capillary driven imbibition.

The initial conditions for saturation $s(t,x)$ are

$$s(0,x) = \begin{cases} s_{max}, & x = 0; \\ s_{min}, & x > 0. \end{cases} \tag{21}$$

The boundary conditions are

$$s(t,0) = s_{max}, \quad \left(\frac{\partial s}{\partial x}\right)_{x=L} = 0. \tag{22}$$

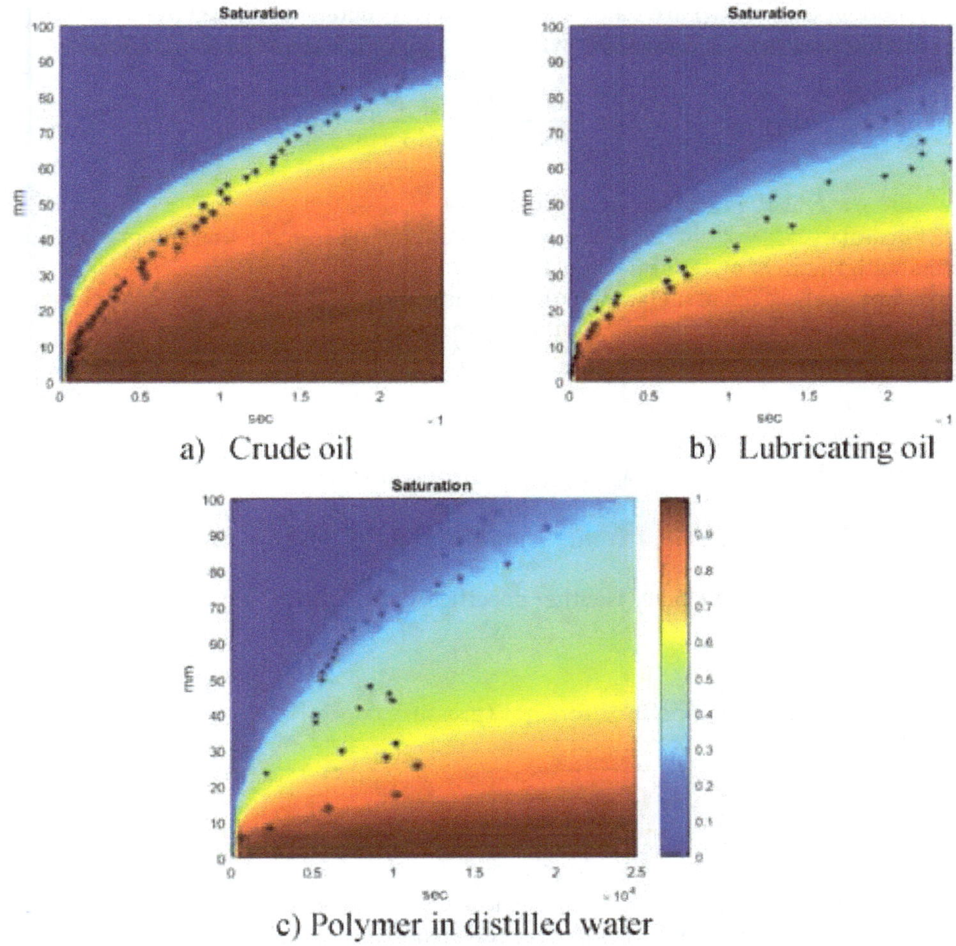

Figure 12.
Results of numerical modeling of the imbibition process of a liquid into a porous medium under the influence of capillary forces compared to experiment results.

where s_{min} and s_{max} are respectively the minimum and maximum possible saturations.

The convection-diffusion equation with initial and boundary conditions describes the displacement of a quantity s propagating as a step along the x-axis from the left to the right, if the convection flux is positive ($F > 0$) [7, 8]. The diffusion coefficient D is always positive; it blurs the front of propagation.

The effective saturation S is

$$S = \frac{s - s_{min}}{s_{max} - s_{min}}. \tag{23}$$

Below, the index "O" corresponds to the displacing fluid ("oil"). The index "G" corresponds to the displaced liquid ("gas"). Dimensionless relative mobilities M_O and M_G are calculated by the Brooks and Corey model:

$$M_O = k_O S^{n_O}, \quad M_G = \frac{\mu_O}{\mu_G} k_G (1 - S)^{n_G}, \tag{24}$$

where n_O, k_O, n_G, k_G are empirical constants. When simulating two-phase seepage, the expression for F is as follows:

$$F = \tilde{F} = -\frac{KM_O}{\mu_O}\frac{\partial p}{\partial x} = \frac{M_O}{M_O + M_G}u(t) \tag{25}$$

where $u(t)$ is the total seepage velocity.

Formula (25) was obtained as a result of averaging in modeling integral parameters with the help of 1D Equation [12].

Additional effect associated with the instability of capillary affected flows is modeled by the application of an auxiliary term to F:

$$F = \tilde{F} + F_0((s - s_{min})(s_{max} - s))^{\alpha} = \frac{M_O}{M_O + M_G}u(t) + F_0((s - s_{min})(s_{max} - s))^{\alpha}. \tag{26}$$

The details of such method for 1D approximation for 3D results are discussed in [13].

This term is proportional to the empirical coefficient F_0. It intensifies the dispersion of the displacement front. The sign of F_0 is positive since the instability intensifies the flow.

The total seepage velocity is either specified (flux control)

$$u(t) = U_0, \tag{27}$$

or calculated using a definite integral (pressure control)

$$u(t) = \frac{K\Delta P_0}{\mu_O}\left[\int_0^L \frac{dx}{M_O + M_G}\right]^{-1}. \tag{28}$$

where ΔP_0 is constant pressure drop.

The diffusion coefficient D is constructed from the sum of the diffusion coefficients due to the capillary forces d_C and the dispersion, which simulates, in the one-dimensional approximation, the blurring of the front due to other effects.

Diffusion formed by capillary forces d_C looks according to the Leverett model:

$$d_C = \frac{C_J a_J |\cos\theta| \sigma \sqrt{K\varphi} M_O M_G}{\mu_O (s_{max} - s_{min})(M_O + M_G)} \bullet \begin{cases} S^{-1-a_J}, & \cos\theta > 0; \\ (1-S)^{-1-a_J}, & \cos\theta < 0; \end{cases} \tag{29}$$

where θ is the wettability angle on the side of the displacing liquid, and C_J, a_J are parameters of the Leverett function. Similar procedure was described in [1].

The dispersion depends on the total flux of the phase and is proportional to its magnitude. Thus, considering dispersion and capillary effects, the diffusion coefficient D is modeled [10, 11] as

$$D = d_C + \left|F - d_C\frac{\partial s}{\partial x}\right|. \tag{30}$$

We note that the additional factor which is proportional to the coefficient F_0 also enters into the dispersion structure.

Processing the mathematical model described above makes it possible to select the coefficients responsible for the blurring of the front on the basis of experimental data. We have obtained that for the three experiments described above, the optimal coefficients are as follows (**Table 1**):

	C_J	a_J	F_0
Crude oil	$3 \cdot 10^{-2}$	0.5	$3 \cdot 10^{-8}$
Lubricating oil	$1 \cdot 10^{-2}$	0.5	$1.5 \cdot 10^{-7}$
Polymer in distilled water	$5 \cdot 10^{-3}$	0.5	$4.5 \cdot 10^{-7}$

Table 1.
Optimal coefficients for the three experiments.

The pressure drop or the specified flux was absent, so that the motion was carried out only by capillary forces.

The distribution of the saturation of the penetrating liquid in the calculated region over time is shown in **Figure 12**. The parameters of each of the three calculations corresponded to the actual experiment. The experimental data are marked with black asterisk markers.

Thus, as can be seen from **Figure 12**, the results of numerical calculations based on the proposed mathematical model are in good agreement with the experimental data. The blurring of the displacement front for each of the liquids is different: the smallest blurring is observed in crude oil (**Figure 12(a)**), and the front is most strongly blurred in the case of imbibition of polymer (**Figure 12(c)**). For each fluid, the empirical constants C_J, a_J, and F_0 are selected. The value of a_J for all liquids is chosen to be 0.5. The coefficients C_J and F_0 are different for each case, but they are of the same order. The coefficients were chosen in such a way that the blurring of the front in the numerical calculation would have the least difference from the experimental data.

6. Conclusions

Experiments on the imbibition of a fluid into an inhomogeneous artificial porous medium under microgravity conditions during parabolic flights are considered. A mathematical model for multiphase seepage considering inertial effects is developed. The results of the experiment are compared with the results of numerical simulation of the flow of a viscous liquid in a porous medium, taking into account capillary and inertial effects. It is shown that when the transition from a more permeable medium to a less permeable one takes place, the velocity of the imbibition front increases. This is because the capillary forces in a medium with a low permeability are larger. On further imbibition into the low permeable medium, the velocity of imbibition decreases much faster than that in highly permeable medium. It is also shown that the dispersion of fluid during its imbibition into a less permeable medium is much higher. A model describing the displacement front blur due to diffusion, capillary effects, and flux instability was developed; the model uses two empirical constants. This model can describe the displacement front peculiarities even in 1D problem setup. To estimate those empirical constants, the results of experiments in microgravity were used; those conditions were needed to exclude the gravity factor affecting the flux and diminishing the capillary effects.

Acknowledgements

Authors express warmest gratitude and high appreciations of the results by Professor Jean Claude Legros, who was initiator and principle investigator in these researches. The experimental hardware and methodology were developed by

Dr. Eric Istasse and became the base of his thesis [15]. The authors with financial support of Russian Academy of Sciences performed theoretical investigations and numerical simulations.

Author details

Nikolay Smirnov[1,2], Valeriy Nikitin[1,2] and Evgeniya Kolenkina (Skryleva)[1,2*]

1 Moscow M.V. Lomonosov State University, Moscow, Russia

2 Scientific Research Institute for System Analysis of the Russian Academy of Sciences, Moscow, Russia

*Address all correspondence to: jennyne@yandex.ru

References

[1] Barenblatt GI, Entov VM, Ryzhik VM. Theory of Fluids Flows through Natural Rocks. Dordrecht, Boston, London: Kluwer Academic Publishers; 1990

[2] Bear J, Bachmat Y. Introduction to Modelling of Transport Phenomena in Porous Media. Dordrecht, Boston, London: Kluwer Academic Publishers; 1990

[3] Bear J. Dynamics of Fluids in Porous Media. New York: Dover Publications Inc.; 1988

[4] Kaviany M. Principles of Heat Transfer in Porous Media. 2nd ed. New York: Springer-Verlag; 1995

[5] Nield DA, Bejan A. Convection in Porous Media. New York/Berlin/Heidelberg/London: Springer-Verlag; 1992

[6] Nigmatilin RI. Dynamics of Multiphase Media. Moscow Science Publication; 1987

[7] Smirnov NN, Dushin VR, Legros JC, Istasse E, Boseret N, Mincke JC, et al. Multiphase flows in porous medium—Mathematical model and microgravity experiments. Microgravity Science and Technology. 1996;**IX/3**:222-231

[8] Smirnov NN, Nikitin VF, Norkin AV, Kudryavtseva OV, Legros JC, Istasse E, et al. Capillary driven filtration in porous media. Microgravity Science and Technology. Hanser Publications, Munich, Germany. 1999;**XII/1**:23-35

[9] Smirnov NN, Legros JC, Nikitin VF, Istasse E, Schramm L, Wassmuth F, et al. Filtration in artificial porous media and natural sands under microgravity conditions. Microgravity Science and Technology. Bremen, Germany: Z-Tec Publishing. 2003;**XIV/2**:3-28

[10] Smirnov NN, Nikitin VF, Ivashnyov OE, Maximenko A, Thiercelin M, Vedernikov A, et al. Microgravity investigations of instability and mixing flux in frontal displacement of fluids. Microgravity Science and Technology. 2004;**XV/3**: 3-28

[11] Smirnov NN, Dushin V, Nikitin V, Philippov Y. Two phase flows in porous media under microgravity conditions. Microgravity Science and Technology. 2008;**20**(3–4):155-160

[12] Smirnov NN, Dushin VR, Nikitin VF, Phylippov YG, Nerchenko VA. Three-dimensional convection and unstable displacement of viscous fluids from strongly encumbered space. Acta Astronautica. 2010;**66**:844-863

[13] Smirnov NN, Nikitin VF, Maximenko A, Thiercelin M, Legros JC. Instability and mixing flux in frontal displacement of viscous fluids from porous media. Physics of Fluids. 2005; **17**:084102

[14] Dushin VR, Nikitin VF, Smirnov NN, Skryleva EI, Tyurenkova VV. Microgravity investigation of capillary driven imbibition. Microgravity Science and Technology. Springer. 2018;**30**(4): 393-398

[15] Istasse E. Determination of capillary characteristics in porous media, Free University of Brussels, Faculty of Applied Sciences, Dissertation presented by Eric Istasse with a view to obtaining the Diploma of Advanced Studies in Applied Sciences, Director: Prof. Legros J.C., Academic Year 2000- 2001, Brussels. 2001

[16] Smirnov NN, Nikitin VF, Skryleva EI. Microgravity investigation

of seepage flows in porous media. Microgravity Science and Technology. Springer. 2019;**31**(5):629-639. DOI: 10.1007/s12217-019-09733-7

[17] Larry WL. Enhanced Oil Recovery. Englewood Cliffs, New Jersey: Prentice-Hall; 1989

9

Preparation, Implementation and Execution of Human Cardiovascular Experiments in Space

Peter Norsk

Abstract

There are eight steps in the preparation, implementation and execution of a human spaceflight experiment: (1) writing a proposal, (2) being selected by a space agency, (3) finding funding, (4) flight feasibility assessment for flight, (5) implementation into a specific space platform (e.g. the Space Shuttle in the past and now the International Space Station), (6) experiment execution, (7) analysis of collected data and (8) publication. The unique features about spaceflight experiments are steps 4–6 because of the limitations of conducting experimental procedures in space. Furthermore, all of the associated equipment have to be developed and approved for spaceflight with all the safety aspects taken appropriately into consideration. In this chapter, two specific experiments from the Spacelab D2 mission in 1993 are used as illustration of these steps as well as describing the use of parabolic flights as a preparatory platform. It is important to have data collected of such a quality that they can be published in science journals with external peer review. It is also important that the data not only have operational spaceflight applications but also can advance knowledge for terrestrial science purposes.

Keywords: cardiac output, cardiovascular, central venous pressure, parabolic flight, weightlessness

1. Introduction

On 22 March 1993, I was closely watching the big screen in the German Space Control Centre in Oberpfaffenhofen near Munich, Germany. The Space Shuttle with the Spacelab D2 mission was just about to lift off, and on board were seven astronauts and equipment for a host of experiments of which four were from Denmark. I was the responsible investigator for these four experiments. I was nervous. In the airplane from Copenhagen to Munich the day before, I had told myself that this would either be my stepping stone to a further career in space physiology and medicine or simply my big Waterloo. The Danish government had payed millions of Kroners in preparation of the studies, and failure was not an option. The problem, though, was that anything random, which was totally out of my and my colleagues' control, could happen at any time and jeopardise our experiments.

The four experiments had been under preparation for some 5 years and supported by Danish space research grants. Eight people had been full time involved in my laboratory: four engineers and four medical doctors. Equipment had been developed, tested and adapted for spaceflight, and many nervous moments had been overcome. It was exciting times and now—finally—everything was about to be launched into space to orbit the Earth at a speed of 28,000 km per hour in a free fall condition inducing chronic weightlessness for 10 days. I felt tense as the launch countdown approached zero.

A few seconds before launch, two of the three Space Shuttle engines ignited. The third one did not. Immediately we knew in the control room that something was wrong. One of the engines did not blast, and within a few seconds, all engines stopped. As water vapour evaporated away from the shuttle, we waited anxiously. What would happen now?

The ignition had failed, and within the next hour or so, the astronauts were let out, and the launch postponed until the end of April. In that interim period, we had to repeat some ground tests on the astronauts in collaboration with several other international teams from Europe and the United States. Finally, after some additional postponements of the launch, the Space Shuttle Columbia ignited on 26 April and went to space, where it completed a 10-day mission with great success. Almost 90 experiments were expertly conducted. Concerning the four Danish human physiology experiments, the data were successfully collected, which 2–3 years later led to three scientific papers [1–3] from our Danish space medicine and physiology laboratory, DAMEC Research Inc. (Danish Aerospace Medical Centre of Research Inc.), at the Copenhagen University Hospital.

The successful completion of our experiments on the Spacelab D2 mission in 1993 with the publications in 1995 and 1996 led to the later expert preparation, implementation and conduction of seven additional studies in space on the Russian Mir station, Space Shuttle Columbia and International Space Station. All studies had cardiovascular adaptation aspects, and parabolic flights were used—not only for preparation of the spaceflight experiments but also to obtain basic science knowledge of human physiological responses to very acute weightlessness of 20 s.

2. Experimental science background

During changes in posture, blood pressure in humans is continuously and acutely regulated by pressure reflexes originating from receptors (baroreceptors) in the aorta close to the heart and in the two carotid (neck) arteries at the base of the skull. In addition, there are pressure sensors in the heart, from which reflexes also originate to participate in blood pressure control. This blood pressure regulation is for the central nervous system to make sure that the perfusion pressures to the brain and other organs are optimal despite the displacement of blood caused by the posture. As an example, the upright posture displaces blood downwards towards the lower body and the legs away from the head and heart. To counteract that so that the blood pressure at heart and head level does not fall too much, which would impede blood supply to the brain, the blood pressure reflexes sense the decrease in pressure and within a few heart beats initiate an increase in heart rate and constriction of the small arteries in the lower body. The opposite occurs, when the posture changes from upright to recumbent or supine.

We have for many years investigated, which blood pressure sensors are the most important for adjusting blood pressure to posture changes. In a whole host of investigations using a combination of various models as well as short-term weightlessness during parabolic flights, we have aimed at isolating the effects of some receptors versus others and found that in order for blood pressure and heart rate to adapt to either the supine or upright posture, the low pressure reflexes originating from the heart and major central veins are pivotal. Without the inputs from these low pressure heart and venous reflexes, the new steady state cannot be achieved. These findings have changed our previous understanding of how the human cardiovascular system adapts to changes in postures and the effects of gravity.

Since blood pressure is also determined by the amount of fluid in the body, we have used the human head-out water immersion model to investigate how the volume of fluid and amount of salt is controlled. By immersing humans in the seated posture to the level of the neck for hours, the fluid- and salt-excreting mechanisms through the kidneys are stimulated, because the headward shift of blood and fluid from the lower body to the heart through various mechanisms informs the central nervous system that the upper body vessels are being overloaded with blood and fluid so that this excess volume must be excreted. The mechanisms for this are still only partly understood, but the main opinion is that there is a connection between the heart and kidneys through what is termed the cardio-renal link so that when the heart chambers are stretched, it initiates a reflex response to the kidneys to excrete more salt and fluid in the urine.

Our research using the seated head-out water immersion model has shown that not only the cardio-renal link and associated hormones control the urinary excretion rates of sodium during shifting of blood and fluid to the upper body but also dilution of the blood with fluid from the tissues plays an important role. During shift of blood from the lower to the upper body caused by the surrounding water pressure, fluid is pushed into the circulation from the lower body tissues. This dilution can affect the kidneys directly as well as release of some kidney-regulating hormones.

Weightlessness in space is a unique condition that cannot be replicated on the ground and where bodily functions can be studied without the intervening effects of gravity. In space, blood and fluids are chronically displaced towards the upper body segments (heart and head), and the daily fluctuations induced by posture changes do not occur. This gives us a unique opportunity to utilise weightlessness in space for exploring the cardiovascular and fluid volume-regulating systems in the human body.

3. Experiments: the spacelab D2 mission in 1993

The experiments on the Spacelab D2 mission in 1993 aimed at understanding whether the blood and fluid shift into the heart during prolonged weightlessness would augment the urinary output of water and salt, just as we usually observe in ground-based simulation models. In one experiment we aimed at measuring how much the fluid pressure leading into the heart (central venous pressure, CVP) increases, which we at that time thought to be the stimulus for control of the urinary fluid and salt excretion. In another experiment we aimed at monitoring the urine production. It was the hypothesis that an increase in CVP induced by weightlessness would augment the excretion rate of fluid and salt.

3.1 CVP experiment

The CVP experiment was originally planned to be done on two of the Spacelab D2 astronauts, but for some technical reasons, it was only accomplished in one. The experimental plan was to shortly before launch insert a long catheter with a pressure transducer at the end into the vein directly leading into the right atrial chamber of the heart through a peripheral arm vein and connect it to a preamplifier and a recording system. The test astronaut would thus be inserted with the catheter and wear the CVP monitoring system until 3 hours into the mission following launch. Thereafter the catheter would be withdrawn. Before the launch of the Space Shuttle, control measurements in different body postures were performed.

The equipment used for measurements of central venous pressure during the Spacelab D2 mission in one astronaut is depicted in **Figure 1**. The central venous catheter (1) with a pressure transducer placed at one end and a connector box at the other was plugged into a preamplifier (2) that was connected to a recording unit (3). A calibration piston (4) could be connected to the reference opening of the catheter and thus induce predetermined pressure changes on the backside of the transducer membrane. In this way, it was tested to what degree the calibration characteristics of the pressure transducer might have changed over time after being inserted into the astronaut. Following the spaceflight, the catheter was brought back to the investigators and tested for change in drift of the tip transducer.

3.2 Urinary excretion experiment

Before the spaceflight, four test astronauts would, over a 4.5-h period, empty their urine bladders in either supine or seated posture, on an hourly basis after being infused through a peripheral vein with isotonic saline in an amount of 2% of their body weight. About a week into the flight, the same would be done following a similar infusion while they thus would be free floating in the space vehicle. Blood was sampled on ground and during flight for determination of water-, salt- and blood pressure-regulating hormones. The volume of urine was measured after each void and samples taken for determination of salt (sodium and potassium) concentrations. In space, the volume of urine was determined by a urine monitoring system delivered by NASA, which was connected to the toilet. If the voids were felt by the subjects to be small, the bladders were emptied into bags and returned to

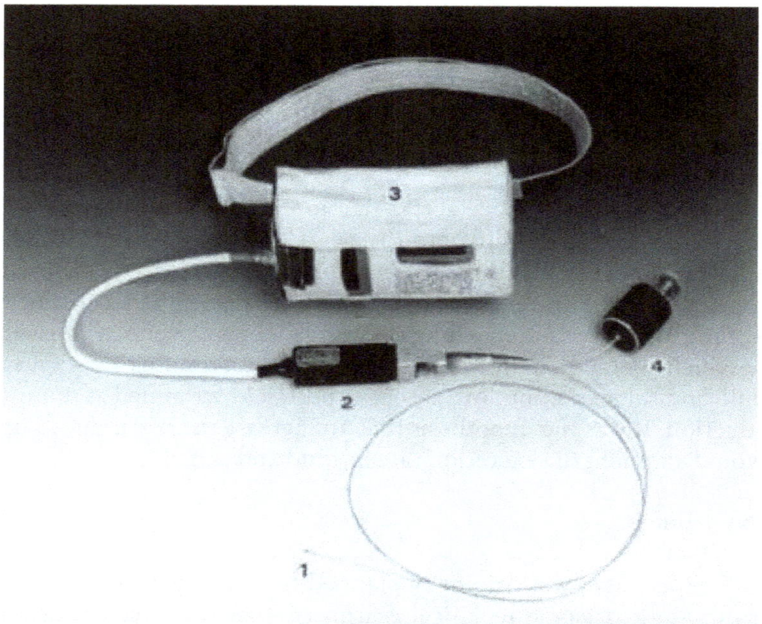

Figure 1.
Central venous pressure equipment, which flew on the Spacelab D2 mission in 1993 [3]. (See text for detailed explanation).

Earth through the trash system on the Shuttle. The saline infusion was conducted over some 20 min by manually inflating a cuff-pressure system, and the amount varied on an average between 1.7 and 1.8 litres.

4. Experiments during subsequent missions (1995–2012)

After the Spacelab D2 mission in 1993, our research team in Denmark conducted additional experiments on various space platforms with one on the Russian space station, Mir, in 1995, where we monitored urine excretion rates over longer flight periods of up to 6 months by having three test astronauts collect urine into bags and the volumes measured by a scale system. The idea was to test whether urine production in weightlessness after intake of an oral water load would be enhanced—just like we tested the same hypothesis regarding urinary salt excretion on the previous Spacelab D2 mission.

During later space missions, we have conducted several additional studies on the Space Shuttle and the International Space Station focusing on the cardiovascular adaptation to short (<30 days) and long (>30 days) duration flights. In particular, we have for this purpose conducted cardiac output measurements by a non-invasive rebreathing technique, which has been developed for spaceflight. Cardiac output is the amount of blood injected by the heart into the body per minute, and this variable is important for understanding the effects of weightlessness on not only cardiac function but also the vascular condition in general. The hypothesis was tested that the weightlessness-induced increase in central blood volume would increase cardiac output and at the same time through the cardiovascular reflexes dilate the arterial resistance vessels to counteract an increase in blood pressure.

Normally, accurate cardiac output estimations require insertion of catheters into the veins and arteries, which makes the measurement difficult on a routine basis in normal healthy humans. With the non-invasive foreign gas rebreathing technique developed for spaceflight, such estimations can be done anytime and anywhere with no harm to the test subject. As indicated in **Figure 2**, the tested person breathes

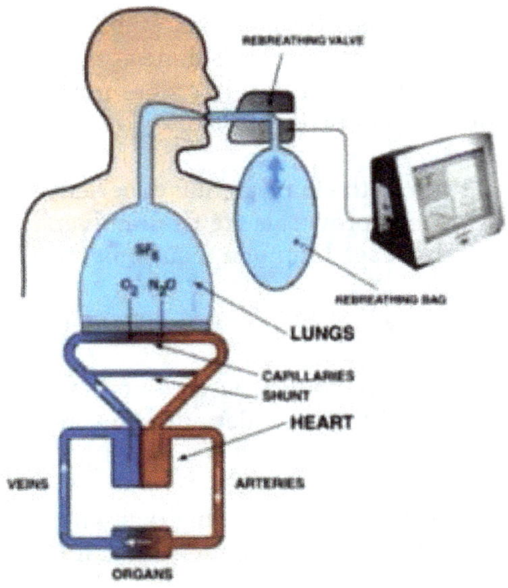

Figure 2.
The principle of foreign gas rebreathing for measurement of cardiac output [4]. (See text for explanation).

back and forth into a rebreathing bag with a gas mixture, in which a tracer gas (e.g. N_2O) is taken up by the blood flowing through the lungs. The disappearance rate of the tracer gas from the lungs to blood is detected by an infrared photoacoustic gas analyser connected to the mouthpiece of the rebreathing person. By knowing the solubility of the gas in the blood, the amount of blood flowing through the lungs per unit of time can be calculated. This amount is equal to cardiac output. The measurements take less than 30 s and are pivotal for understanding how cardiac output adapts to various conditions such as weightlessness in space [4].

This methodology is currently being used on the International Space Station for various research projects and has been developed from a mass spectrometry detection technique for measuring gas concentrations to using infrared photoacoustic gas detection. This has made it possible to have a much less voluminous and more user-friendly equipment on board the space station. Also, this technique has been tested against golden standard invasive clinical techniques, where excellent correlations have been found [5].

5. Microgravity experiment implementation and execution

The cardiovascular experiments that we have conducted in space through the past three decades entail the following steps: (1) proposal, (2) selection, (3) funding, (4) feasibility assessment, (5) implementation, (6) execution, (7) data analysis and (8) publication. The unique features about spaceflight experiments are steps 4–6, because of the limitations of conducting experimental procedures in space. Furthermore, all of the associated equipment have to be developed and approved for spaceflight with all the safety aspects taken appropriately into consideration.

5.1 Proposal

The first step for performing experiments in space is to develop a proposal and respond to a space agency solicitation from, e.g. the European Space Agency (ESA) or National Aeronautics and Space Administration (NASA). ESA's topics are usually broader than NASA's, because the latter are mostly focused on operational aspects. In both cases the proposal formats are rather similar and the subsequent selection process very much like the way it is done at the national levels with peer review panels and scientific merit scorings. In order for a proposal to be successful, the following criteria must generally be fulfilled:

1. Qualified research team with experimental experience from a university or a recognised company. Most proposals are from universities, and experience in space research is an advantage but not a requirement.

2. Clearly written proposal, which fulfils the requirements stated in the call.

3. Adherence to all of the rules and instructions—otherwise the proposals will be rejected upon receipt, and this includes adhering to the stated deadlines for submission.

4. Relevancy for utilising spaceflight meaning that what is proposed to be measured should be responsive to the flight environment (e.g. weightlessness).

One thing to keep in mind when writing a proposal is to formulate it so that non-experts in the field can also get something meaningful out of it, because at the time of selection and thereafter, decision-makers who may not be scientists might make a judgement as to the appropriateness of spending resources in space for this particular experiment.

5.2 Selection

When a proposal is submitted in response to a research announcement, the first step is that it will be evaluated by scientists appointed by the space agency or a group of space agencies. The scientists—or peer reviewers—are usually experts within the field of the topic of the solicitation, who are not involved in collaborations with the proposers. The peer review panels are usually led by an agency representative, and the panel will score each proposal on a scale between zero and 100. Scores between 90 and 100 are categorised as "Excellent" or "Outstanding", 80 and 89 as "Very Good", 70 and 79 as "Good", 50 and 69 as "Fair" and below 50 as "Poor". The score threshold for selection may vary between space agencies, but usually no proposals are selected with a score lower than 70.

The next step for the space agency representatives is to—based on the peer review scores—perform final selections. In this case, not only the scientific merit scores play a role but also the relevancy of the proposal for the agency. Usually a subset of the highest scientifically scoring proposals are selected, but sometimes even proposals with the highest scores may not be finally selected because of less relevancy for the operational purposes of the agency. In this regard, there are different policies between the space agencies. For NASA, deep space explorations are the main drivers, while ESA usually focuses almost entirely on the scientific merit and the proposal's ability to produce new fundamental knowledge to the scientific community.

5.3 Funding

In order for a selected proposal to be executed in space, funding has to be obtained. This can either be done by grants from local and national authorities or from the space agency itself. The problem in particular for European researchers is that for ESA to consider selection of a proposal, it is advantageous to have obtained national funding or a declaration of intention of funding already before submission. In many cases, national authorities will only indicate intention of funding, should the proposal be selected, but they usually will not guarantee it. This can sometimes create a hen and an egg problem: The space agency will—before it actually finally selects a proposal—require guaranteed funding from a national authority, whereas the national authority requires that the agency selects the proposal. The proposers usually obtain intention for funding in letters from the national funding authorities, and usually the space agency will accept that. In our case, when the research team was supported for selected experiments, we referred to an existing grant that could overlap with selection of a new proposal.

Funding of a grant for experiments in space usually only covers the expenses incurred by the experimental research team. The space infrastructure, such as access to a space vehicle and its astronauts (e.g. the Space Shuttle), is delivered by the space agencies.

5.4 Feasibility

When selection and funding are obtained, the space agency will conduct a feasibility study to evaluate whether the experiment can be implemented in space and whether there are technical or other obstacles. If these cannot be overcome, the proposal will be de-selected. Usually the experiment is modified, if obstacles are detected. An obstacle for implementation may not just be technical such as lack of availability of a technique or equipment but may also be lack of astronaut crew time for execution of the experiment. In that case, the experiment is usually modified in close collaboration with the research team. If the experiment changes considerably, the space agency may require an additional scientific peer review to evaluate whether the scientific quality is still high enough, but this is not the usual process.

5.5 Implementation

When the space agency feasibility assessment has been successfully completed, meaning that the experiment can be conducted in space, the implementation process begins. All of our previous research team's selected experiments (10 in total) have been somehow modified during the feasibility assessment and implementation processes. The renal experiment on the Spacelab D2 mission in 1993 was changed by decreasing the number of inflight sessions from two to one, because of limited crew time. The purpose was as previously described to evaluate the effects of applying an intravenous saline fluid load to the test astronauts on renal excretion rates of fluid and sodium. We had originally planned a session with infusion and a session without, but only the infusion session could be implemented in space. Otherwise the experiment was kept intact except that it was actually improved by changing our proposed saline loading from an oral saltwater load to intravenous infusion. This was done because an experimental complement was created for the space mission, whereby several experiments were to be executed in an integrated fashion, and a US experiment had suggested to use saline loading by infusion. Thus, this intravenous infusion of isotonic saline was planned to be done for the first time ever in space.

Although the urinary experiment was not difficult to implement from a technical point of view, it was totally different regarding the CVP experiment. After selection of this experiment, many managers within ESA and NASA thought it would be impossible to be allowed to conduct such an invasive study. We succeeded anyway, which was because of one important thing: the backing of the appointed payload commander. Without this support for the experiment, it would never have been executed. The reason for the astronaut support was because of thrust in our research group's ability to conduct the study, which we obtained by always being well prepared for pre-flight briefings of the astronauts as well as for the pre-flight control studies. We always made sure that as much of the data that had been collected were properly analysed between the different ground tests and that the data were presented to the astronauts during the subsequent tests so that they together with the investigators could follow the progress of the study throughout the pre-flight period.

5.6 Execution

During the Spacelab D2 mission, I was standing in the mission control centre in Oberpfaffenhofen near Munich in Germany holding my breath and watching the big screen. All investigators followed the executions of their experiments from the mission control centre, and just before initiation of our renal experiment, which as described earlier was integrated into one complement, a valve was stuck in some of the equipment. If the valve problem was not solved, we would all risk that none of the experiments in the complement would be conducted. It was a tense moment, when the payload commander after directions from mission control finally got the valve to work and initiated the flow of measurements. What a relief!

During execution of our urinary excretion experiment a mistake did happen, whereby urine bags, which were to be collected after flight directly from the trash can in the Space Shuttle, were not correctly labelled. It meant that we could not readily identify, from which crew members the urine in the bags derived. The way the problem of identification was solved was by measuring the concentrations of five different substances in each urine bag and comparing them to samples that had been collected inflight from each bag before trash storage. Each of the samples had been correctly labelled. Since each crew member had a unique pattern of concentrations of the selected substances, the matching and identification of the bags with the samples were possible.

Spaceflight experimentation often requires creative solutions to unexpected problems.

The CVP experiment went well in one astronaut. Originally, we had planned for two astronauts to be instrumented, but unfortunately the catheter broke in one during an extended prelaunch period, where the catheters had been successfully inserted into two astronauts, but where the prelaunch period was extended for 2 days over a weekend because of a failure in one of the shuttle's navigation systems that had to be changed. During some leisure activities, it broke and was withdrawn before launch.

During execution of the experiment on ground before the flight, which is called the baseline data collection to which all inflight data were to be compared, the biggest obstacle occurred during the execution process of one of the pre-flight test sessions on the ground. The obstacle demonstrated that it is not only a challenge to implement and execute an experiment during the flight phase but that ground tests can be limiting factors too. What happened is that the gas analyser used for rebreathing experiments to determine cardiac output and respiratory variables (**Figure 3**) for some reason did not work. Even though these measurements were not directly involved with our urinary experiment, it sent our experiment to jeopardy, because it was totally integrated into an experimental complement to be executed in concert. The breakdown happened at the Aerospace Medical Institute at the German space centre, DLR, and since the astronauts' test time was extremely limited with many other obligations, it was made clear to the experimental team that the astronauts would withdraw from the experiments, if the equipment was not working the next morning. We knew then that we were in trouble!

We were otherwise all ready to conduct the baseline data collections the next day, and if the gas analyser could not be fixed, it would mean that all of our experimental efforts would go down the drain. What could we do?

Ingenuity, imagination and thinking out of the box are usually essential in solving unexpected problems associated with spaceflight. In fact, this is what characterises this discipline. To my disappointment most of the officials and investigators gave up immediately. It was late afternoon, and the experiments were to be commenced early next morning at 07.00 am. The payload commander had left with a statement that he and his astronaut team would show up on time the next morning, and if the equipment did not work, the experimental complement would be deleted from the mission.

I and one of our ESA representatives soon conferred with each other, and we promised that we would demonstrate to the payload commander that this problem could be fixed in time. The only question was how? At the time we did not know that the problem was a burned capacitor, so we planned to have a technician immediately flown down from the company, Innovision A/S, in Denmark, which had developed and built the equipment.

What we did was risky, unusual and not according to the normal rules and regulations, but we were running out of time. We had to rent a private airplane

Figure 3.
*This rack was called Anthrorack and developed for the spacelab D2 mission. It consisted of several pieces of equipment of which one was a mass spectrometer gas analyser used for respiratory analysis and cardiac output determinations by rebreathing (**Figure 2**). During the ground-based data collections, a capacitor was burned and had to be replaced at a very critical time before the mission.*

within a few hours, because there were no commercial flights at that time. We had to establish an ESA guarantee for payment to the airline company, and we—above all—had to get in contact with the technician in Denmark. It soon turned out that he was available and the money for renting the airplane could be secured (after tough negotiations with ESA) and everything seemed possible.

The technician was late at night transported by taxi to a nearby local airport some 200 km away, from where he lived, entered the plane and came to Cologne around 4 am in the morning. We picked him up at the airport in Cologne and brought him to the aerospace facility, and by a miracle, he quickly identified the problem to be a burned capacitor and substituted it by another from a similar equipment.

At exactly 07.00 am before the baseline data collection was to begin, we were ready. The astronauts entered with the expectation that the tests would be cancelled. We could inform them otherwise, and with a rare expression of facial recognition, the payload commander and his fellow astronauts professionally moved ahead to be ready for the tests.

Everything went smoothly!

This as well as the inflight incident with the stuck valve were pivotal obstacles for the outcome of many of the physiological experiments during the Spacelab D2 mission. Had they not been overcome, I would probably not have been able to continue my space physiology career for the next 30 years.

5.7 Data analysis

It is pivotal to make sure that the data collected are correct. One basic rule is for the principal investigator to always be present or to have proper representation at each of the pre- and post-flight experimental sessions and to be monitoring how the data collections are done during flight—preferably from a mission control centre. If that is not made sure of by the investigators of a study, one cannot be sure that the circumstances surrounding the collections are fully understood and that handling of blood samples is done correctly and according to specifications. Furthermore, the investigators have to make sure to be readily available during executions of their experiment should inquiries from space agency representatives need acute responses and interventions. Otherwise, it is unlikely that the data can be trusted.

The investigators must also be proactive and tenacious in obtaining the collected data in space that usually are stored on inflight computers. One way to make sure that the data are correctly handled is to push the mission controllers to download as much as possible from space to ground as soon as the data are collected, because no one knows what could happen afterwards to the storage. Usually, it is not a problem should downloading not be possible, but in this case, it can also delay the post-flight analysis of the data because of bureaucratic impediments to obtain it.

During the very sad and unfortunate accident of the STS-107 mission on February 1, 2003, where the Space Shuttle, Columbia, and crew were lost during re-entry in the atmosphere, I was in charge of an experiment. We conducted inflight cardiac output rebreathing experiments as well as blood pressure monitoring. The data were downloaded to the mission control centre during the mission, which made it possible for us to publish the data so that the experiments—despite the very sad circumstances—were not done in vain.

During the Euromir 95 mission, which was a long-duration mission on the Russian space station Mir, where the ESA astronaut, Thomas Reiter, stayed for 179 days in space, I was in charge of a urinary collection experiment following an oral water load. I obtained the inflight data directly from Thomas Reiter himself shortly after his flight, which is unusual, but we did so to bypass the bureaucracy. Otherwise it could have taken weeks to obtain it.

5.8 Publication

The most important part of all investigations including those in conjunction with human spaceflights is to publish the results in as widely distributed scientific journals as possible. The whole purpose of obtaining the data is to gain new knowledge, and by publishing in science journals with external peer review, there is a certain guarantee for data quality and interpretation. It usually takes 2 years after the end of a spaceflight mission to have the data published, but many times, it takes longer. The investigators, however, owe it to everybody involved as well as society in general to produce a scientific publication as fast as possible.

From the Spacelab D2 mission in 1993, our research team succeeded in publishing three papers within 3 years of the mission [1–3]. During later missions on the Russian space station Mir, the Space Shuttle Columbia (STS-107) and the International Space Station, we conducted five additional experiments focusing on how the human cardiovascular system adapts to short- and long-duration flights [6–10]. This is important for understanding the long-duration health effects of future deep space missions that may last up to 3 years on a mission to Mars.

6. Parabolic flights

In preparation of the CVP experiment for the Spacelab D2 mission, we in 1991 participated in a series of ESA-funded parabolic flights at an air base in Bretigny-sur-Orge, near Paris in France. The purpose of participating in these flights was not only to test the technical feasibility of the equipment in weightlessness but also to obtain short-term data during this condition and compare them to longer effects of spaceflight (see **Figure 4**). At that time, the parabolic flights were conducted by a Caravelle, which flew in a Keplerian trajectory, thereby creating a free fall condition (0 g) symmetrically around the top of the trajectory for 20 s. Some 20 s before and after the 0 g period, the plane underwent a period of increased g's from 1 up to 2. Thus, it is a very short period of weightlessness that is created in this way, but it is the only way to induce real weightlessness in humans without going into actual space.

The CVP equipment was also tested during longer weightless periods (some 60 s) in a fighter airplane (Draken) in Denmark in one of the investigators. This test was supported by the Royal Danish Air Force. All of these tests were conducted in seven subjects (Caravelle) and in an additional one subject (Draken) and demonstrated that the equipment worked during short-term variations in g's between 0 and 4. In addition we obtained data on effects of short-term changes in g-loads on CVP including effects of weightlessness for comparisons with spaceflight.

For further interpretation of the data, we later after completion of the D2 mission performed another series of CVP experiments during 20 s of weightlessness during parabolic flights [11]. In that context, we added measurements of oesophageal pressures through an air tube that was swallowed by the test subject

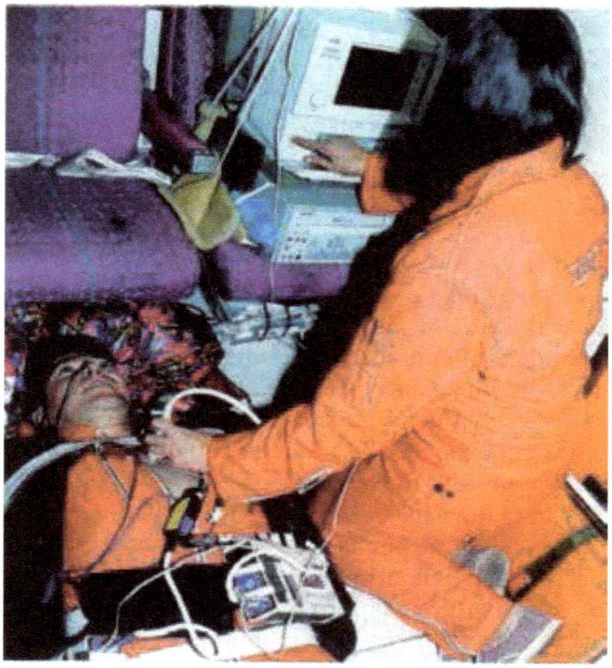

Figure 4.
Dr. Regitze Videbaek measuring the size of the heart chambers in a subject during an ESA parabolic flight campaign. The airplane ascents into a parabolic (Keplerian) trajectory to create weightlessness for 20 s. The subject is also instrumented with invasive monitoring equipment for estimating central venous pressure (CVP), which was also used for the D2 spacelab mission in 1993 on board the space shuttle Columbia [3].

through the nose for obtaining intrathoracic pressures. Intrathoracic pressures are the pressures surrounding the heart. Those pressures were not measured during the Spacelab D2 mission, so the parabolic flight data helped us interpret the CVP data from space.

The process of getting access to parabolic flights is not very different from getting access to spaceflight. Investigators must usually respond to solicitations put forward by a space agency and go through the scientific selection and feasibility assessment processes. The space agency will supply the investigators with the infrastructure such as the flights, but investigators must find their own funding, which usually also applies for adjustments of the equipment to fit into the airplane. In some cases, investigators will have more direct access to the parabolic flight venue, if their experiments concern technical feasibility assessments for a spaceflight. Obtaining experimental baseline data from these flights for comparisons with space data can also be allowed at the discretion of the relevant space agency.

7 Results and conclusions

7.1 Spacelab D2 mission

From the Spacelab D2 mission, our CVP and urinary experiments showed us a new mechanism as to how blood and fluid are shifted from the lower to the upper portions of the body in weightlessness and that the excretion rate of a saline load is not faster than on Earth. Both results were surprising and revealed new insight. Likewise, it was a surprise that the agitating (sympathetic) part of the autonomous nervous system was stimulated during weightlessness and that it was not—as expected—supressed. In ground-based simulation studies using 6^0 head-down bed rest or acute seated head-out water immersion, the opposite is usually seen. Thus, there is a difference in effects of weightlessness in space and the simulation models on the ground.

Despite the upward blood shift to the heart and head, CVP was measured to decrease in space compared to being horizontal supine on the ground (see **Figure 5**). We had expected it to be increased. The data we obtained were only from one astronaut, but a US-led team during two other missions also measured CVP directly with invasive catheters and found decreases. We thereafter performed a parabolic flight study and measured CVP with same technology as during the D2 mission and found similar acute decreases during the 20 s of weightlessness [11]. However, we also observed that the heart was expanded despite the decreased CVP, because simultaneously the oesophageal pressure also decreased and even more so. From ultrasound images taken of the heart during the parabolic weightless period, we observed an expansion of the cardiac chambers, so the ostensible discrepancy between the decrease in CVP and the expanded heart could be reconciled by the expansion of the thorax that further stretches the heart and gives an erroneous impression of the change in its feeding pressure (CVP, **Figure 6**).

7.2 Subsequent space missions

From our later inflight experiments [6–10] following the Spacelab D2 mission, the main conclusions can be briefly summarised as follows:

- Cardiac output and stroke volume increase by some 35–40% during months of flight in space, which is caused by the weightlessness-induced upward fluid shift.

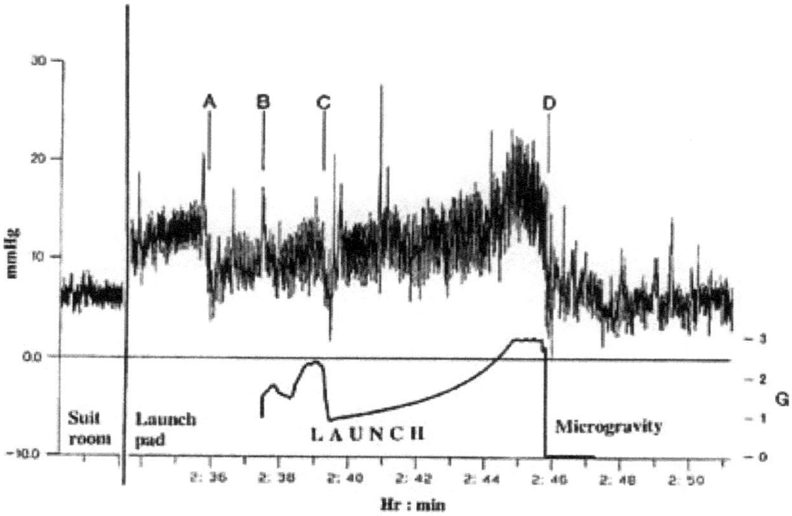

Figure 5.
Central venous pressure (in mm Hg) as a function of time in one astronaut before launch of the space shuttle in the suit room with the space suit on and in the shuttle on the launch pad in the supine leg-up position. (A) Closing of the helmet visor. (B) ignition. (C) Release of the solid rocket boosters from the ascending shuttle. (D) Entering weightlessness. The g-load (G) is indicated at the bottom [3].

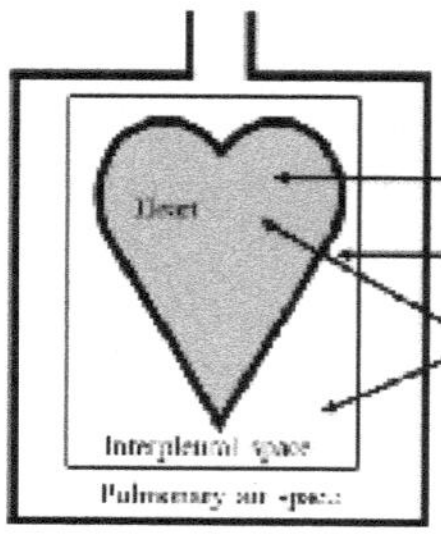

	1 G	0 G	Delta
CVP =	5.8	4.5	- 1.3
IPP =	1.5	- 4.1	- 5.6
tCVP = CVP – IPP =	4.3	8.6	4.3

Figure 6.
The parabolic flight experiment which helped us interpret the Spacelab D2 mission data. Central venous pressure (CVP) was measured directly with long catheters with transducers at their tip placed near to the heart chambers in supine subjects during the parabolic manoeuvre. Simultaneously, the intrathoracic pressures (IPP) were also measured through long air-filled tubes with balloons at the end in the oesophagus. By subtracting IPP from CVP, the transmural heart distension pressure (tCVP) can be estimated. As can be seen, the tCVP increased in weightlessness (0 G) by 4.3 mm Hg (Delta) even though CVP fell by 1.3. Thus, parabolic flight data could help interpret those obtained during spaceflight [3, 11].

- The arterial resistance vessels dilate and decrease the circulatory resistance by some 40% and blood pressure by 10 mm Hg.

- The sympathetic nervous activity is not decreased in space and is at the level of being upright on ground, which is supported by the attenuated urinary excretion rates of fluid and sodium.

- The dilatation of the arteries and the high sympathetic nervous activity is in contradiction to each other, and the mechanism is not yet known (**Figure** 7).

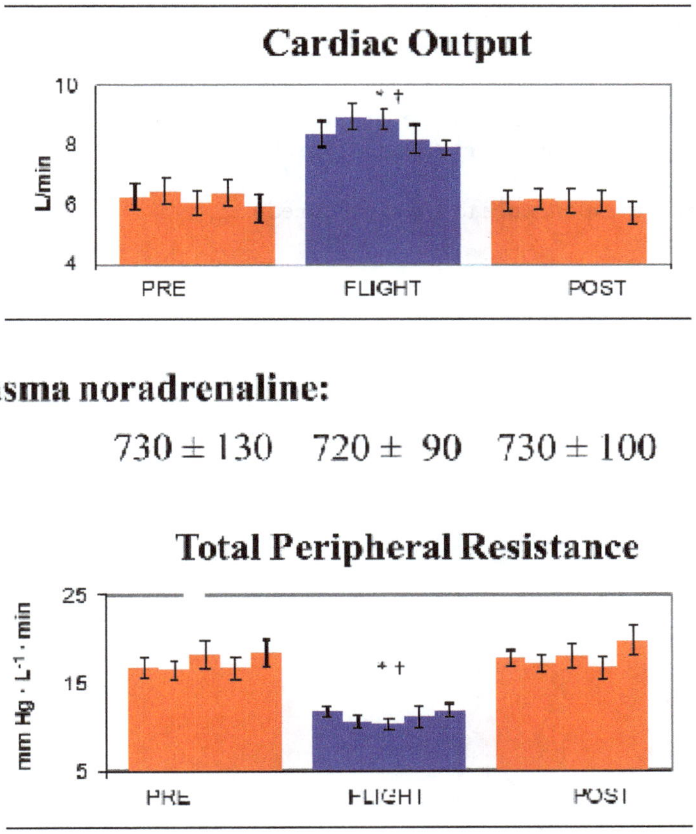

Figure 7.
During missions to the international Space Station in the period of 2006 to 2012, we conducted measurements of cardiac output by a non-invasive rebreathing method in eight astronauts and found a clear-cut increase in some 35% between the 3rd and 6th month in space [10]. At the same time, blood pressure is decreased, which indicates that the total vascular resistance is decreased by almost 40%. In contradiction to this, noradrenaline levels are not suppressed but maintained unchanged from ground-based upright levels. Thus, the mechanism of chronic peripheral vasodilatation in space is still unknown [10, 12].

Thus, to our experience, experiments in space have revealed some new insight and mechanisms into human physiology, which could be of importance in interpreting the health consequences of long-duration flights in the future. By comparing the effects of long-duration (3–6 months) spaceflight on the International Space Station [12] with those of short-term shuttle flights [8], there are at least two important and surprising observations: (1) The shift of blood and fluid from the lower body segments into the heart, which increases cardiac output, is even bigger, and (2) blood pressure is more decreased by a more pronounced peripheral vasodilation (**Figures 6** and 7).

Acknowledgements

This paper was supported by the NNX16A069A, NASA Cooperative Agreement to Baylor College of Medicine for the Translational Research Institute for Space Health (TRISH).

The help and dedication of Mr. Poul Knudsen, technician at Innovision, is deeply appreciated for the repair of the faulty component described in Section 5.6.

Author details

Peter Norsk
Center for Space Medicine and Department Molecular Physiology and Biophysics, Baylor College of Medicine, Houston, Texas, USA

*Address all correspondence to: norsk@bcm.edu

References

[1] Norsk P, Drummer C, Rocker L, et al. Renal and endocrine responses in humans to isotonic saline infusion during microgravity. Journal of Applied Physiology. 1995;**78**:2253-2259

[2] Gabrielsen A, Norsk P, Videbaek R, Henriksen O. Effect of microgravity on forearm subcutaneous vascular resistance in humans. Journal of Applied Physiology. 1995;**79**:434-438

[3] Foldager N, Andersen TAE, Jessen FB, et al. Central venous pressure in humans during microgravity. Journal of Applied Physiology. 1996;**81**:408-412

[4] Clemensen P, Christensen P, Norsk P, Groenlund J. A modified photo- and magnetoacoustic multigas analyzer applied in gas exchange measurements. Journal of Applied Physiology. 1994;**76**:2832-2839

[5] Gabrielsen A, Videbaek R, Schou M, Damgaard M, Kastrup J, Norsk P. Non-invasive measurement of cardiac output in heart failure patients using a new foreign gas rebreathing technique. Clinical Science. 2002;**102**:247-252

[6] Norsk P, Christensen NJ, Bie P, Gabrielsen A, Heer M, Drummer C. Unexpected renal responses in space. The Lancet. 2000;**356**:1577-1578

[7] Christensen NJ, Heer M, Ivanova K, Norsk P. Sympathetic nervous activity decreases during head-down bed rest but not during microgravity. Journal of Applied Physiology. 2005;**99**:1552-1557

[8] Norsk P, Damgaard M, Petersen L, et al. Vasorelaxation in space. Hypertension. 2006;**47**:69-73

[9] Gabrielsen A, Norsk P. Effect of spaceflight on the subcutaneous venoarteriolar reflex in the human lower leg. Journal of Applied Physiology. 2007;**103**:959-962

[10] Norsk P, Asmar A, Damgaard M, Christensen NJ. Fluid shifts, vasodilation and ambulatory blood pressure reduction during long duration spaceflight. The Journal of Physiology. 2015;**593**:573-584

[11] Videbaek R, Norsk P. Atrial distension in humans during microgravity induced by parabolic flights. Journal of Applied Physiology. 1997;**83**:1862-1866

[12] Norsk P. Adaptation of the cardiovascular system to weightlessness: Surprises, paradoxes and implications for deep space missions. Acta Physiologica (Oxford, England). 2020;**228**(3):e13434. DOI: 10.1111/apha.13434

10

Red Blood Cell Dynamics: The Contribution of Microgravity in the BIOMICS Project

Thomas Podgorski, Gwennou Coupier and Christophe Minetti

Abstract

The complexity of blood flows is an intense subject of research since the pioneering works of Poiseuille and through studies over a wide range of scales, from the single red blood cell to dense suspensions in capillary networks, using a large variety of techniques. The subtle effects that are responsible for the structure of blood flows in vessels can benefit from the use of microgravity platforms in order to suppress sedimentation that prevents precise measurements of red blood cell dynamics. The BIOMICS experiment was performed in the MASER11 and MASER12 sounding rockets and was preceded and followed by several parabolic flight experiments in which two important phenomena were investigated using red blood cells and biomimetic model systems like lipid vesicles: the lift forces that push red blood cells away from walls and hydrodynamic interactions between cells that contribute to spreading, mixing, and segregation of different cell types. Parabolic flights played a crucial role in the definition of the scientific questions, preliminary experiments, hardware development and testing, as well as the definition of protocols, and were central in an experimental program combining ground and flight experiments on different platforms.

Keywords: blood cells, vesicles, suspensions, hydrodynamic interactions, holography

1. Introduction

The flows and rheology of suspensions of particles in a fluid are a longstanding topic in physics and mechanics, with a wide spectrum of applications from geophysics (sediment transport, mud flows), chemical engineering or biological fluids. Owing to the complexity of the physics involved in the hydrodynamics of these systems, which couple fluid and solid mechanics, chemical physics or biology and are spread over several scales from microscopic to macroscopic, they are an intense subject of research in different fields. Blood is a particular example of these complex fluids that has attracted a renewed interest in the past 15–20 years thanks to the rapid development of numerical simulation and new opportunities for experiments that appeared with the rise of microfluidics and new optical techniques of investigation, both in vivo and in vitro.

Blood is indeed a suspension composed of around 40–50% by volume of red blood cells and 1% of leukocytes and platelets in plasma, an aqueous solution of salts and proteins which behaves as a Newtonian liquid under normal conditions.

The main function of red blood cells, which are deformable cells made of a membrane encapsulating a haemoglobin solution, is the transport of oxygen. Their complex dynamics [1–4] and the hydrodynamic and mechanical interactions between cells and with the vascular walls have decisive consequences on the rheology of blood, the circulation of which covers a large range of vessel sizes and flow rates. Pathologies associated with blood flow and cardiovascular function, most of which linked to modifications of red blood cell mechanical properties and flow conditions, are a primary cause of mortality worldwide, and specific disorders have also been identified among astronauts in relation to long spaceflight and exposure to microgravity. Besides heart rate alteration and changes in plasma volume which lead to modification of the shear stress experienced by vessel walls and possible endothelial dysfunction, biochemical perturbations have also been reported, such as an increase of amylase activity [5] as well as with variations in red blood cell membrane phospholipid composition [6]. These phenomena have a possible impact on red blood cell mechanics and interactions and are a motivation in the space domain and more generally at the physics-medicine interface for a better understanding of the influence of mechanical properties of red blood cells on their dynamics. While blood dynamics has always been a topic of interest for physicists since the pioneering work of Poiseuille[1] who studied haemodynamics in veins and capillaries in the 1830s, there has been a new surge of interest for interdisciplinary research on blood in recent years thanks to the development of new experimental techniques and the increasing performance of numerical simulation which allow to tackle the difficulty of studying red blood cell dynamics at the microscale.

Blood flows through a complex system of vessels with dimensions that span over several orders of magnitude from the main arteries (several cm in diameter) down to capillaries with diameters as small as 5 μm that are comparable to the size of cells. At the smallest scales (arterioles, capillaries and venules), blood can no longer be considered a homogeneous fluid, and the complex mechanics of red blood cells lead to the formation of structures, inhomogeneous cell concentration and variations of the local effective viscosity of blood and heterogeneous distribution of red blood cells in capillary networks. These phenomena are directly related to mechanical and hydrodynamic interactions between flowing cells and with vessel walls. In the experiments that we report here, we focused on two basic mechanisms that are fundamental for red blood cell dynamics in small vessels: the lift and migration phenomena that move cells away from vessel walls and the hydrodynamic interactions between flowing cells that give rise to a shear-induced diffusion phenomenon in vessels (see **Figure 1**): when deformable objects such as red blood cells or giant lipid vesicles (which we used as a model for red blood cells) flow near a solid surface or in a channel, hydrodynamic forces tend to push them away from walls (**Figure 1a** and **b**), leading to the formation of a cell-free layer that lubricates the flow and decreases the overall flow resistance. Similarly, when cells meet in the flow, they experience a cross-streamwise displacement (**Figure 1c** and **d**), which balances the migration towards the centreline and leads to an equilibrium cell concentration profile in the section of the channel.

These phenomena, although they are ubiquitous in blood flows in vivo where they have direct consequences in pathological situations, and in vitro where they can be exploited for lab-on-chip applications such as cell sorting and analysis, had seldom been quantified experimentally when the biomimetic and cellular systems (BIOMICS) project started.

[1] Jean Léonard Marie Poiseuille (1797–1869) who was a physicist and a physician actually derived the more general Hagen-Poiseuille law (which describes the laminar flow of viscous liquids in tubes) from his studies on blood flow.

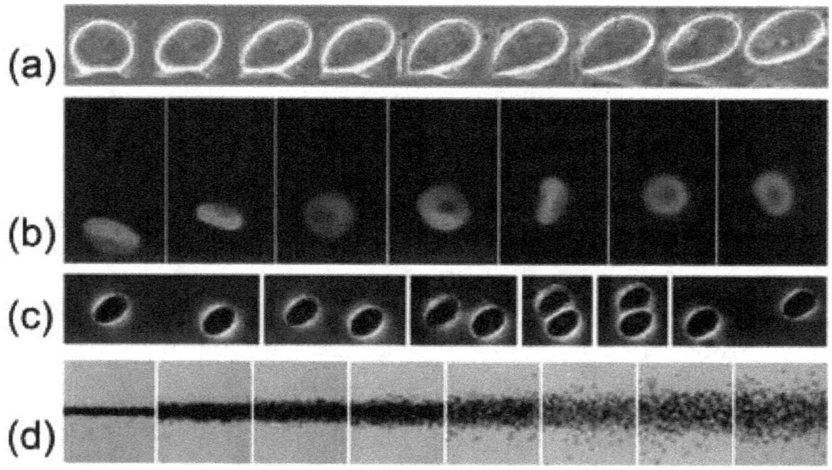

Figure 1.
Basic phenomena of vesicle and red blood cell dynamics in confined flows. (a) Unbinding and lift of a vesicle in shear flow, (b) migration of a red blood cell in channel flow (from [7]), (c) repulsive hydrodynamic interaction in a pair of vesicles in shear flow (from [8]) and (d) shear-induced diffusion of red blood cells in a channel (from [9]).

An experimental difficulty faced by scientists when trying to make a quantitative evaluation of migration forces and velocities at the single cell level is the rapid sedimentation of cells. Indeed, the density difference between red blood cells and plasma (or any suspending buffer with physiological osmolarity and viscosity) is about 0.1 g/cm^3, which leads to sedimentation velocities of several μm/s for cells with a diameter of 7–8 μm. This sedimentation velocity is indeed of the same order of magnitude as lift velocities, which prevents from making precise and quantitative measurements of these subtle hydrodynamical effects under gravity. Thanks to the access to microgravity platforms granted by the *Centre National d'Etudes Spatiales* (CNES, French Space Agency) and the European Space Agency (ESA) since 2004, we have been able to perform several reference experiments which constitute a solid basis for the understanding and modelling of red blood cell dynamics in flow and the mechanics of the microcirculation.

In the following, we will review the preparatory steps that led to the BIOMICS experiments on the dynamics of vesicle suspensions in shear flow that took place during the MASER 11 (2008) and MASER 12 (2012) sounding rocket flights as well as different experiments that were performed in parabolic flights for the preparation of the sounding rocket experiments, equipment validation and preliminary results. Additional and complementary experiments that gave a significant amount of additional results using the equipment that was developed for these opportunities will also be summarized. Besides the exciting challenges that are inherent to microgravity and space-related projects, the development of equipment and experimental techniques for these missions was a formidable opportunity for spinoffs in the form of many other experiments, both on the ground and in microgravity.

2. The BIOMICS experiment

The experimental collaboration between the teams of the Laboratory for Interdisciplinary Physics (LIPhy) in Grenoble, France, and the Microgravity Research Center (MRC) in Brussels, Belgium, was initiated in the framework of the BIOMICS Topical Team organised by ESA. After the initial meetings in 2004 and

2005, the teams gathered during the 2005 biennial symposium of the European Low Gravity Research Association (ELGRA) in Santorini, Greece. The team of Grenoble had started working on the dynamics of giant lipid vesicles in flow a few years earlier. Giant vesicles, which are lipid bilayer membranes enclosing a fluid with a characteristic size between 5 and 50 μm, are a convenient tool for physicists to study the dynamics of soft objects in flow and can be considered as a simple model for red blood cells. Interests raised in developing experimental systems that would allow precise characterization of the structure and dynamics of suspensions in confined flows while at the same time the team of Prof. Frank Dubois at MRC had developed holographic microscopy techniques that had great potential for the study of the dynamics and blood cell suspensions. In addition, the technique, which we will briefly describe later, had already been implemented in space experiments, notably the protein crystallization experiment PromISS [10] aboard the International Space Station, which was an indication of its robustness in microgravity and flight environment.

From these discussions, a project emerged aiming at a quantitative characterization of the structure (concentration profile) of a vesicle suspension in simple shear flow between plates. This configuration was chosen as one in which the two fundamental mechanisms (lift force and shear-induced diffusion) compete, providing means to measure the parameters of these processes and test theoretical models. Besides, the 3D reconstruction capabilities of digital holographic microscopy (DHM) [11] developed by MRC appeared to be a powerful tool to study dynamical phenomena in suspensions with statistical significance thanks to the large quantity of data generated by the processing of DHM images. Although there was at that time little available data on vesicles or red blood cell migration velocities or on shear-induced diffusion coefficients, a rough estimation of time scales in the involved phenomena was made based on preliminary experiments and related literature, from which it was suspected that at reasonable shear rates (50–100 s^{-1}), a time of tens of seconds or minutes would be required to achieve a steady state in a shear flow between plates with a gap of tens or hundreds of micrometres. As these experiments were to be performed in microgravity to avoid sedimentation over these time scales, sounding rockets were quickly identified as the appropriate platform.

3. Preparatory steps and preliminary experiments

3.1 Designing the experiment

The first draft of the sounding rocket experiment proposal was written quickly after the ELGRA symposium and the first meeting between the scientific team, ESA and the Swedish Space Corporation (SSC) took place on November 7, 2005 in Paris, France. During this meeting, the general objectives of the experiment were defined as well as a general schedule and a preliminary design was presented.

While the main optical instrument, the digital holographic microscopy, was already available at MRC, the first step in the project was to design a prototype of shear flow chamber for several purposes which include (i) the dimensioning of the final system to obtain proper flow conditions, (ii) testing and tuning the optical system and the 3D-reconstruction capabilities of the DHM with vesicle samples and (iii) testing the experiment principle on ground and in parabolic flights in order to progress in the definition of experiment parameters such as vesicle sample characteristics, flow conditions and time scales.

The requirements for this shear flow chamber were rather strict and constraining: we needed to produce a constant shear flow between plates with shear rates in the range of 1–100 s^{-1} during a period of up to several minutes. The gap between plates had to be in the range of 100–200 µm in order to be large compared to vesicle sizes in our samples (typically 5–30 µm) but still sufficiently small to allow reaching a steady state of the vesicle suspension distribution at the time scale of the experiment. There were constraints on the parallelism of the plates to ensure constant and uniform shear rate in the chamber, on water tightness of the system and the possibility to evacuate air bubbles when filling. Finally, for compatibility with the optical technique and the configuration of the DHM, plates had to be transparent and of optical quality, and the bottom one needed to be thin compared to the working distance of objective lenses.

Thanks to the skills of engineers at LIPhy, the shear flow chamber of **Figure 2a** and **b** was designed. In this design, a transparent plexiglass disc (diameter 12 cm, thickness 1 cm) rotates about the central axis inside a chamber made of aluminium for the top part (except for a small glass window allowing the passage of the incident laser light) and closed at the bottom by a 2-mm-thick glass disc. This produces a shear flow between the bottom glass plate and the rotating disc. The rotating axis, which is rather long for mechanical stability and alignment, is entrained by a cog wheel linked to a stepping motor and gear box. This flow chamber was built in early 2006 and tested on ground and in parabolic flights (2006) in order to set up filling procedures, check optical compatibility and obtain the first images that would help refine the experiment plan, especially regarding time scales.

As for all prototypes, this first design had a number of flaws that were highly instructive for the design of the final experiment by SSC, who were in charge of developing the experimental module for MASER. These included difficulties to fill the chamber without entrapping air bubbles, insufficient mechanical alignment of the rotating disc, some loss of optical quality due to the materials (plexiglass), and a relative fragility of the large and thin bottom glass disc. Nevertheless, it provided the first digital holographic images of vesicles in shear flow, which allowed to start developing sophisticated software for their treatment. On the scientific side, a first rough curve of the evolution of the distance between sheared vesicles and the bottom plate was obtained.

In parallel, during the development phase of the cell, a first draft of the experiment requirements document was produced, and the final version of the flow chamber was designed by SSC, as shown in **Figure 2c** and **d**. Gaining from the experience of the prototype, significant improvements were made: an excellent mechanical precision ensuring parallelism of the shearing surfaces, mechanical stability and a compact design, optical quality thanks to the use of glass only in the optical path; a clear view of the interior from the top that allows to monitor the proper filling and bubble removal and a much smaller dead fluid volume of the chamber that allows to optimize sample use.

3.2 Testing sample compatibility and limits

Space-related experiments that are performed or launched on sites that are far from the research laboratory require to think about details that usually do not come up to the scientist mind when entering the domain of gravity-related research. Sounding rocket experiments have specific constraints that have to be taken into account, especially when dealing with biological or biomimetic samples. Indeed, doing the experiment requires preparing samples in advance, travelling with them to Esrange (Kiruna, Sweden) for the flight campaign, staying there for pre-flight

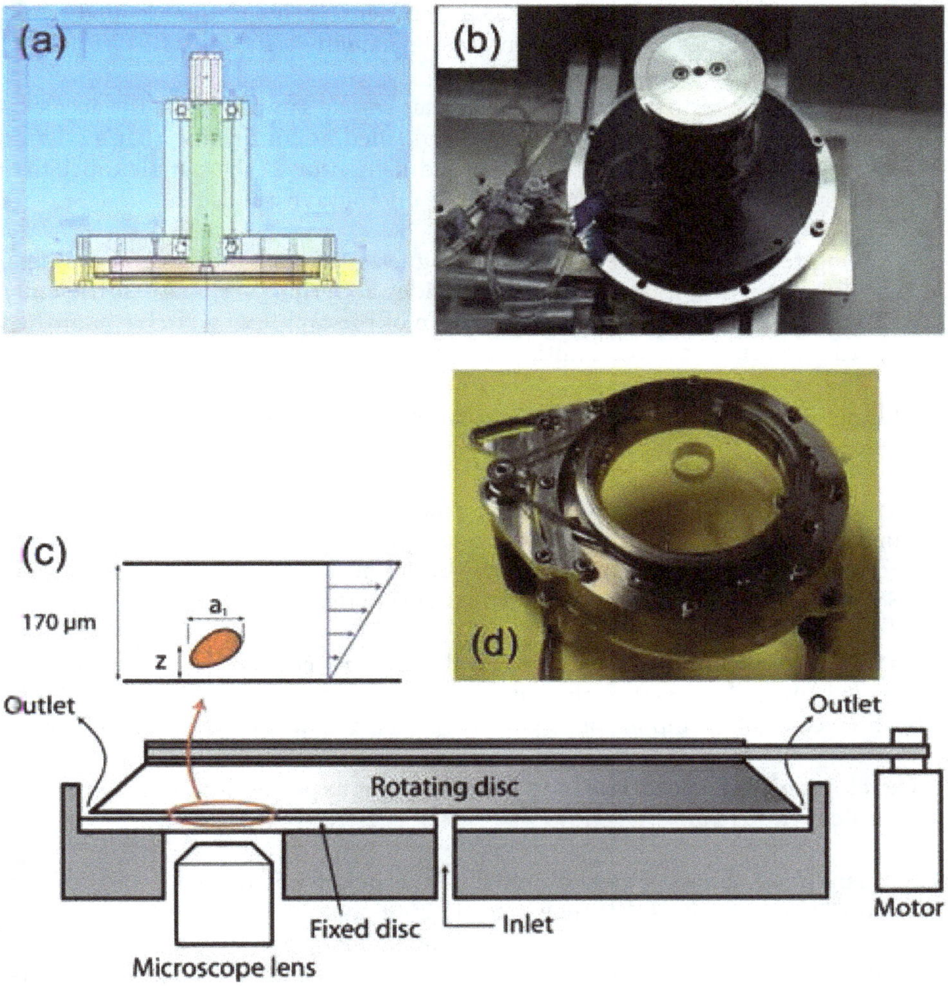

Figure 2.
Two versions of the shear flow chamber developed for microgravity experiments. (a) and (b) Sketch and picture of the prototype designed at LIPhy and tested in the parabolic flights campaigns of spring 2006 (ESA) and September 2006 (CNES); (c) and (d) sketch and picture of the shear flow chamber designed and developed by SSC and used in the MASER 11 (2008) and MASER 12 (2012) sounding rocket flights and in ESA and CNES parabolic flight campaigns since 2007 for various experiments on vesicles and red blood cell dynamics (from [12]).

tests and wait until weather conditions are favourable to start a countdown, which may take up to several weeks on site. Then samples need to withstand a waiting time of up to several hours in the experimental module before launch and resist the strong acceleration levels experienced when the motor ignites and in the ascending phase of the rocket (up to 12 g).

These conditions and constraints are of course extremely different from what happens in the home laboratory where the scientist can prepare samples and experiment with them within hours in ideal environmental conditions. While doing experiments in parabolic flights, which we had done previously, is already a challenge (samples have to be transported to Bordeaux 1 week before the first flight, wait for about an hour in the experimental rack before the first parabola comes and withstand vibrations and accelerations of the plane's take-off and of hyper gravity phases (1.8 g) before and after each parabola), sounding rocket campaigns represent a significant step forward in complexity. The requirements for the MASER 11 campaign, the first one for the BIOMICS experiment, included then to qualify our samples through a series of environmental tests, namely:

- Centrifugation tests that would ensure that vesicle samples resist the acceleration levels of the rocket. They were also aimed at confirming that a stirring system would be necessary to prevent sedimentation of the samples once they are installed in the module and before the microgravity phase.

- Stirring tests: samples need to resist magnetic stirring over several minutes, the stirring (by a magnetic bar inside a syringe) should sufficiently homogenize the samples, and degradation of the samples had to be quantified in order to tune the needed initial concentration.

- Lifetime tests to quantify the degradation of samples over times of several weeks during the campaign.

- Travel tests to check the possible degradation of samples during travels that may last a day or two, during which samples are no longer at 4°C and their containers may be shaken. Some samples would also need to be sent to SSC in Solna, Sweden, by mail for various tests during the preparation.

As trivial as they may seem, these simple but time-consuming and sometimes long tests were indeed extremely critical for the feasibility of the project. They were all successful, an important step to move on to the final development phase.

3.3 Developing a full prototype for parabolic flight experiments: a decisive technical validation

As mentioned above, a parabolic flight experiment was developed for several ESA and CNES campaigns that took place in 2006 and 2007 before the MASER 11 campaign, with the purposes of testing equipment and procedures as well as obtain scientific results that would serve defining the optimal parameters for the sounding rocket campaign. The setup, which was developed in close collaboration between MRC and LIPhy, is shown in **Figure 3**. The experiment consists in injecting a vesicle suspension in the shear flow chamber. A constant shear flow is created, and the dynamics of the suspension is recorded with the digital holographic microscope. Vesicles initially sedimented on the bottom wall are lifted up thanks to a hydrodynamic lift force and drift towards the middle plane of the shear flow chamber.

The experiment setup is made of a shear flow chamber, mounted on a digital holographic microscope. The instrument allows, after processing the recorded holo-graphic information, to obtain information on the 3D position and shape of vesicles in the chamber during the flow. An injection system (**Figure 4**) consisting in several syringes containing samples and filling and rinsing fluids (vesicle suspensions and glucose-sucrose solutions) allows to sequentially inject samples in the flow chamber through the bottom glass plate. The first two campaigns (spring and fall 2006) were performed with the prototype shear flow chamber, and a final preparation parabolic flight campaign took place in September 2007 with the final flow chamber designed and built by SSC, with the presence of a SSC representative during the campaign to build know-how and procedures that would be useful for the sounding rocket campaign.

From a technical viewpoint, these three campaigns were decisive for the team as they provided important information for hardware development and procedures. As the project emerged from a new collaboration that started only 1 year earlier

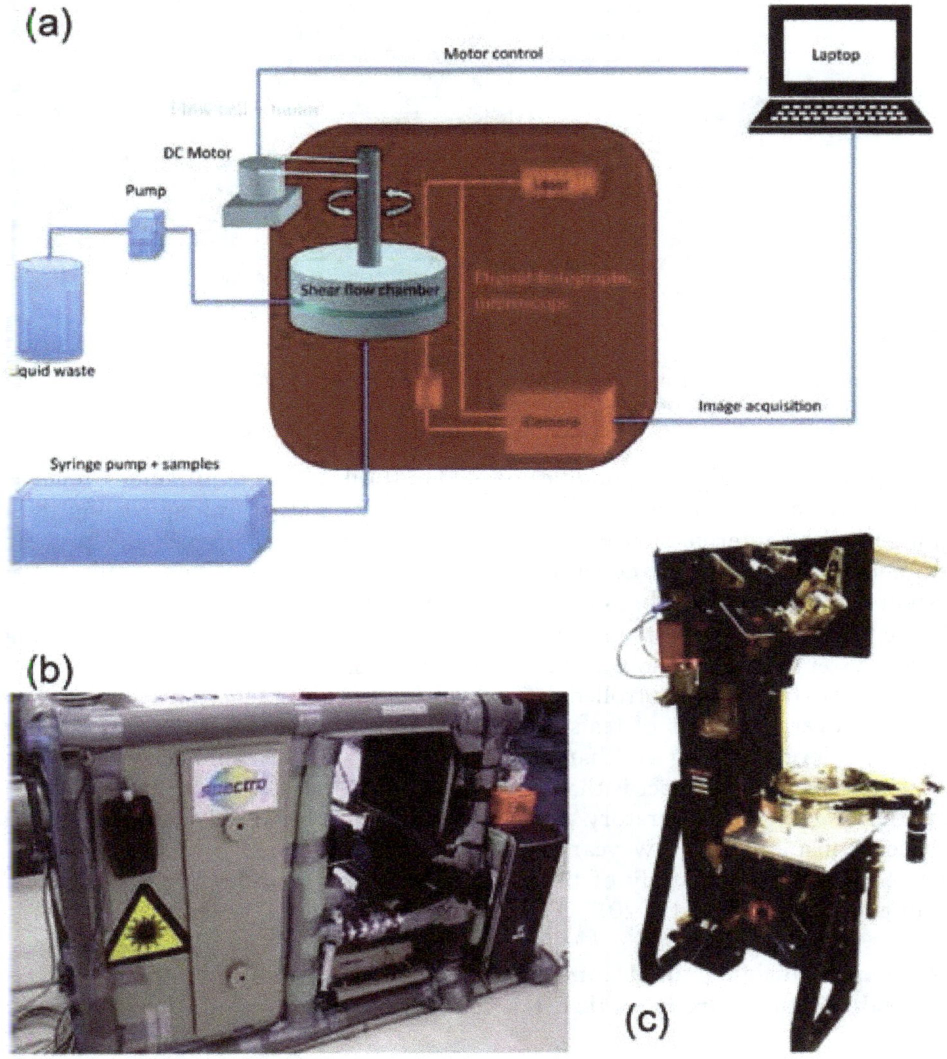

Figure 3.
(a) Design principle of the parabolic flight setup used for BIOMICS experiments. The shear flow chamber is installed on a digital holographic microscope that allows 3D reconstruction of vesicle and red blood cell suspensions in flow. Samples are injected thanks to a multi-channel syringe pump; (b) parabolic flight experimental rack, 2006 version; (c) digital holographic microscope designed and developed at MRC, ULB.

thanks to the networking opportunities offered by ESA Topical Teams, it was also a precious time for the scientific team to know each other's skills, strengths and weaknesses as well as human and social qualities. A good team spirit is an asset in long-term projects in which groups of collaborators spend long periods of intense work time in different environments that are far from home. The 2006 and 2007 campaigns allowed to forge what was called the "Moutchic Spirit" and allowed to maintain a continuous collaboration that is still alive today.

Technical definition of the MASER experiment moved on during this period, especially with the development of two important sub-systems: a sample storage and injection device inspired by the rudimentary initial parabolic flight system and a sounding rocket version of the digital holographic microscope to suit the geometrical and mechanical constraints of MASER rockets. The injection system

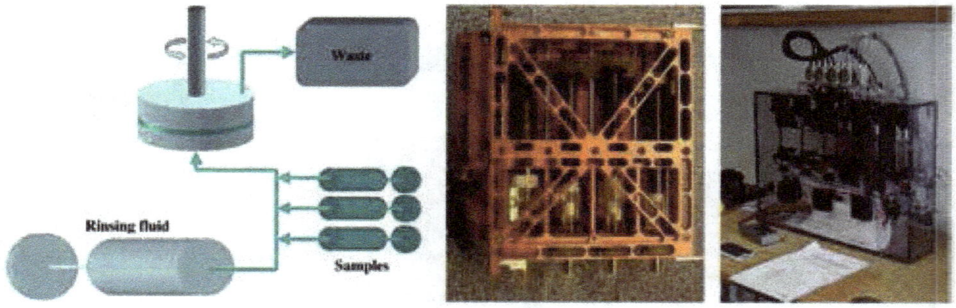

Figure 4.
Multichannel injection system. Left, sketch of the system; middle, multichannel syringe pump with embedded stirring system designed by SSC for the BIOMICS module in MASER rockets; right, similar system developed at LIPhy for recent parabolic flight campaigns (2017–2019).

developed by SSC (**Figure 4**, middle) needed to fulfil several requirements such as having three samples in separate syringes as well as a bigger filling/rinsing fluid syringe and an internal stirring system to homogenize the samples and prevent sedimentation. On the practical aspect, ergonomic considerations required the system to be easily filled before late access with an easy removal of air bubbles. Unlike parabolic flight systems that used disposable plastic syringes, the sounding rocket system includes more rigid and precise glass syringes enclosing a small magnetic stirring bar controlled by external electromagnets. This device also required several sessions of tests at SSC premises in Solna, Sweden, in order to have well-defined filling and handling procedures. A noticeable achievement of the whole BIOMICS project is that the technology of this device that was built on experience gained in laboratory and parabolic flight experiments became itself a source of inspiration a few years later for the development of a new injection device needed for the study of the aggregation of red blood cells in a parabolic flight experiment started in 2017 (**Figure 4**, right).

3.4 Scientific insight gained from parabolic flights: original and important results on the hydrodynamic lift of vesicles and cells

On the scientific side, it was necessary to gain a better insight on one of the main mechanisms of structuration of vesicle suspensions under shear: the hydrodynamic lift of vesicles due to their interaction with walls. This phenomenon, in which vesicles are pushed away from walls in shear flow towards the centre of the flow, sets the time scale of the establishment of a steady spatial distribution in conjunction with hydrodynamic interactions between vesicles or cells. The first parabolic flight campaigns in 2006–2007 allowed to estimate this time scale for the MASER experiment timeline at the same time that equipment and procedures were tested, and the study that had fundamental interests in itself was pursued during additional parabolic flight campaigns in 2009. The results of the pre-MASER 11 parabolic flight campaigns led to the publication of a first research article on the hydrodynamic lift of vesicles [12] that has received a strong attention from the community and has become a reference article. It provided the first experimental validation of earlier theoretical predictions of scaling laws governing this phenomenon [3].

The experimental principle and sample results are shown in **Figure 5**. The experimental procedure is a good example of the way the particular succession of different g-levels in parabolic flights can be exploited to give original results: during the normal and hyper-gravity phases, the flow is stopped, allowing sedimentation of

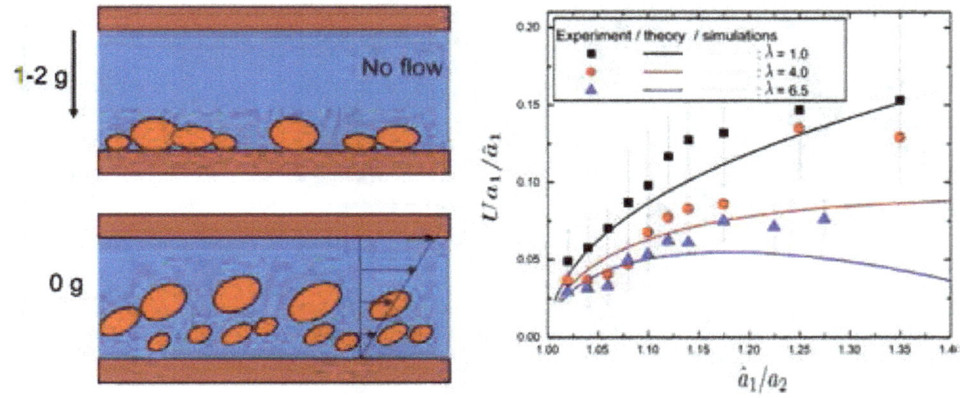

Figure 5.
Experiments on the lift of vesicles in shear flow near a wall performed in parabolic flights for the determination of lift coefficients. The motor of the shear flow chamber is stopped during normal and hypergravity phases (1–2 g) to allow sedimentation of vesicles on the bottom wall, and the fluid is sheared during microgravity phases only to study the resuspension of vesicles. Results show a strong dependency of the dimensionless lift velocity U *on vesicle parameters such as viscosity ratio* λ *and elongation* a_1/a_2 *(from [13]).*

vesicles on the bottom wall of the shear flow chamber, providing a reproducible initial condition for all parabolas. Then, when the microgravity phase starts, the shear flow starts, leading to progressive lift of vesicles. The dynamics is recorded by digital holographic microscopy, which allows after reconstruction to analyse the evolution of the distance between vesicles and walls as a function of time. This phenomenon is strongly dependent on the size of vesicles, their geometric characteristics (surface to volume ratio or equivalently, their reduced volume, a dimensionless parameter that compares the internal volume to the volume of a sphere having the same membrane area) and mechanical properties (viscosity ratio λ between internal and external fluids). We showed that the lift velocity follows a scaling law in which the velocity is proportional to shear rate, vesicle diameter and $1/z^2$ where z is the distance to the wall. The proportionality constant (a dimensionless lift velocity) could be extracted as a function of vesicle parameters and numerical simulations performed later quantitatively matched experimental results: the dimensionless lift velocity U is stronger when the vesicles are more deformable and thus more asymmetric with respect to the flow direction. This asymmetry is favoured by a low viscosity ratio λ between the internal and external fluids and by a higher surface to volume ratio of the vesicles that gives them more flexibility. This is reflected by a higher apparent aspect ratio a_1/a_2 of the long and short axis of the quasi-elliptic recorded shape of vesicles (see **Figure 5**). From the theoretical and numerical perspective, this experiment was a benchmark that allowed a quantitative validation of numerical models [13].

Results from parabolic flight campaigns were also the opportunity to establish reliable image processing methods for the DHM images [11, 14]. A powerful reconstruction and segmentation technique were indeed needed to process the large quantity of data generated by these experiments (up to thousands of flowing objects for a typical parabola). An interesting outcome was also that this technique could be used to measure 3D shapes of vesicles and concentration profiles in other types of flow [15], capabilities that could be useful for other applications.

3.5 Developing new sample preparation techniques thanks to microfluidics

The second BIOMICS sounding rocket flight (MASER 12) required the development of new sample preparation techniques in order to study the dynamics of

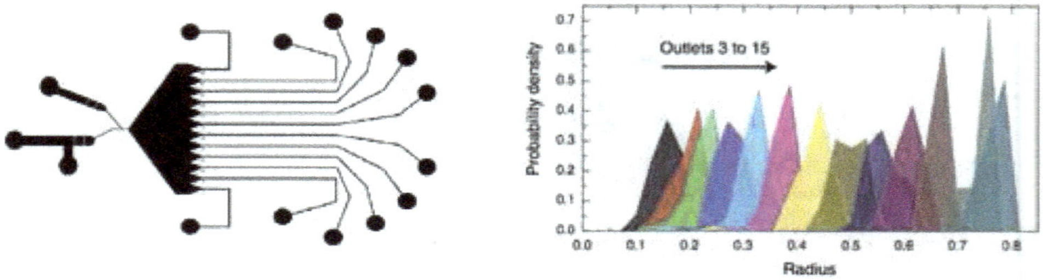

Figure 6.
Sketch of the microfluidic sorting device developed at LIPhy for the production of monodisperse vesicle samples for the MASER 12 experiment (overall size 50 × 30 mm) and example of vesicle size distributions at the different outlets of the system [17].

well-controlled monodisperse or bidisperse vesicle samples. The aim was to be able to control the size distribution of vesicles in samples, in order to study more specifically the interaction between the lift- and shear-induced diffusion phenomena in monodisperse samples, as a function of vesicle size, and to get a quantitative understanding of the role of heterogeneous pair interactions in suspensions (i.e. collisions or hydrodynamic interactions between objects of different sizes). This was motivated by segregation or margination phenomena that had been reported in blood flow (platelets or white cells tend to be closer to vessel walls while red blood cells flow in the centre) [16] and whose mechanisms still deserved quantitative description and modelling.

The classic production method for giant lipid vesicles, electroformation, produces samples that are highly polydisperse in size. While other methods (e.g. using micro-emulsions as an intermediate) had been proposed for a better control of vesicle size, they were not suitable for the production of rather large samples (>10 mL) with significant vesicle concentration. We therefore decided to keep the efficient electroformation method and developed a microfluidic sorting device to produce monodisperse samples (**Figure 6**).

The system is based on the principle of pinch-flow fractionation, and its performance is described in [17]: the initial vesicle suspension is introduced through one inlet, while pure suspending fluid is introduced at through the other inlet with a much higher flow rate. This effectively squeezes vesicles against one wall after the converging bifurcation, leading to a configuration in which the distance between the centre of mass of vesicles and the wall is equal to their radius. A diverging chamber then amplifies this ordering and distributes vesicles in several outlets. As shown in **Figure 6**, we were able to separate vesicles in to up to 16 different fractions with a size dispersity that is of order 10% only in each fraction, which met the requirements of the experiment we wanted to implement in MASER 12. This system which was developed in the framework of microgravity experiments has been used many times since then for various experiments on different types of samples at LIPhy and other laboratories.

4. Sounding rocket flights

4.1 Experimental module

The BIOMICS module that was developed by SSC for the sounding rocket flights was directly inspired from the experimental setup that we had developed and tested in parabolic flights, by adapting it to the geometrical, mechanical, electrical and environmental constraints of MASER rockets.

The module, approximately 60 cm in height, is built around a central optical breadboard on which the holographic microscope is assembled (**Figure** 7). This version of the digital holographic microscope was developed by the Lambda-X company from Belgium in order to meet specifications (compact design, robustness of the alignment of elements and insensitivity to strong accelerations and vibrations). Compared to the parabolic flight version, the sounding rocket module also includes thermal control, an overview camera that allows the team to check that correct filling of the chamber is achieved and that rotation takes place as scheduled during the experiment and of course specific electronics for experiment management, data storage and transmission. The module housing is equipped with a hatch for late access before the flight: to avoid sample degradation, the injection unit (red part in **Figure** 7) and its syringes are filled with samples after countdown has started and installed in the module only minutes before take-off.

After a series of environmental and functional tests in Solna, Sweden, as well as the first flight simulation tests which included setting up an efficient protocol for the filling of the syringe unit without entrapping bubbles, the module was ready and approved for the flight.

4.2 The campaign: beyond the far north to outer space

The weeks before the MASER 11 and 12 campaigns were hectic. After months of hardware testing, protocol definition and parabolic flight experiments, the team's attention was largely devoted to the preparation of samples. As we were aware of the significant risk of countdown delays and interruptions, and the need to perform several flight simulation tests on site before the launch, a large quantity of samples was needed to deal with the expected and unexpected. It took several days and nights in the weeks preceding the campaigns to prepare the necessary volume of vesicle samples by electroformation, followed by the time required to pass them through the microfluidic sorting device (for MASER 12) and finally check their quality.

This intense, almost monastic, activity was a premise of the polar environment in which we were going to be immersed for several weeks. Although scientists do not have the chance to go and conduct their experiment in space themselves, all the more so when it comes to automated and remotely controlled sounding rockets, the journey to the Esrange launch site near Kiruna in Swedish Lapland is in itself an immersion in a scenic, unusual and dramatic environment. And although we do not experience the degree of confinement that astronauts do in spacecrafts and orbital

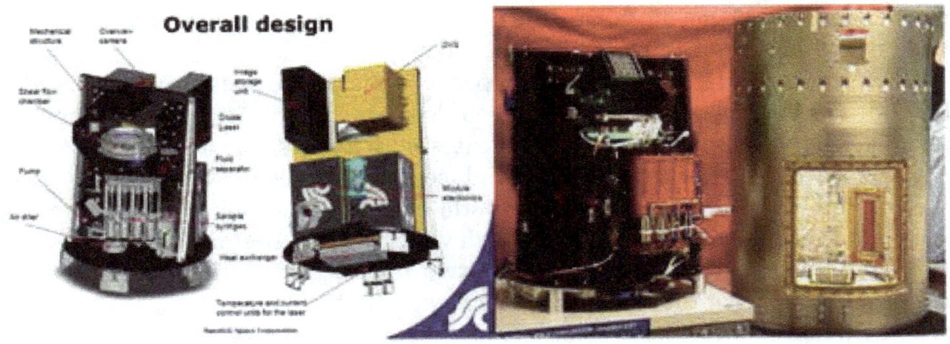

Figure 7.
The BIOMICS module developed by SSC with a contribution of lambda-X for the holographic microscope. Left, overall design. Right, picture of the module and its payload housing with the late access hatch open.

platforms, the serene wide open spaces of Lapland and its harsh climate give a small taste of what it is to be confined at the edge of the common world. The succession of days and nights was indeed confusing—the MASER 11 campaign took place in May 2008, with almost no dark night, and MASER 12 in February 2012, with rather short days and temperatures that reached −40°C—and on countdown days, our schedules were not punctuated by the usual beacons of the clock but by launch opportunities and the organisation around foreseen launch times that were dictated by weather conditions. The MASER 11 campaign was especially long due to almost permanent windy conditions that led to an unusual number of interrupted countdowns. A strange feeling, which work and social activities actually did not make unpleasant, was the impression of being in a universe halfway between the village of the old TV series "The Prisoner" and the classic film "Groundhog day", doomed to repeat the same procedure that lasts several hours until the wind sends its white bouncing ball that sends everyone to bed. Despite the large quantity of samples that has been prepared, these repeated countdown stops had seriously eroded the stock when finally a successful launch took place.

4.3 Results

From a technical viewpoint, the two campaigns were a success for the BIOMICS experiment: the whole flight sequence took place nominally, with an excellent image quality that would ensure ideal conditions for data processing. Experience gained from the first flight (MASER 11) also leads to hardware improvement (modifications of the laser diffuser in the holographic microscope) that ensured even better interferometric contrast in MASER 12.

From a scientific viewpoint, we have been able to show that the suspension reaches a steady and symmetric distribution centred between plates in shear flow, even at low volume fraction, a configuration that is impossible to reach under normal gravity (**Figure 8**). As this distribution is the result of the balance between the lift force (already well characterized in parabolic flights) and shear-induced diffusion, it allows us to determine a dimensionless diffusion coefficient f_2 that only depends on vesicle mechanical properties, from the height and width of the distribution (see, e.g., **Figure 9**, left panel). From different sets of data and independently of vesicle size, we found $f_2 = 0.063 \pm 0.024$, a value that is consistent with theoretical predictions, suggesting that it can be indirectly derived from the analysis of individual pair interactions [8, 13, 18].

For the MASER 12 campaign, we considered three samples of vesicles of different sizes and a viscosity contrast of $\lambda = 1$. Suspensions of large, medium and small vesicles were successively injected. The radii of these populations lie, respectively, in the ranges 22–35 μm, 15–20 μm and 5–15 μm. The flight sequence included the study of the three separate populations, which clearly showed that large vesicles produce sharp, centred concentration peaks while small ones produce broader distributions. While it was known from studies on the lift phenomenon in parabolic flights that the lift force increases with vesicle size and deflation, the influence of these parameters on shear-induced diffusion coefficients as well as the interplay between lift and diffusion was still unclear at this point, especially since shear-induced diffusion also increases with vesicle size and deformability.

Finally, when injecting a mixture of two populations, as in **Figure 9** (right panel), we found that the small vesicles are expelled from the centre, as a result of both the weaker lift forces they experience and the asymmetry in their interaction with large ones: small particles are indeed pushed away from the centre zone by large ones. This segregation phenomenon is similar to platelet or leucocyte margination in blood flow and has strong implications for the structure of blood flow,

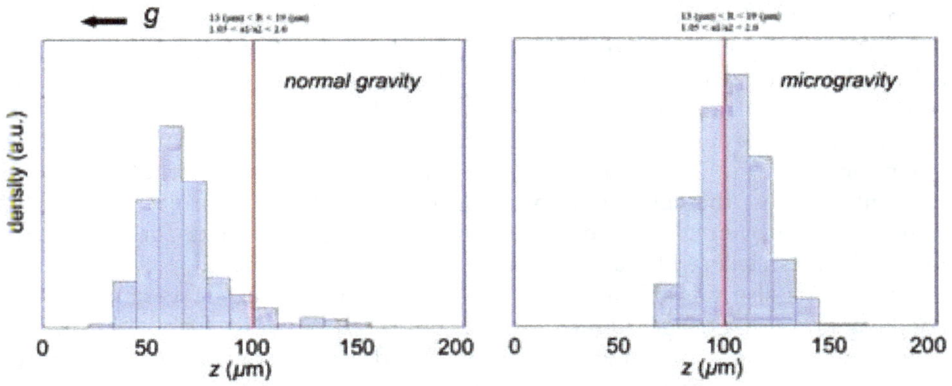

Figure 8.
Comparison of vesicle distributions in the gap of the shear flow chamber on the ground (normal gravity) and during the MASER 11 flight. The red line corresponds to the centre of the gap. In microgravity, the distribution is centred and symmetric, allowing an evaluation of vesicles dynamic parameters (diffusion coefficients and lift forces), while the distribution is off-centred and asymmetric under normal gravity.

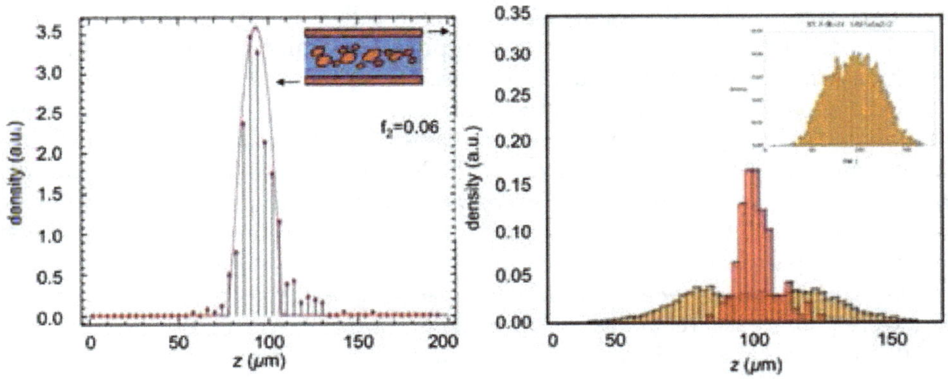

Figure 9.
Some results from MASER 11 and MASER 12 [13, 18]. Left, the steady distribution of a monodisperse vesicle sample allows the determination of shear-induced diffusion coefficients (inset, sketch of the experiment). Right, steady distribution for a bidisperse suspension of vesicles in MASER 12 (yellow, small vesicles with average radius 10 μm; red, big vesicles with average radius 20 μm). The double peak for small vesicles (to be compared to the unique, smooth peak when they are alone as shown in the insert) reveals segregation phenomena in the flow of polydisperse cellular suspensions.

blood rheology in small vessels and the immune response. As such, it is a topical issue in the blood flow community which has received renewed interest over the past decade through advances in theoretical modelling and numerical simulation. Our experimental results provide quantitative data for the fine-tuning of models which are still a topic of interest today.

5. Experience and follow-up

Like all space-related projects, the preparation and realization of these sounding rocket experiments have had multiple repercussions in terms of experience, technical developments and inspiration for a large number of works that have taken place since then in the research carried out in the laboratory and during many subsequent parabolic flight campaigns.

The parabolic flight setup and the shear flow chamber that were developed in this framework were subsequently used for several other experiments which include the lift of red blood cells [9], the complex dynamics of red blood cells in shear flow [19] or more recently the dynamics of aggregation of red blood cells. More experiments are indeed still planned today with different samples such as polymeric capsules. In addition to large quantities of experimental results on different systems, the collaboration that was initiated in this project also led to significant improvements in the processing of holographic data for suspension flows, which can be used for a wide variety of purposes [11, 14, 15].

The results also raised many questions on various aspects of vesicle and red blood cell flows that triggered many studies in the lab. Notably, for a better understanding of the phenomenon of shear-induced diffusion, a study of the elementary mechanism behind it was performed in microfluidics, namely, the hydrodynamic repulsion in a pair of interacting vesicles in a shear flow [8], while shear-induced diffusion was characterized in channel flows of red blood cells [9]. The lift phenomenon that we fully characterized for vesicles [12, 13] and red blood cells [9] in simple shear flow thanks to parabolic flight experiments led to the question of validity of the established scaling laws in channel flows where the velocity profile is parabolic instead of linear. In that case, we proposed empirically modified scaling laws for both vesicles [20] and red blood cells [7] that are now a reference for the development of cell-sorting applications and the refinement and benchmarking of numerical simulation codes.

These projects have had a strongly structuring effect on our collaborative network at the European scale, beyond the central collaboration between MRC and LIPhy, with ramifications of the topic with German and Italian collaborators, for instance. A strong expertise in the preparation and manipulation of samples was gained, which led to the establishment and sharing of reference protocols with several other partners.

On management aspects, dealing with space-related projects as all team members were young researchers when the collaboration started was a decisive opportunity for building experience and know-how on the management of large-scale projects in a complex environment, combining the characteristics of experiments on large instruments with the specificities of the space domain (technical constraints, security, interactions with many different actors between space agencies, contractors and technical staff).

6. Conclusions

Developing a space experiment is usually a long process with specificities that are often unknown to the average scientist. Besides specific procedures and constraints related to security and quality control, these experiments require an unsuspected number of tests and validation as well as extremely precise protocols that have to be defined in advance and do not withstand improvisation. As strange or caricatural as they may seem to the experimentalist who usually works on his laboratory bench, this collection of many different tests is indeed a precious key to a successful experiment by ensuring that no detail, which may result in experiment failure, is overlooked.

In many cases, parabolic flights are a recommended or even compulsory step for the precise definition of the scientific question, tuning of parameters and protocols and testing of hardware. For the BIOMICS project, it was the opportunity to develop a full prototype of the experiment and directly interact with it, in microgravity conditions (see **Figure 10**), before the development of an automated and

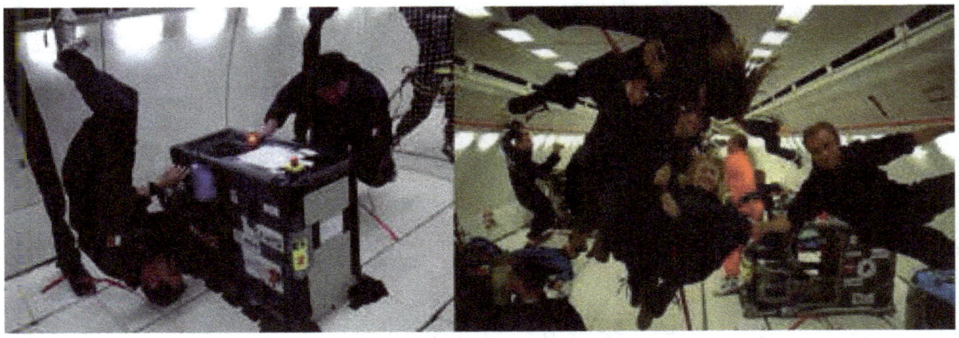

Figure 10.
Team at work and group picture in parabolic flights (photos, CNES).

remotely controlled module for the sounding rocket flights. A decisive advantage of parabolic flights is that, despite rather short microgravity periods (22 s), the number of repetitions is large and the interval between campaigns is short, of order weeks or months only. In addition, the possibility to get important scientific results during this preparation phase should not simply be viewed as a side product of the final space experiment but rather as a part of a multi-faceted project that involves several experimental platforms.

The road to space is long but intense and extremely well-paced by the organising agencies and contractors. Beyond the fact that successful launches felt like a relief and achievement in themselves, the whole process of a well-conducted preparation and campaign itself is a source of scientific, technical and humane benefits and spin-offs that go far beyond the sole sounding rocket experiment.

Acknowledgements

The authors acknowledge the support and help of Jérôme Giraud and Patrice Ballet, engineers at LIPhy, for their skills in the design, development and testing of the shear flow chamber; of Christian Lockowandt from SSC for his involvement in the pre-MASER parabolic flight campaign and of all SSC and Lambda-X members involved in the project and of Dr. Olivier Minster from ESA-ESTEC for his organisation of the Topical Teams. Continuous support of the teams by space agencies, ESA and CNES is acknowledged. The authors especially thank Wolfgang Herfs and Antonio Verga, project managers of the MASER 11 and 12 campaigns, as well as Dr. Vladimir Pletser of ESA and Jean-Baptiste Behar and Sébastien Rouquette of CNES, who were responsible of the parabolic flight campaigns of this project. The project was only made possible thanks to the strong involvement of Dr. Natacha Callens (now at ESA) in the genesis of BIOMICS along with Pr Frank Dubois (MRC) and at all stages of the MASER 11 preparation and campaign.

Author details

Thomas Podgorski[1,2]*, Gwennou Coupier[1] and Christophe Minetti[3]

1 Laboratory for Interdisciplinary Physics (LIPhy), UMR5588 CNRS, Université Grenoble Alpes, Grenoble, France

2 Laboratoire Rhéologie et Procédés (LRP) – UMR5520 CNRS, Université Grenoble Alpes, Grenoble-INP, Grenoble, France

3 Microgravity Research Center, CP-165/62, Université Libre de Bruxelles (ULB), Brussels, Belgium

*Address all correspondence to: thomas.podgorski@univ-grenoble-alpes.fr

References

[1] Popel AS, Johnson PC. Microcirculation and hemorheology. Annual Review of Fluid Mechanics. 2005;**37**:43-69

[2] Minetti C, Audemar V, Podgorski T, Coupier G. Dynamics of a large population of red blood cells under shear flow. Journal of Fluid Mechanics. 2019;**864**:408-448

[3] Vlahovska PM, Podgorski T, Misbah C. Vesicles and red blood cells in flow: From individual dynamics to rheology. Comptes Rendus Physique. 2009;**1**:775-789

[4] Mader M, Misbah C, Podgorski T. Dynamics and rheology of vesicles in a shear flow under gravity and microgravity. Microgravity Science and Technology. 2006;**18**(3/4):199-203

[5] Markin A, Strogonova L, Balashov O, Polyakov V, Tigner T. The dynamics of blood biochemical parameters in cosmonauts during long-term space flights. Acta Astronautica. 1998;**42**(1-8):247-253

[6] Rizzo AM, Corsetto PA, Montorfano G, Milani S, Zava S, Tavella S, et al. Effects of long-term space flight on erythrocytes and oxidative stress of rodents. PLoS One. 2012;**7**(3):e32361

[7] Losserand S, Coupier G, Podgorski T. Migration velocity of red blood cells in microchannels. Microvascular Research. 2019;**124**:30-36

[8] Gires PY, Srivastav A, Misbah C, Podgorski T, Coupier G. Pairwise hydrodynamic interactions and diffusion in a vesicle suspension. Physics of Fluids. 2014;**26**:013304

[9] Grandchamp X, Coupier G, Srivastav A, Minetti C, Podgorski T. Lift and down-gradient shear-induced diffusion in red blood cell suspensions. Physical Review Letters. 2013;**101**:108101

[10] Dubois F, Joannes L, Legros JC. Improved three-dimensional imaging with a digital holography microscope with a source of partial spatial coherence. Applied Optics. 1999;**38**:7085-7094

[11] Dubois F, Yourassowsky C, Callens N, Minetti C, Queeckers P, Podgorski T, et al. Digital holographic microscopy working with a partially spatial coherent source. In: Ferraro P, Wax A, Zalevsky Z, editors. Coherent Light Microscopy. Springer Series in Surface Sciences. Berlin: Springer; 2011. pp. 31-59

[12] Callens N, Minetti C, Coupier G, Mader MA, Dubois F, Misbah C, et al. Hydrodynamic lift of vesicles under shear flow in microgravity. Europhysics Letters. 2008;**83**:24002

[13] Bureau L, Coupier G, Dubois F, Duperray A, Farutin A, Minetti C, et al. Blood flow and microgravity. Comptes Rendus Mécanique. 2017;**345**:78-85

[14] Minetti C, Podgorski T, Coupier G, Dubois F. Fully automated digital holographic processing for monitoring the dynamics of a vesicle suspension under shear flow. Biomedical Optics Express. 2014;**5**:1554-1568

[15] Minetti C, Callens N, Coupier G, Podgorski T, Dubois F. Fast measurements of concentration profiles inside deformable objects in micro flows with reduced spatial coherence digital holography. Applied Optics. 2008;**47**(29):5305-5314

[16] Goldsmith HL, Spain S. Margination of leukocytes in blood flow through small tubes. Microvascular Research. 1984;**27**(2):204-222

[17] Srivastav A, Podgorski T, Coupier G. Efficiency of size-dependent separation by pinched flow fractionation. Microfluidics and Nanofluidics. 2012;**13**:697-701

[18] Podgorski T, Callens N, Minetti C, Coupier G, Dubois F, Misbah C. Dynamics of vesicle suspensions in shear flow between walls. Microgravity Science and Technology. 2011;**23**:263-270

[19] Minetti C, Audemar V, Podgorski T, Coupier G. Dynamics of a large population of red blood cells under shear flow. Journal of Fluid Mechanics. 2019;**864**:408-448

[20] Coupier G, Kaoui B, Podgorski T, Misbah C. Non-inertial lateral migration of vesicles in bounded poiseuille flow. Physics of Fluids. 2008;**20**:111702

11

GRIP: Dexterous Manipulation of Objects in Weightlessness

Jean-Louis Thonnard, Laurent Opsomer, Philippe Lefèvre, Vladimir Pletser and Joseph McIntyre

Abstract

The aim of the GRIP experiment is to investigate how gravity impacts the kinematics and dynamics of the upper limb during dexterous manipulation of objects and how the central nervous system adapts to long-term exposure to microgravity and subsequently back to Earth gravity. Hence, we proposed to conduct a set of experiments on healthy human subjects, involving the manipulation of an instrumented object during exposure to normal and microgravity, and to study how the central nervous system adapts motor control in order to cope with the new physical environment. More particularly, the coordination between the grasping force (or grip force, GF) and the load force (LF) is studied, as well as the adaptation of the movement dynamics and kinematics and the interaction between cognitive and sensory cues that establish a reference frame for the human brain. Here we describe the background motivation, the parabolic flight tests that initiated the scientific hypotheses and the technical and scientific process that led to the implementation of the GRIP experiment currently on board the International Space Station (ISS).

Keywords: manipulation, grip, weightlessness, gravity, force, kinematics, space, cognitive, human, brain

1. Introduction

1.1 Project background

A stable grip on handheld objects is of primary importance to lifting and moving actions particularly when such objects are used as tools. During object manipulation, predicting the consequences of one's own movements is necessary to avoid unwittingly dropping the object. Studies of the forces employed in the dexterous handling of objects have found that the grip forces are tuned to prevent accidental slips and yet are not so excessive as to crush a fragile object or to cause muscle fatigue [1]. Flanagan and Wing [2, 3] examined grip force modulation as subjects performed either point-to-point or cyclic arm movements with a handheld load. They found that variations in inertial forces caused by the subjects' own arm movements over a range of accelerations produced synchronous changes in grip forces that rose and fell with the changes in the tangential load forces on the fingers while also taking into account the friction between the fingers and the object. A tight temporal coupling between the grip force (GF) and the load force (LF) has been documented in a large variety of tasks engaging different kinds of objects, grips, loads or mode of transport and locomotion. In other words, grip forces were controlled in anticipation of the fluctuations in inertial forces.

Changes in gravity can be considered as major perturbation for these tasks, which must be handled by the motor system. This is particularly a challenge for loads applied by gravity to the body because the muscle activities used to compensate these loads appear to be programmed in a highly predictive manner, probably based on a lifetime experience in a normal 1 g environment. It has been shown that during exposure to microgravity in parabolic flights, the control of interaction forces adapts at least partially to the lack of gravity [4–9], yet evidence indicates that anticipation of gravity's effects persists in the short term and that adaptation is not fully complete [10–12]. The motivation for the GRIP experiment, to be performed in long-duration space flight, is to understand how the central nervous system adapts to an environment without gravity's effects and what will be the consequences of long-term adaptation when an individual returns to a normal (Earth) or partial (Moon or Mars) gravitational field.

1.2 Hypothesis

We hypothesized that the central nervous system anticipates the effects of gravity in a normal Earth environment but that after long-term exposure to 0 g, grip force/load force coordination and control of upper limb trajectories will adapt to the particularities of the 0 g environment by means of re-optimization of the motor control policy.

1.3 Research objectives

The experiment described here targets specific questions about the effects of gravity on dexterous manipulation, questions that cannot be addressed in the normal terrestrial environment. The research carried out in this project contribute to our understanding of how the human nervous system controls movement both on the ground and in the microgravity environment of spaceflight. Data from these experiments may also be used to identify potential hazards for astronauts as they move between different gravitational environments. These studies could also contribute to the design and control of intelligent haptic interfaces to be used in challenging environments such as space.

More specifically, GRIP addresses the following questions:

- Can the central nervous system dissociate static (due to gravity) and inertial (due to acceleration) loads in the control of precision grip?

- Is grip force control for the manipulation of objects more sensitive to the friction between the fingers and the object in microgravity?

- Do up/down asymmetries observed in dynamics of precision grip and kinematics of the upper limb disappear during long-term exposure to microgravity?

- What visual and kinaesthetic cues are used by the central nervous system to establish an "up and down" reference frame for the control of limb trajectories and applied forces in the absence of gravity effects?

- How does grip force/load force coordination evolve during long-term exposure to microgravity?

- What after-effects influence grip force control on return to a gravity environment?
- Does the learning of kinematics parallel the learning of dynamics in microgravity?
- How does gravity alter decisional processes as measured through the reaction time during fast upper limb movement?

2. Preparation of grip during parabolic flights

Our first parabolic flight campaign participation was during the 26th campaign organized by the European Space Agency (ESA) that took place in Merignac (France) in 1999. Then between 1999 and 2014, we participated in 21 parabolic flight campaigns. This allowed us to properly prepare our GRIP experiment before launching it to the ISS. During this preparation period, the GRIP project allowed us to train 11 PhD students and publish 32 scientific papers. We present hereafter as examples some interesting results of these experiments in parabolic flights.

2.1 Gravity influences the rhythm of our movements

We investigated humans' ability to sustain rhythmic movements in different gravity environments [13]. By analogy with a simple pendulum system, the self-generated pace should grow as a function of gravity. However, because the natural period would be infinite in 0 g, this simplistic model fails in microgravity. Therefore, we hypothesized that the movements are partly driven by a central pattern generator (CPG) in a closed loop with the arm. Since neuronal control systems cooperate with the physical constraints imposed by the dynamics of the body and the environment, we expected that a change of gravity would induce a change of frequency to perform efficient rhythmic movements that have adapted to the resonant frequency as a result of gravity level changes. Since there is no resonant frequency in 0 g, however, we predicted that the pace adopted by the participants would rely mainly on the intrinsic frequency of the neural oscillator.

Twelve right-handed volunteers participated in the study. Subjects were successively confronted with periods of normal gravity (1 g), hypergravity (1.8 g) and microgravity (0 g) during parabolic flights. They were instructed to perform rhythmic arm movements with a handheld object of approximately 200 g mass around two virtual obstacles situated 3 m in front of them, following an "infinity-sign-shaped" trajectory. The first group of six participants (self-paced) were instructed to perform the movement at a self-generated pace. The second group of six volunteers (metronome-paced) followed the rhythm dictated by a metronome (one cycle every 1.5 s) (**Figure 1**).

We showed that the frequency of a rhythmic movement of the upper limb was systematically influenced by the different gravity conditions created in parabolic flights. The period of the arm movement shortened with increasing gravity levels. In weightlessness, however, the period was more dependent on instructions given to the participants, suggesting a decreased influence of resonant frequency. Our results are in agreement with a computational model of a CPG coupled to a simple pendulum exposed to gravity. We demonstrated that the innate modulation of rhythmic movements by CPGs is highly flexible across different gravity contexts. This further supports the involvement of CPG mechanisms in the achievement of efficient rhythmic arm movements. Our contribution is of major interest for the study of human rhythmic activities, both in a normal Earth environment and during microgravity conditions in space.

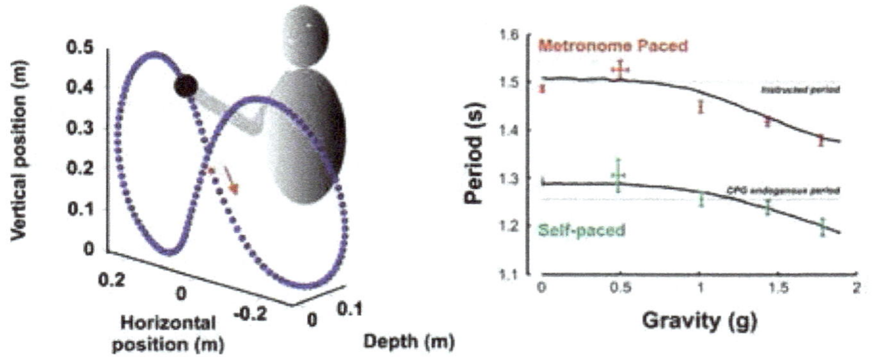

Figure 1.
(Left) One typical cycle trial of a participant in 0 g. The blue dots represent the object position sampled every 10 ms. The arrow marks the path to follow. (Right) Period as a function of gravity: experimental (points) and simulation (lines) data. The data points are the mean periods adopted by the participants in the self-paced group (green) and in the metronome-paced group (red). The bottom dotted line indicates the central pattern generator endogenous period (1.25 s). The top dotted line indicates the instructed rhythm for the metronome-paced group (1.5 s). Self-paced and metronome-paced data were fitted by the CPG pendulum model (solid lines). Error bars represent 95% confidence intervals on both axes (adapted from Ref. [13]).

2.2 Gravity influences the kinematics of our movements

In another study [14], we investigated the effect of hypergravity induced by parabolic flights on the trajectory of vertical pointing movements to test the hypothesis that motor commands are optimized with respect to the effect of gravity on the limb. The subjects sat in front of three visual targets that were aligned vertically with respect to the aircraft floor and that were separated by 18 cm. The centre target was in front of the subject's shoulder and defined the horizontal arm position. They were asked to grasp a manipulandum (mass 250 g, grip aperture 4.5 cm) with the right hand and to perform visually guided pointing movements toward the current target with arm-straight rotations around the shoulder. Upward movements (from the centre to the top) and downward movements (from the centre to the bottom) were randomly interleaved to avoid anticipatory movements. All subjects performed control experiments in normal gravity conditions (1 g) prior to the in-flight experiment. The subjects performed the task during the 0 g and subsequent 1.8 g phases of each parabola (they did not perform the task under normal gravity conditions during the flight). Each subject performed from 60 to 80 trials in each direction and in each gravity condition. The analysis reported in **Figure 2** focuses on the data acquired during the hypergravity phases.

First, the simulations in normal gravity reproduced the asymmetry in the velocity profiles (the velocity reaches its maximum before half of the movement duration), which typically characterizes the vertical pointing movements performed on Earth, whereas the horizontal movements present symmetrical velocity profiles. Second, according to the simulations, the optimal trajectory in hypergravity should present an increase in the peak acceleration and peak velocity despite the increase in the arm weight. In agreement with these predictions, the subjects performed faster movements in hypergravity with significant increases in the peak acceleration and peak velocity, which were accompanied by a significant decrease in the movement duration. This suggests that movement

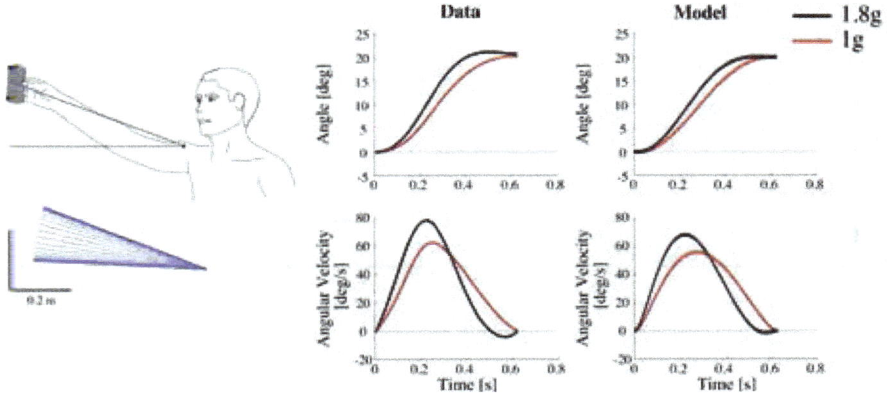

Figure 2.
(Left) Illustration of a subject performing the task. (Right) Trajectories and angular velocity as a function of time in normal gravity (red) and hypergravity (black) conditions for upward movements. Experimental data (left) and model simulations data (right) (adapted from Ref. [14]).

kinematics change in response to an increase in gravity, which is consistent with the hypothesis that motor commands are optimized and the action of gravity on the limb is taken into account. These results provide evidence for an internal representation of gravity in the central planning process and further suggest that an adaptation to altered dynamics can be understood as a reoptimization process.

2.3 Gravity influences the dynamics of our grip

Grip force-load force coordination was studied during cyclic vertical arm movements with a handheld instrumented load across different gravity conditions (0, 1, 1.8 g) induced by parabolic flight maneuvers [4].

Eight adult subjects (including two women) participated in the study. Four subjects had no previous experience in microgravity (the non-experienced subjects, NES), whereas the other four had more than 100 parabolas each to their credit (the experienced subjects, ES).

The results showed that the grip force was modulated in parallel with the load force fluctuations due to the arm movements, regardless of the gravity condition. At new gravity levels, the phase shift between the grip force and the load force was equivalent to that observed in 1 g, even on the first trials of the non-experienced subject. By contrast, the level of the grip force modulation was dependent upon the gravity level, and its adjustment seemed to require some adaptation for the non-experienced subjects. This is illustrated by the more vari-able grip force-load force coordination in the phase diagrams of the first parabola for non-experienced subjects (see P1 in **Figure 3**). The experienced subjects (ES) adjusted the level of the grip force modulation to the new gravity level as soon as they executed the task in the aircraft. The non-experienced subjects had some difficulty in maintaining the imposed movement and applied unnecessarily high safety margins in their grip force during their first trials at 0 and 1.8 g. In the subsequent trials, they progressively decreased their grip force, and no further evolution in the grip force-load force coupling was seen after the fifth parabola (see P5 in **Figure 3**).

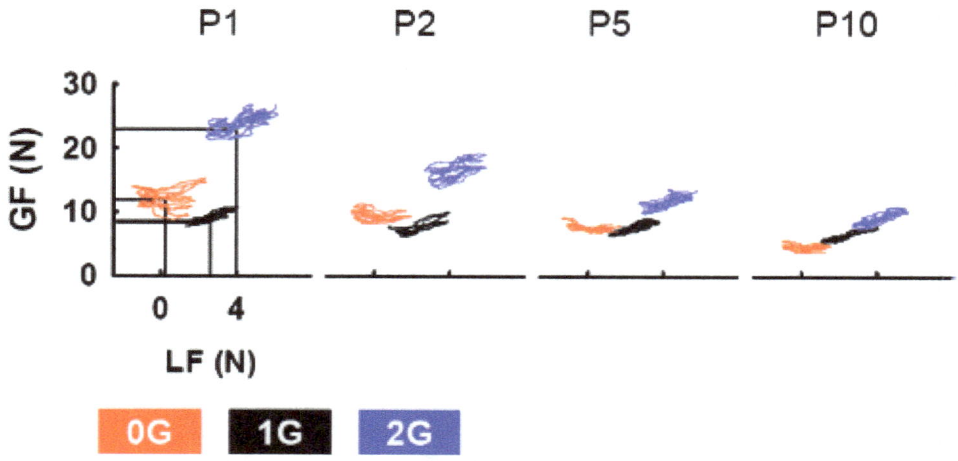

Figure 3.
Phase diagrams illustrating grip force-load force (GF-LF) coordination during vertical cyclic arm movements for non-experienced subject (NES). By decreasing both the level of GF and the variance of GF-LF coordination throughout the 10 parabolas (from P1 to P10), non-experienced subjects progressively tend toward a single GF-LF relationship across the different gravity levels. Identical load force ranges were obtained by varying separately the weight and inertial components of the load force across three different gravity levels (red, 0 g; black, 1 g; and blue, 2 g), and the same coupling GF-LF was observed after the fifth parabola (P5) (adapted from Ref. [4]).

2.4 Restraint system preliminary tests in parabolic flights

Weightlessness is a specific environment where minor limb movements may result in other parasitic body movements. In order to avoid such parasitic body movements when performing the arm movements for the GRIP experiments, the subject will have to be adequately restrained. Indeed, the restraint system will have to maximize the freedom of arm movements while minimizing the trunk, pelvis and legs movements. At the same time, the restraint system must be comfortable, allow a quick egress in case of emergency and avoid inducing pain. During the 53rd ESA parabolic flight campaign of October 2010, different concepts of restraint system were tested, and the subjective perception was recorded for various subjects (see **Figure 4**). The different

Figure 4.
Test of the GRIP restraint system using a GRIP mock-up chair during the 53rd ESA parabolic flight campaign of October 2010 (photo: ESA).

concepts were tested for the sitting and supine posture, for different arm movements (vertical and horizontal oscillations and vertical collisions) and for different feet positions. It has been shown that the pelvis restraint was strongly recommended and that shoulder straps were important. The main observation was that the weightlessness environment provided by the parabolic flights was required to allow evaluating the impact of the parasitic movements on the subject stability.

3. Technical development of grip prior to the launch to the ISS

Before launching any equipment to the International Space Station (ISS), it must follow a technical development answering several categories of stringent requirements regarding mass, size, used materials, electrical design, power consumption, electromagnetic compatibility, structural integrity, fracture control, overall safety of utilization, human interfaces, etc. for ground usage, for interfacing with the launch vehicle, for transfer phases and for final utilization on the ISS. It is a long but essential process to send a scientific experiment in space.

The industrial consortium was chosen after a phase of tendering. The prime contractor was Qinetiq Space nv (previously known as Verhaert D.D.) from Kruibeke, Belgium, with Arsalis, a spin-off of the University of Louvain in Belgium, as main subcontractor in charge of the manipulandum design and electronics. Charnwood Dynamics, from the United Kingdom, was a main supplier of the CODA motion visualization system.

The technical development approach decided to design, develop, build and test the GRIP instrument was what is called the *prototype* approach in which two models are developed. The first model is called the Engineering Model and serves as the main development model. The second is called the Flight Model and is developed after the Engineering Model to be the unit that will eventually fly in space. It was also decided to combine the two initial design phases A and B into a single-phase A/B and the last two technical development and test phases C and D into a single-phase C/D. Due to delays in the definition of some interfaces, it was decided to add an intermediate bridging phase between the end of the Phase A/B and the beginning of the Phase C/D.

Phase A/B was devoted to first, the detailed definition of the scientific requirements of the GRIP experiment and their translation into technical requirements; second, a conceptual design definition where several solutions were proposed by the industrial team and discussed with the scientists to come to a commonly agreed concept; and third, a more detailed definition to bring the chosen concept to a more evolved design with definition of dimensions, subsystems and overall masses, power and interfaces estimations, etc. This Phase A/B was initiated in May 2009 and lasted until March 2010. During this phase, several project reviews and subsystems reviews took place. This Phase A/B culminated in a major review, called the preliminary design review (PDR), whose goal was to finally agree on a detailed design of the GRIP instrument.

In parallel to this design phase, a series of experiments and tests were taking place during parabolic flight campaigns to support the development of concepts and design of some subsystems, like subject restraining straps, supine position support, GRIP chair interfaces etc., as all these "taken-for-granted" design elements that work well on ground in a 1 g environment were not necessarily efficient or optimal in a 0 g environment.

The bridging phase took place from April 2010 till December 2011 and allowed to conduct additional definition tests of various subsystems on ground and during parabolic flights, mainly for the GRIP chair, stretcher, reference frame and manipulandum; requirements for experiment control software were consolidated; the manipulandum was bread boarded for testing its calibration and three-dimensional (3D) tracking accuracy; further bread boarding of printed circuit boards assembly, of the finger humidity sensor and of the lock mechanism was conducted. **Figure 5** summarizes the different steps taken during the Phase A/B and the bridging phase.

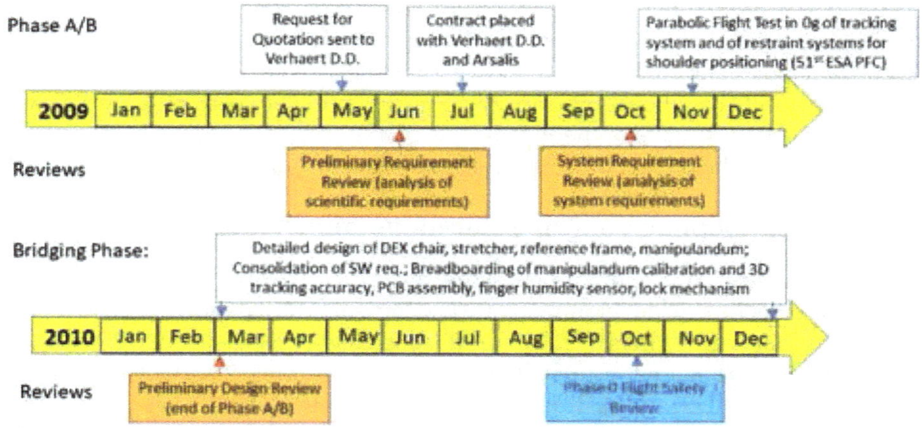

Figure 5.
Timeline of phase A/B and bridging phase (ESA PFC: ESA parabolic flight campaign).

Phase C/D was aimed at first, completing the definition work to incorporate all the results of separate development and test of GRIP subsystems, in parallel to the development of the GRIP hardware and, second, formally agreeing on the design of each subsystem regarding mechanics, electrical design, electronics, software, etc. through dedicated specific technical reviews. A major change occurred in the middle of the technical development when the launch vehicle was replaced. The initially foreseen vehicles (ESA's Automated Transfer Vehicle and Russian Progress cargo) were replaced by newly available launch vehicles (Orbital Cygnus and SpaceX Dragon) resulting in a formal Contract Change Notice that yielded a major verification of structural load requirements and test programme definition. The part of the Phase C/D that corresponds to the classical Phase C ended with a major design review, called the critical design review, during which the final design was agreed upon and frozen, i.e. changes after this review would no longer be possible or would cost a lot of efforts and could result in a non-readiness for launch. This CDR lasted 4 months and ended in September 2012, with a view to a launch in October 2014 (**Figure 6**).

The part corresponding to a normal Phase D, i.e. the final building, assembly and testing was then formally initiated (although it is commonly a good practice to anticipate the parts procurement, assembly and building). In parallel to technical procurement and development, several reviews took place, not only for technical aspects but also safety reviews and crew reviews (**Figure 7**).

In addition to the Engineering and Flight Models' development, additional Science Models and Training Models had to be built to support on one hand the development of all scientific protocols for the experiment execution with the ground support centre and, on the other hand, to allow the astronaut trainers at the European Astronaut Centre to develop and validate crew procedures before experiments could be conducted on board the ISS. The Phase C/D ended after all tests were successfully completed, at the final Flight Acceptance Review in August 2014, and the Flight Model was delivered for launch vehicle integration. In total, 18 project reviews and four safety reviews were conducted during the technical development Phases A/B and C/D. It was essential to have a good and regular communication between the scientific and engineering teams, to conduct preliminary scientific and technical tests in 0 g during parabolic flights and to be supported by the various space agencies (ESA, CNES, NASA) and the Belgian Science Policy Office.

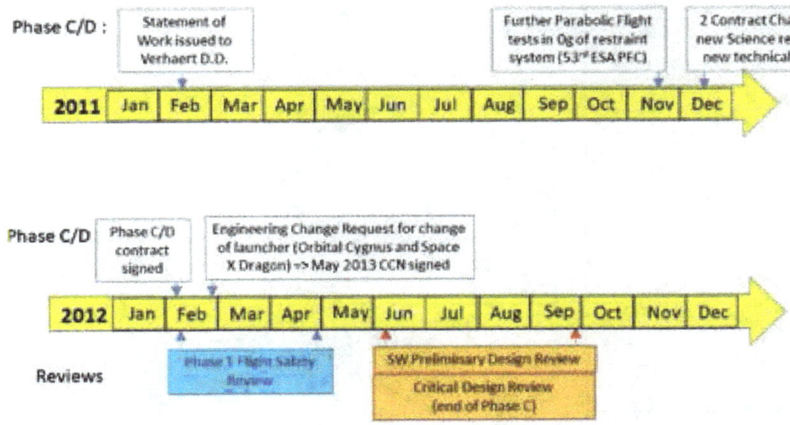

Figure 6.
Timeline of phase C/D until CDR (ESA PFC: ESA parabolic flight campaign).

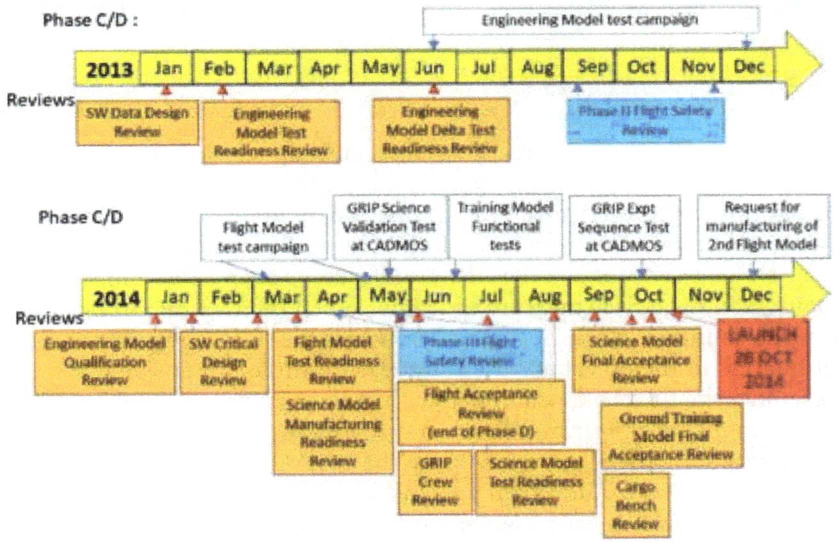

Figure 7.
Timeline of phase C/D from CDR until first launch.

4. Grip instrumentation

The equipment is similar for pre-, in- and postflight sessions. Four similar models (see **Figure 8**) of the GRIP equipment exist: the Engineering Model (for sessions performed in the European Astronaut Centre, Cologne), the Science Model (for sessions performed at the Johnson Space Centre, Houston), the Ground Model (for testing sessions performed at the CADMOS) and the Flight Model (for in-flight sessions performed on-board the ISS).

All models are similar and are equipped with the same hardware items listed below:

The chair box (see **Figure 8**) acts as a transport box and as a restraint system when performing the experiment in the sitting position. The chair box accommodates back and foot support, restraint belts, the manipulandum and wrist box,

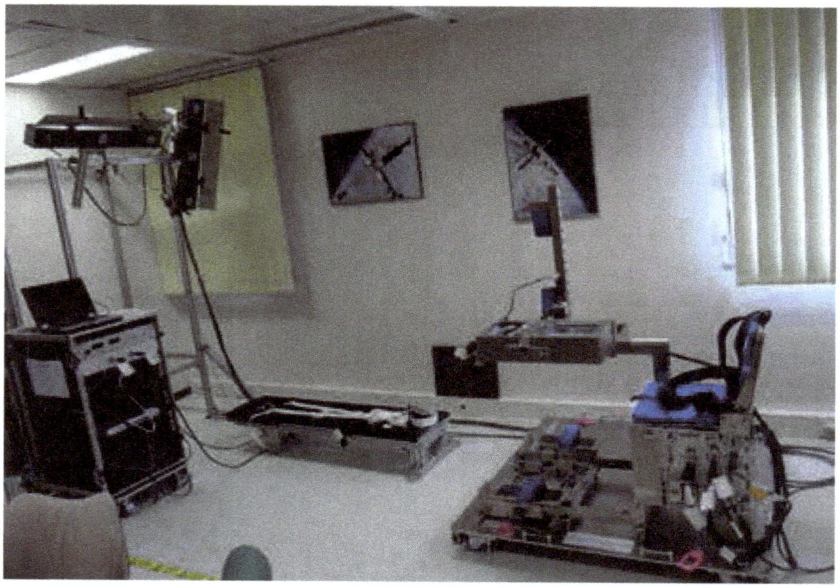

Figure 8.
The GRIP set-up in the seated configuration.

electronics unit for power, data interfaces and signal conditioning, mounting interface for the reference frame, laptop and headphone.

The manipulandum (**Figure 9**) is an instrumented device that can be held in a precision grip between the thumb and index finger. The manipulandum comprises force and torques sensors, accelerometers, gyroscopes and two moisture sensors. Force sensors in the device measure six-dimensional interaction forces and torques between the thumb and the object and between the index finger and the object. Accelerometers in the object are used to measure three-dimensional acceleration. Surfaces under the fingers are equipped with sensors to measure skin moisture. The object is equipped with infrared light-emitting diode (LED) markers to allow for the measurement of the 3D movement by the CODA tracking system (see CODA tracking system below). Additional weights can be added to the manipulandum to change its mass. A system is implemented to provide for a test of the coefficient of friction of the contact between the fingertips and the manipulandum.

The wrist box (see **Figure 9**) accommodates eight additional infrared LED markers (see CODA tracking system below). The wrist box is attached to the wrist of the subject in order to track the movement of the arm.

The CODA tracking system (**Figure 10**) consists of two motion-tracking units (CODA units, Charnwood Dynamics Ltd.). Each unit encompasses three cameras. These cameras are able to track the position of infrared LED markers placed on the manipulandum and on the wrist box. Each unit can then reconstruct the 3D position of each marker by triangulation. The CODA system is therefore used to measure the movement of the hand and object in 3D space. One unit is equipped with an additional webcam assembly.

The reference frame (**Figure 11**) is a mechanical frame that includes the following:

1. The target frame that provides the upper and lower tapping surfaces and the target LEDs that are illuminated under automatic control to define the target position for point-to-point movements. The target frame also accommodates two CODA markers for the spatial reference frame.

2. The utility box that consists of a cover with audio generator, touchscreen user interface, electronics and fan and a cradle for the manipulandum retainer and the additional masses. It accommodates two CODA markers for the spatial reference frame as well. It includes control computer and associated electronics to automatically run the experiment (provide instructions to the subject, generate sequence of targets) and to measure and store data.

The GRIP set-up can be reconfigured to perform the experiment in supine position. In that case, the back support of the chair box is folded, the reference frame position is changed, and the subject lies on a bed equipped with restraint belts.

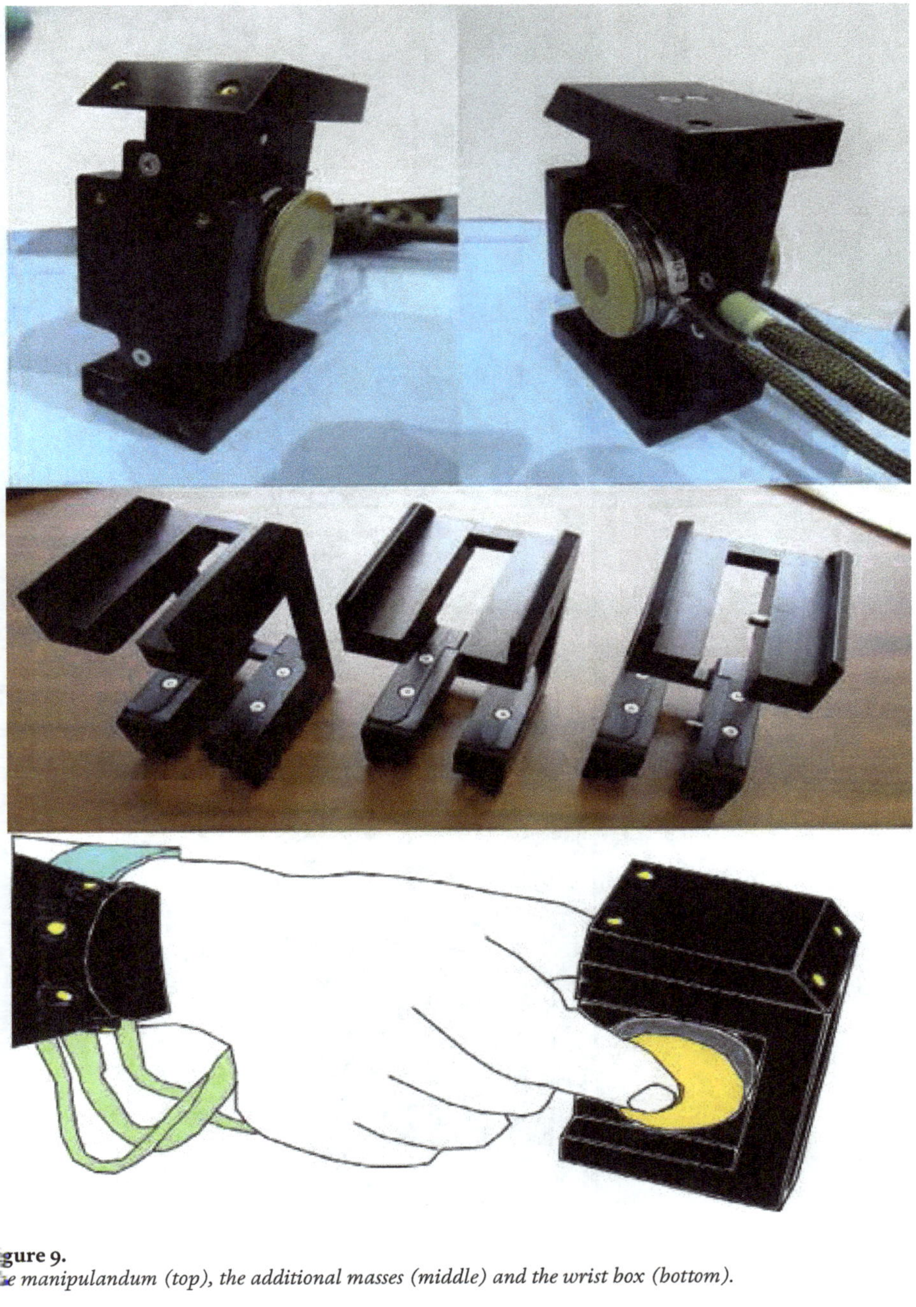

Figure 9.
The manipulandum (top), the additional masses (middle) and the wrist box (bottom).

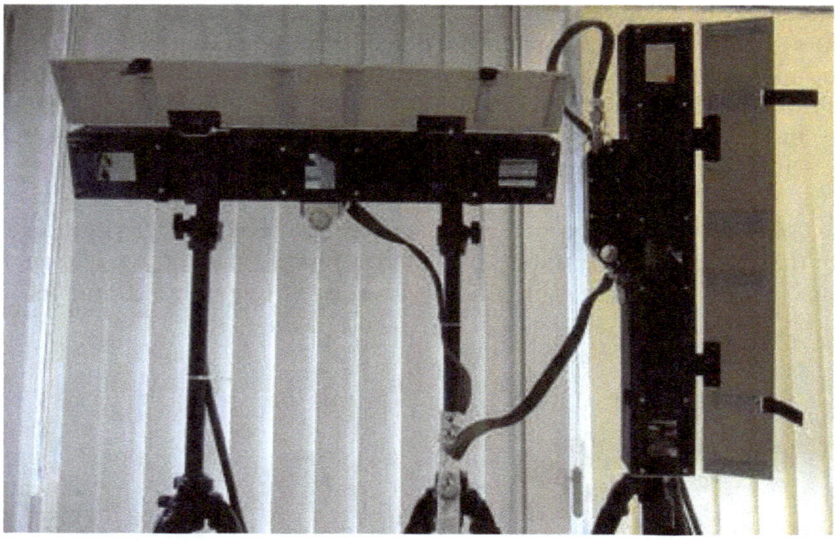

Figure 10.
The tracking system (CODA cameras).

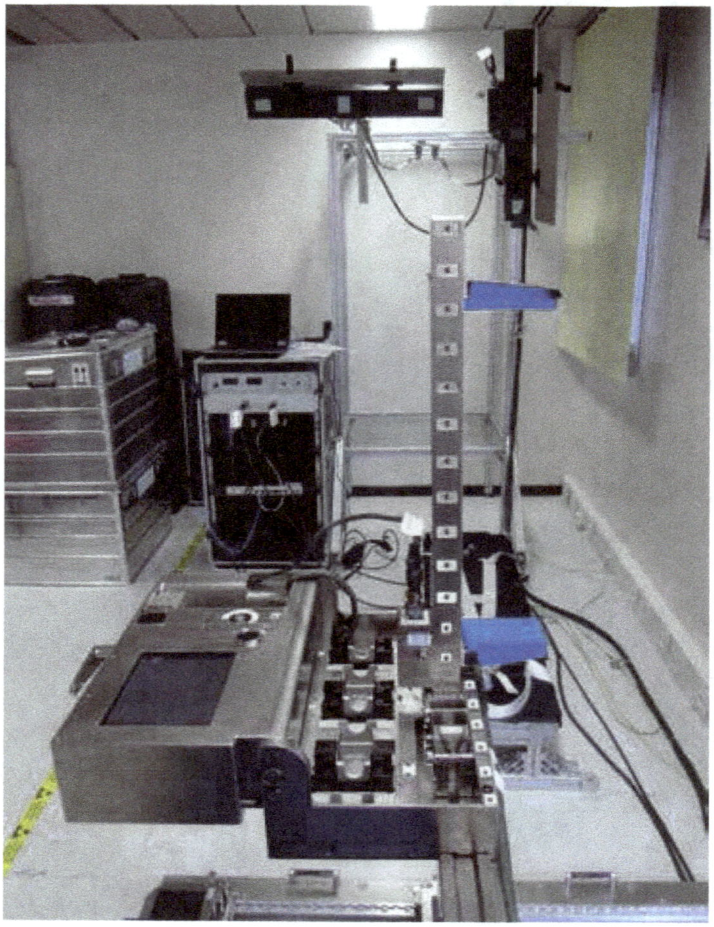

Figure 11.
Astronaut view of the setup in the seated position.

5. Grip in the International Space Station

The launch took place in October 2014, and unfortunately, the rocket exploded a few seconds after take-off (**Figure 12**). It resulted in the complete destruction of the GRIP Flight Model. It was decided very rapidly by the ESA and with the support of the Belgian Science Policy Office to rebuild the GRIP Flight Model during an additional extension phase from December 2014 till 2017.

The second launch took place successfully in February 2017 and the first commissioning of GRIP in May 2017. The first on-board 6-month experiment took place with the first subject in 2018 (**Figure 13**). The second and third subjects successfully performed GRIP experiments on board the ISS in 2019, and two additional subjects are planned in 2020.

During every task of the experiment, the astronaut holds the manipulandum between the thumb and index finger of the right hand. He/she performs a variety of movements (oscillations, point-to-point movements and controlled collisions) while holding the manipulandum in the hand and being either in sitting or supine position. The movements of the manipulandum and of the hand are measured, as well as the forces acting between the fingers and the manipulandum and the moisture of the fingers. A typical full session is broken into three sub-sessions: dynamics seated, references seated and references supine protocols.

Figure 12.
Explosion at launch of the Antares rocket with orbital sciences Cygnus CRS-3 vehicle on 28 October 2014 with the first flight model of the GRIP instrument (credit: NASA).

Figure 13.
The first on-board 6-month experiment took place with the German ESA astronaut Alexander Gerst in 2018 (credit: ESA/NASA).

5.1 Protocols

5.1.1 Dynamics seated protocol

Participants perform a set of experimental trials in the seated posture.

First, they perform a test of the coefficient of friction. We have developed a simple and reliable method to estimate the coefficient of friction in weightlessness. Our method is based on active, back-and-forth movements of an astronaut's finger on the manipulandum six-axis force sensor. The static coefficient of friction is computed as the ratio of the tangential to the normal force at slip onset. (For more details see Barrea et al. [15]).

Then, they perform oscillatory movements of the arm with three different masses for the manipulandum, at three different frequencies. The sequence of masses and frequencies is instructed by the control computer and may be varied from session to session. Thereafter, the participant performs targeted point-to-point movements to visual targets presented in random order in both the vertical and horizontal directions (see **Figure 14**). At the end of this protocol, they repeat the test of the coefficient of friction.

5.1.2 References seated protocol

Participants perform another set of experimental trials in the seated posture. First, they perform a test of the coefficient of friction. Then, they perform rapid discrete movements between two target positions located along a vertical line with eyes open. After that they perform rapid, discrete movements to the remembered position of the same vertically arranged targets, with eyes closed. Rapid, discrete movements are subsequently

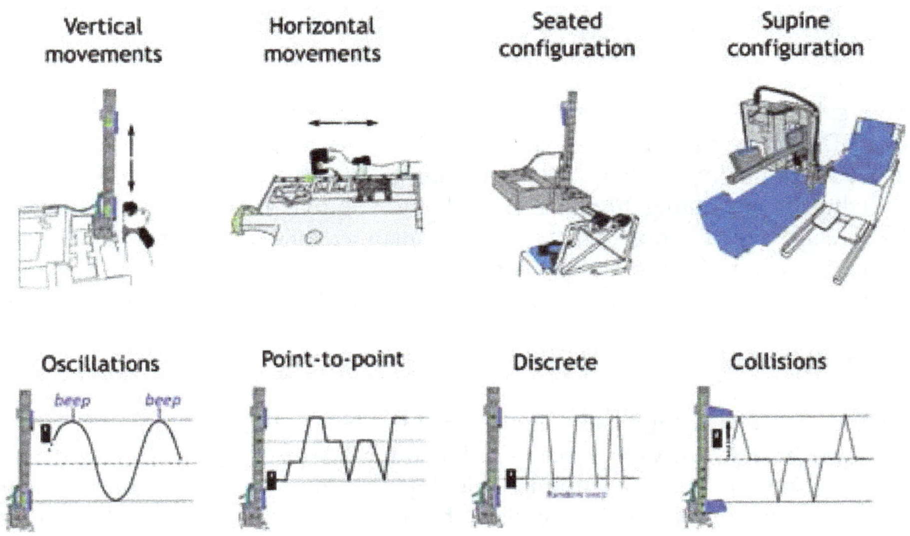

Figure 14.
The astronaut performs a variety of movements (oscillations, point-to-point, discrete movements and controlled collisions) while holding the manipulandum in the hand and being either in sitting or supine position.

performed with eyes open and with eyes closed, as above, with the targets located along a horizontal line. Finally, participants perform a set of tapping gestures in which the manipulandum is moved rapidly upward or downward to collide with surfaces that are rigidly attached to the target structure (see **Figure 14**). Audible tones are used to indicate to the participant which surface (above or below) must be tapped on each trial, in random order. At the end of this protocol, each participant repeats the test of the coefficient of friction.

5.1.3 *References supine protocol*

Participants perform the same set of trials as described for the "references seated" protocol, but now in the supine posture (lying down on his/her back).

5.1.4 *Sensor check*

In addition to the science protocols described above, a "sensor check" protocol has to be performed at least once during each increment for which the GRIP experiment is performed on orbit and shall be performed at least once after the final GRIP test session on orbit. Specific oscillatory movements are performed with the manipulandum in order to verify the integrity of the force sensors and accelerometers. This calibration procedure is performed on both ground and flight equipment to ensure that any differences in measured values between experimental trials performed on the ground and during flight cannot be attributed to differences in the hardware, rather than to differences in sensorimotor processing by the central nervous system in weightless conditions.

5.2 Session planning

A full session consists of the dynamics seated, references seated and references supine sub-sessions. There are two preflight Baseline Data Collection (BDC) sessions. BDC1 may be performed any time 360 days prior to launch. BDC2 must be scheduled not more than 180 days prior to launch. There are three reduced postflight sessions lasting 30 min each, in between 1 to 6 days post return on Earth. Additionally, three full postflight sessions are planned: one early session 12 days after return and two late sessions 30 days after return, with a minimum of 1 week between sessions. Pre- and postflight BDC sessions are taking place in Cologne or in Houston

5.3 Commissioning

Before the first in-flight session on-board the ISS, a commissioning session was performed by a surrogate subject in order to verify the equipment integrity. The commissioning included a complete deployment of the equipment, a "sensor check" protocol and a reduced version of the dynamics seated, references seated and references supine protocols.

6. Support of ground control centers

The development of the GRIP experiment and its implementation in the ISS would not have been possible without a strong support of experts, which was provided by the CADMOS (*Centre d'Aide au Développement des Activités en Micropesanteur et des Opérations Spatiales*) from the CNES (*Centre National d'Etudes Spatiales*, Toulouse, France). The central role of CADMOS is to help scientists prepare their experiments, to provide support and monitoring and to downlink the data from the ISS. In particular, CADMOS' team provided support for the Scientist Validation Test and the Experiment Test Sequence that were both performed at CADMOS facilities on the ground to test the hardware and the smoothness of the operations and coordinated the Commissioning that was run on-board the ISS by ESA French astronaut Thomas Pesquet to test the equipment. CADMOS also acts as a crucial link between the scientific team and the Columbus Control Centre (COL-CC) which is based in Munich and which controls all experiments performed on-board the Columbus module. CADMOS also coordinates with the European Astronaut Centre in Cologne for the training of the astronauts and can support the scientists during the Baseline Data Collection when it is judged necessary. BDCs performed in Houston are supported by the Johnson Space Centre from NASA.

7. Conclusions

The GRIP project investigates object manipulation in microgravity. It is an epic journey to the ISS that is nicely illustrated and motivated in this short (5 min) video [16].

This project required a lot of essential ingredients. Indeed, it relies on excellent scientific competences, an outstanding support from the different agencies (Belspo/Prodex, ESA, NASA, CADMOS) and contractors (QinetiQ, Arsalis). A lot of patience, professionalism and determination were also needed to reach a long-term goal and to overcome many obstacles over more than 20 years. Indeed, GRIP was not even spared a rocket explosion. One of the key elements of this success is the valuable trust among the different partners that was often supported by a true

Figure 15.
The GRIP badge in front of the ISS porthole with the earth in the background (courtesy of Luca Parmitano).

friendship that built up over the years. This is truly a team work that is a reflection of the immense success of the ISS project built on international collaborations.

The experiments could recently be performed on board the ISS thanks to all the preparatory scientific work of excellent quality conducted by all our PhD students in our laboratories and in parabolic flights. Finally, it is very important to give special thanks to our experimental subjects on board the ISS. Indeed, it is important to stress that we were often stunned by the dedication of all our astronaut subjects who conducted GRIP experiments in conditions that were sometimes difficult.

We are convinced that GRIP experiments will make a very significant contribution to our understanding of dexterous manipulation and adaptation to long-term microgravity. This should also shed some light on motor learning and hopefully guide clinicians who aim at helping patients suffering from motor control pathologies (**Figure 15**).

Author details

Jean-Louis Thonnard[1,2*], Laurent Opsomer[1,2], Philippe Lefèvre[1,2], Vladimir Pletser[3,4,5] and Joseph McIntyre[6,7,8]

1 Institute of Neurosciences (IoNS), University of Louvain, Belgium

2 Institute of Information and Communication Technologies, Electronics and Applied Mathematics (ICTEAM), University of Louvain, Belgium

3 Blue Abyss, Space Training Operations, Liverpool, United Kingdom

4 Chinese Academy of Sciences, Beijing, China

5 European Space Agency, Noordwijk, The Netherlands

6 Centre National de la Recherche Scientifique, France

7 TECNALIA, Basque Research and Technology Alliance (BRTA), Spain

8 Ikerbasque Research Foundation, Spain

*Address all correspondence to: jean-louis.thonnard@uclouvain.be

References

[1] Johansson RS, Westling G. Roles of glabrous skin receptors and sensorimotor memory in automatic control of precision grip when lifting rougher or more slippery objects. Experimental Brain Research. 1984;**56**(3):550-564

[2] Flanagan JR, Tresilian J, Wing AM. Coupling of grip force and load force during arm movements with grasped objects. Neuroscience Letters. 1993;**152**(1-2):53-56

[3] Flanagan JR, Wing AM. The stability of precision grip forces during cyclic arm movements with a hand-held load. Experimental Brain Research. 1995;**105**(3):455-464

[4] Augurelle A-S, Penta M, White O, Thonnard J-L. The effects of a change in gravity on the dynamics of prehension. Experimental Brain Research. 2003;**148**(4):533-540

[5] Crevecoeur F, Thonnard J-L, Lefèvre P. Forward models of inertial loads in weightlessness. Neuroscience. 2009;**161**(2):589-598

[6] Crevecoeur F, McIntyre J, Thonnard J-L, Lefèvre P. Movement stability under uncertain internal models of dynamics. Journal of Neurophysiology. 2010;**104**(3):1301-1313

[7] Hermsdörfer J, Marquardt C, Philipp J, Zierdt A, Nowak DA, Glasauer S, et al. Moving weightless objects. Grip force control during microgravity. Experimental Brain Research. 2000;**132**(2000):52-64

[8] Nowak DA, Hermsdörfer J, Philipp J, Marquardt C, Glasauer S, Mai N. Effects of changing gravity on anticipatory grip force control during point-to-point movements of a hand-held object. Motor Control. 2001;**5**(3):231-253

[9] White O, McIntyre J, Augurelle A-S, Thonnard J-L. Do novel gravitational environments alter the grip-force/load-force coupling at the fingertips? Experimental Brain Research. 2005;**163**(3):324-334

[10] McIntyre J, Zago M, Berthoz A, Lacquaniti F. Does the brain model Newton's laws? Nature Neuroscience. 2001;**4**(7):693-694

[11] Opsomer L, Théate V, Lefèvre P, Thonnard J-L. Dexterous manipulation during rhythmic arm movements in Mars, moon, and micro-gravity. Frontiers in Physiology. 2018;**9**(July):1-10

[12] Papaxanthis C, Pozzo T, McIntyre J. Kinematic and dynamic processes for the control of pointing movements in humans revealed by short-term exposure to microgravity. Neuroscience. 2005;**135**(2):371-383

[13] White O, Bleyenheuft Y, Ronsse R, Smith AM, Thonnard J-L, Lefèvre P. Altered gravity highlights central pattern generator mechanisms. Journal of Neurophysiology. 2008;**100**(5):2819-2824

[14] Crevecoeur F, Thonnard J-L, Lefèvre P. Optimal integration of gravity in trajectory planning of vertical pointing movements. Journal of Neurophysiology. 2009;**102**(2):786-796

[15] Barrea A, Bulens DC, Lefèvre P, Thonnard J-L. Simple and reliable method to estimate the fingertip static coefficient of friction in precision grip. IEEE Transactions on Haptics. 2016;**9**(4):492-498

[16] Cetiner E. Dexterous Manipulation in Microgravity [Internet]. 2017. Available from: https://youtu.be/_783SofVVs8 [Accessed: 18 May 2020]

12

Challenges to Airway Management in Space

Preethi Reddy

Abstract

Human interest in space exploration is boundless. We are driven to investigate the unknown and push the limits of our understanding of our universe. Given that space flights are for extended periods of time —in the hazardous environments of space and the growth of the space tourism industry is credibly anticipated; the incidence of medical and surgical events is bound to increase during space travel. Airway management becomes an essential skill in such situations. Microgravity, shortage of medical personnel, inability of the crew to return to earth expeditiously or access real time assistance from earth are some of the reasons that warrant training and preparation of the crew, towards this end. The purpose of this chapter would be to explore the challenges and the various recourses available for airway management during space travel.

Keywords: space travel, airway management, space medicine, space flight, anaesthesia

1. Introduction

There is expanded access to space. Government agencies and private companies alike have planned manned missions to the moon and to Mars, for the coming years. Currently, we are also witnessing a meaningful growth in the space tourism sector.

It is now more important than ever to increase our understanding of human physiology and pathology in space. Furthermore, it is of prime importance that astronauts can manage medical and surgical emergencies, that may arise especially given that a space tourist—as against the typical astronaut—is unprepared for the rigours of space travel and therefore exposed to a higher risk of medical complications [1].

Man's tryst with space began in 1961. The International Space Station (ISS), which has been in orbit for almost 20 years now, has enabled humans to stay in space for long durations. This has provided a swifter and a more profound bioastro-nautics development.

The environment of space is harsh and challenging, with a prolonged exposure to multiple stressful stimuli, radiation, weightlessness, isolation, and confinement to tight enclosed spaces for long periods of time. Microgravity, which affects all organ systems, has the most profound effect on human physiology [2].

Travelling to Mars will require transitioning between three different gravitational fields: being weightless during a six month interplanetary flight, being at about one third of the Earth's gravity on Mars, and re-acclimatising to Earth's gravity upon return [3].

2. Airway management in space

With increasing flight durations, there is an increased prospect that a medi-cal emergency will entail airway management. It is currently estimated that the probability of a medical intervention requiring general anaesthesia, over the course of a 950 day mission to Mars and with six crew members is 2.6%. This speaks to the importance of how even the unlikeliest of events could imperil the mission and lead to loss of life [1]. Airway management is an important skill that is required to man-age a great many medical emergencies. Some of the possible scenarios necessitating airway management in space are listed in **Table 1** [1, 4, 5].

Seventeen medical emergencies were documented during spaceflight between the periods of 1961–1999 [1]. In one instance, in 1962, on the Mercury 7 flight, Scott Carpenter, an American astronaut, aspirated food crumbs in orbit and in 1975 several astronauts on the Apollo-Soyuz mission developed a mild form of chemical pneumonitis after accidentally inhaling propellent fluid during re-entry [4]. Incidentally, none of the seventeen cases have required intubation. Also, no one has required GA in space to date [1].

Medical evacuation is not an option, in case of an airway emergency, owing to the distance as well as the absolute need to maintain oxygenation to avoid brain death. It is therefore necessary that immediate care is provided, while on board [6]. This warrants a crew equipped with emergency care skills as well as continuous training, to prevent skill erosion [1, 6, 7]. Furthermore, communication delays prohibit real time telemedicine support. For example, the communication delay between Earth and Mars is about twenty minutes—one way [1]. Airway manage-ment skills are thus critical to the mission of exploring space safely.

In the first half of this chapter, we will deal with the physiology of airway management and the physiological adaptations of the human body in space. This would provide the essential foundation to understand the challenges associated with airway management in space.

The term 'Airway Management' refers to the maintenance of airway patency and ensuring adequate ventilation and oxygenation. Successful airway management entails that the practitioner anticipates and predicts difficult airway and at the same time devises an airway management plan. The practitioner should also be adequately skilled to execute that plan, with the available resources. In order to enable this plan, anaesthesia is typically required—to provide patient comfort, limit airway reflexes, and to moderate the hemodynamic response to airway instrumentation [8].

3. Physiology of airway management

3.1 Pre-oxygenation

Hypoxaemia can occur on induction of anaesthesia and muscle paralysis on account of hypoventilation and apnea. Pre-oxygenation or denitrogenation helps to replace the nitrogen in the lungs with oxygen. This, consequently, extends the apnea time and allows the anaesthesiologist to secure the airway and resume ventilation.

Pre-oxygenation is achieved by providing 100% oxygen via a face mask, at a flow rate of 10-12 L/min to prevent rebreathing. This can be achieved by asking the patient to breathe for 3 min using tidal volume ventilation; or by taking 8 vital capacity breaths over 60 seconds. During this process, it must be ensured that there are no leaks around the face mask [8].

1. Hypoxic cardiopulmonary arrest
2. Airway obstruction
3. Foreign body aspiration
4. Burns or smoke inhalation
5. Coma
6. General Anaesthesia (GA)

Table 1.
Indications for airway management in space.

3.2 Pulmonary aspiration of gastric contents

Patients are required to have an empty stomach to reduce the risk of regurgitation and pulmonary aspiration of acidic gastric contents. The American Society of Anaesthesiologists task force recommends 4 hours of fasting from breast milk, 6 hours of fasting from infant formula, non-human milk and solid foods; and up to 8 hours or more from fried or fatty food. Clear fluids may be allowed up to 2 hours prior to anaesthesia [9].

Prophylactic drugs may be beneficial in patients with specific risk factors for aspiration. They help in decreasing gastric volume and increasing the gastric fluid pH. The commonly used drugs (alone or in combination) are—non-particulate antacids, promotility drugs and H2-receptor antagonists. These drugs may be used alone or in combination [8].

3.3 Airway reflexes and the physiological response to intubation of the trachea

One of the main functions of the larynx is protection of the airway. Sensory receptors in the glottic and subglottic mucosa are triggered—on airway instrumentation—leading to the adduction of the vocal cords and laryngospasm. Furthermore, foreign body irritation of the lower airway can result in bronchospasm [8].

Airway instrumentation causes an intense noxious stimulus via the vagal and glossopharyngeal afferents. This results in a reflex autonomic activation, manifesting as hypertension and tachycardia. Although this response lasts only for a short duration, it may have serious consequences in patients with significant cardiac disease. Also, CNS activation can occur leading to an increase in the electroencephalographic activity, cerebral metabolic rate and blood flow, which may result in an increased intracranial pressure [8].

3.4 Anaesthesia for airway management

General anaesthesia is the most common technique employed in airway management. A rapid acting intravenous anaesthetic agent is most commonly used for induction of anaesthesia, followed by a neuromuscular blocking agent to provide muscle relaxation [8, 10].

Rapid sequence induction is used when there is an appreciable risk for gastric regurgitation and pulmonary aspiration of gastric contents. In this technique, after pre-oxygenation, cricoid pressure is applied. This is followed by an induction dose of an intravenous anaesthetic and 1–1.5 mg/kg of intravenous succinylcholine. The trachea is then intubated without any attempts at positive pressure ventilation. The cricoid pressure is applied constantly until the airway is secured.

Inhalation induction of anaesthesia with volatile anaesthetics is commonly used in paediatric patients—to provide a needle free experience—and in adults where intravenous access is difficult or when this technique is desirable [8].

Intravenous induction without neuromuscular blocking drugs is used for LMA (laryngeal mask airway) placement. Propofol is the drug of choice for this technique due to its distinct ability to suppress airway reflexes and produce apnea [11, 12].

Awake airway management is indicated, but not limited to, difficult mask ventilation and difficult intubation [13]. In such a case, the pharyngeal muscle tone and patency of the upper airway is maintained. This allows for spontaneous ventilation and acts as a safeguard against aspiration. It also provides an opportunity for a quick neurological examination, if indicated. Awake airway management is achieved by topicalisation of the airway with local anaesthetics [8].

3.5 Equipment for airway management

Equipment for basic airway management includes face masks for pre-oxygenation and delivery of inhalational anaesthetic agents, supraglottic airways are devices that are inserted blindly into the pharynx to provide a conduit for ventilation without requiring tracheal intubation, and endotracheal tubes, that provide maximum protection against the risk of aspirating gastric contents while establishing a defini-tive airway and at the same time, allowing positive pressure ventilation with higher airway pressures.

In patients with known or predicted difficult airway, videolaryngoscopy, rather than direct laryngoscopy is indicated since videolaryngoscopy inherently provides better glottic visualisation as well as effortlessly employed by non-experts [8].

3.6 Laryngoscopy and endotracheal intubation

Endotracheal intubation is established as the gold standard for airway management. It is typically achieved by direct laryngoscopy with patients placed in the sniffing position. A line of sight must be established from the mouth to the larynx. Direct laryngoscopy displaces the hyoid, tongue and epiglottis anterior to a line running from the upper teeth to the glottis.

In this technique, the mouth is opened, the laryngoscope blade is inserted and the tip is positioned to apply a lifting force exposing the glottis. The endotracheal tube is then inserted through the vocal cords into the trachea [8, 14].

4. The human body in space

The Earth's constant gravitational force is an important factor in the evolution of life on this planet. It has determined the development of all forms of life. All biological adaptations on land and water have been influenced by its interactions with gravity forming complex systems for stability, fluid regulation, gravity sensing, and locomotion.

The human body responds to microgravity in the same manner that it responds to senescence (ageing): Both ageing and microgravity produce a decline of biological function [15]. Also, like ageing, microgravity causes a negative calcium balance leading to a loss of bone density, muscle atrophy, cardiovascular and haematic changes, and metabolic, endocrine, and sleep disturbances. In microgravity, astronauts undergo rapid senescence. However, they subside over time on returning to Earth, departing from the typical path of the ageing process. This correspondence of symptoms combined with a prolonged stationing in space requires that we are alive to the issue of accelerated ageing in space [15].

4.1 Cardiovascular system

Changes in the cardiovascular system because of microgravity are paramount from the standpoint of the anaesthesiologist. Gravity influences the equilibrium of the various functional fluid compartments of the vascular system and in particular that of the venous capacitance vessels [16]. In the upright posture, there is higher arterial pressure in the feet (200 mmHg) and lower pressure in the head (70 mmHg) relative to the heart (100 mmHg). In space, this gradient is absent, leading to redistribution of body fluids toward the head [2, 15, 17]. This phenomenon is referred to as “fluid shift”. As a result of this, astronauts develop facial puffiness coupled with reduced volume in the lower limbs [2, 16].

Also, due to this ‘fluid shift’, there is engorgement of the central circulation. Mechanoreceptors sense this blood redistribution activating autonomic offloading and volume regulating reflexes leading to vasodilation and pooling of blood in the viscera and tissues, and initial renal fluid and salt loss. Most of these adaptations occur within 6–10 hours of spaceflight. After one week in space, the plasma volume reduces and the intracellular volume increases [15, 16]. In the same period, the RBC mass drops by about 10%. This “space anaemia” can again be attributed to the fluid shift toward the upper body, which is associated with an increase in kidney tissue oxygen partial pressure leading to the inhibition of erythropoiesis. A second hypothesis explains this as being due to haemolysis of recently formed RBCs [2, 16, 18].

Despite the headward fluid shift, paradoxically the central venous pressure is not increased. Further, a reduction in the intrathoracic pressure and the loss of gravitational force on the cardiac muscle may even reduce it [15, 16].

Microgravity affects the heart rate and blood pressure minimally [16, 19]. The initial headward fluid shift increases the stroke volume and cardiac output. Subsequently, after a few days of adaptation, the resulting hypovolemia and cardiac atrophy, increase the ejection fraction and decrease the stroke volume. The left ventricle mass reduces by 8% due to reduced myocardial load in microgravity [16, 20–22]. The left ventricular systolic function is minimally affected even though diastolic dysfunction has been identified in astronauts [16].

The nitric oxide release as a result of endothelial cell adaptation to microgravity and the loss of tone due to smooth muscle cell deconditioning causes vasodilation. Systemic vascular resistance reduces after 1 week of weightlessness due to this vasodilation [15, 16].

The baroreflex response is weakened by 50% after just 24 hours of being in space. It is constrained after long-duration spaceflight and these changes linger on for up to 2 weeks after returning to Earth. There are changes in adrenergic-receptor sensitivity in microgravity: beta-adrenergic receptors sensitivity is increased and alpha-adrenergic receptors sensitivity is decreased. There is also an increased risk of arrhythmias in space due to catecholamine discharge [16].

In space, aerobic capacity may be either maintained or increased. On return to Earth, there is an orthostatic challenge due to readaptation to gravity. As a result of this, astronauts experience reduced stroke volume and cardiac output which leads to landing-day orthostatic stress [2].

4.2 Endothelial changes

Microgravity and reduced motor activity produces endothelial changes by altering the regional blood flow and vascular transmural pressure, which in turn, produces an adaptation of vasomotor tone and long term vascular remodelling mainly in the endothelium and smooth muscular cells [16].

This microvascular endothelial dysfunction in astronauts, plays a material role in osteoporosis, muscle atrophy and cardiovascular deconditioning considering that the endothelial cells of the microvasculature cover a surface area that is fifty times larger than that of all the large vessels put together [15].

4.3 Space motion sickness

Space motion sickness is a result of neurovestibular disturbance that happens to about two-thirds of astronauts. It occurs within a few minutes of being in space and gradually resolves over a period of 48–72 hours. Nevertheless, it can last up to a few days and can reappear after landing. Some causes that are suggested as a possible hypothesis include: an increase in cerebrospinal fluid and intracranial pressure due to the headward fluid shift, a lowered threshold for vestibular stimulation due to central volume expansion and the absence of gravity triggering an abnormal vestibular activity leading to a parasympathetic overstimulation [15, 23].

Space motion sickness is characterised by an imbalance in spatial orientation, balance, gaze control and autonomous vestibular function. Symptoms include facial pallor, cold sweating, stomach awareness, anorexia, vomiting, nausea, headache and malaise [2, 15, 16].

4.4 Eye

On Earth, venous return from the head, neck and upper trunk is supported by gravity. Unlike the lower half of the body, the veins draining this region do not have valves and lack muscular contraction. In space, there is reduced arterial blood supply and venous flow from the eye. This increases the venous pressure and filtration at the capillaries causing an increase in both intracranial pressure and IOP [15].

4.5 Effects on the musculoskeletal system

Extended exposure to microgravity leads to a loss of bone and muscle mass due to its reduced use and perfusion changes. Inadequate nutrition and stress are additional reasons that lead to muscle atrophy [2]. Weight bearing bones: lumbar spine, pelvis, femoral neck and trochanter, and calcaneus and postural muscles: back, abdominal wall, lower limbs are most commonly affected [2, 24].

In addition to absence of gravitational loading, decreased Vitamin D production—partly due to low levels of sunlight—leads to decreased calcium fixation in bones and reabsorption in kidneys. Higher ambient levels of carbon dioxide, leading to respiratory acidosis also contribute to bone loss [2, 15] Increase in urinary calcium coupled with a reduction in diuresis and decreased fluid intake increases the risk of kidney stones [25–27].

4.6 Effects on the respiratory system

Microgravity induced changes in the lungs have been the subject of much interest for decades. The ventilation to perfusion ratio attains equilibrium in the absence of gravity [15, 16]. There is an increase in the total alveolocapillary surface which in turn improves the lung diffusing capacity [16]. Gas exchange in space does not undergo a substantial change but there is reduced oxygen consumption and carbon dioxide production. This is attributed to a reduced physical activity in space and change in the ventilation to perfusion ratios between the upper and lower lung regions. These factors lead to an overall reduction in the metabolic rate [15].

Changes in the thoracoabdominal compliance is advantageous to the pulmonary function [16].

Microgravity can cause weakening of the respiratory muscles leading to a reduced rib cage expansion. Thus, there is an increased contribution of the abdomen to tidal volume [28]. Intra-abdominal pathology and subsequent intra-abdominal hypertension is important to providing life support and mechanical ventilation. However, it has been demonstrated that intra-abdominal gas insufflation during laparoscopic surgery in space and the subsequent intra-abdominal hypertension is made better by the absence of gravity [15, 16].

It has been suggested that changes in the respiratory system are similar to those that occur to individuals on prolonged bed rest and are anatomical in nature [15]. These typically occur over several weeks. There is a significant reduction in tidal volume [15, 16] and residual volume [29]. Vital capacity and forced vital capac-ity reduce initially followed by subsequent recovery. Functional residual capacity (FRC) reduces by 500 ml and remains at that level for the remainder of the period in space [29, 30]. Peak inspiratory and expiratory flows are also not significantly altered. However maximum inspiratory pressure significantly reduces, while the maximum expiratory pressure (MEP) at total lung volume initially reduces at month 2 and month 4, but recovers by month 6 of being in microgravity. The MEP at FRC however is not affected [15].

4.7 Immune system

Immune system dysregulation occurs in space. High levels of physical and psychological stress—immediately before and after space flight— physiological stress, isolation, confinement, disrupted circadian rhythms are some of the contributing factors to immune system dysregulation [2]. Additionally, increase in levels of glucocorticoids and catecholamines, may also contribute to change in the immune system [3, 31]. Various studies have demonstrated that lack of gravity impairs the signalling pathways that are necessary for early T-cell activation. This leads to changes in the organisation of the cytoskeleton and microtubule organising centres [2].

Immune system dysregulation can lead to an increased incidence of hypersensitivities, autoimmunity, allergies, infectious diseases, latent viral reactivation and even malignancies [3].

Microbes undergo several changes in their characteristics in space. Notably, bacteria cultured on board have increased pathogenicity [32]. The microorganisms present in the human body, are transmitted easily between persons, in such confined habitats [3].

4.8 Gastrointestinal motility

Gastrointestinal motility is reduced in space especially in the first 72 hours. It has also been observed that the gastric content pH decreases [16].

4.9 Weight loss

Astronauts experience a weight loss of up to 5% after a 6 month stay on the International Space Station (ISS). This is explained by a mismatch between caloric intake and caloric expenditure [16, 33, 34].

4.10 Psychological effects

Confinement and isolation in constrained spaces, for extended periods of time, affects one's psychological health. Even with screening, training, and support; behavioural issues, cognitive conditions, and psychiatric disorders among crew members, is to be expected. Decline in mood, cognition and morale can occur. Sleep disorders due to changes in their circadian rhythms is also quite common [2, 3].

Extended exposure to stress, isolation and changes in circadian rhythm can have a psychological impact on astronauts. Cognitive impairment, sleep disorders, psychosomatic symptoms, anxiety and even depression can occur [7].

Personnel skills like team coordination, communication, logistics, etc. and technical skills like troubleshooting equipment, use of safety equipment, orientation, etc. contribute to the health and safety of astronauts. Selection of suit-able crew, training and maintenance of skills during the mission, is important. Therefore, medical and psychological benchmarks for crew-member selection ought to be very high [7].

4.11 Exposure to radiation

Space travel presents the additional risk of exposure to harmful radiation. On Earth, we are shielded from cosmic radiation by the Earth's magnetic field and its atmosphere. However, on a space station astronauts are exposed to up to ten times the radiation they are exposed to while on Earth. Radiation in space can cause radiation sickness and degenerative tissue disease, among many other serious issues [3].

5. Challenges to anaesthesia delivery and airway management in space

Anaesthesia is important for airway management. Blunting airway reflexes and hemodynamic response to airway instrumentation is a chief consideration. This can be achieved by anaesthesia.

Delivery of anaesthesia in space is highly demanding and complicated, and the reasons can be grouped into three main categories: physiological, technical and human as shown in **Table 2** [16, 17, 35].

5.1 Physiological considerations

5.1.1 Challenges related to cardiovascular changes in space

In weightlessness, a new physiological equilibrium is established, adapted to the reduced loading conditions. However, this equilibrium is delicate as is the tolerance to any additional event or even an interventional procedure. Reaction of the human body to blood loss, anaphylaxis or any event that reduces cardiac function, may be further compromised. General anaesthesia and mechanical ventilation may also adversely affect this physiological equilibrium [16].

On landing in gravity environments different from Earth, cardiovascular changes and hypovolemia causes orthostatic intolerance. The aerobic capacity is also impaired as a result of hypovolemia, anaemia and orthostatic intolerance [16, 36]. These fac-tors combined with space motion sickness, limits the crew's ability to perform tasks effectively.

Preloading with intravenous (IV) fluids before the induction of general anaesthesia is important to prevent cardiovascular collapse [37]. Any significant

1. Physiological
   - Cardiovascular changes
   - Fluid shift
   - Gastrointestinal System
   - Pharmacology in Space environment
   - Choice of anaesthetic technique
2. Technical
   - Fluid generation and handling
   - Vascular access
   - Closed cabin pressures
   - Medical equipment
   - Use of restraints
   - Telemedicine and information technology
3. Human
   - Crew Skills
   - Psychological effects on crew

Table 2.
Factors complicating delivery of medical care in space.

hypovolemia should be treated concomitantly with IV fluids and vasopressors. Alpha agonists such as phenylephrine, metaraminol, midodrine, norepinephrine should be at hand; higher doses than usual may be required. Beta-agonists and beta-antagonists should be used with care [16]. For induction of anaesthesia, drugs that preserve cardiovascular stability, such as ketamine, are preferred [37].

5.1.2 Fluid shift

Although not documented, headward fluid shift and facial oedema can complicate the intubating conditions [16]. Drug distribution during spinal anaesthesia may also be altered due to the cephalad fluid shift and is therefore not recommended, in microgravity [38].

5.1.3 Gastrointestinal system

Space motion sickness accompanies gastroesophageal reflux in astronauts, sometimes lasting the entire mission. It may even persist after their return to Earth. The gastroesophageal reflux along with decreased gastrointestinal motility puts the crew at a risk for pulmonary aspiration, both during and after flight [16, 17].

5.1.4 Pharmacology in the space environment

Both the pharmacokinetics and pharmacodynamics of drugs are altered in weightlessness [39].

Cardiovascular changes, weight changes, changes in hormonal, electrolyte and immunoglobulin levels, decrease in the amount of microsomal P-450 as well as its dependent enzymes are some of the factors that cause changes to the pharmacokinetic and pharmacodynamic properties of drugs in space [16, 40]. As a result, the corresponding drug dosages need to be altered as well [39].

Also, long term storage of drugs may render them ineffective or even toxic [16].

A notable mention: the depolarising muscle relaxant succinylcholine is contraindicated due to disuse atrophy of muscles and changes in the neuromuscular junction, and the increased risk of hyperkalemia after prolonged exposure to microgravity [17, 39]. Instead, rocuronium is recommended to be used as an alternative [16, 41].

5.1.5 Choice of anaesthetic technique

One of the limiting aspects of the anaesthesia protocol for microgravity is that it should be carried out by a small crew of non-medical personnel, with limited training. In several low-income countries, anaesthetic procedures are regularly performed by non-medical personnel, with relatively low complications. Simplified versions of the protocols—one which can easily be followed by non-physicians—must be developed.

The worst case scenario approach should be the basis for making the choice of the anaesthetic technique. It must be borne in mind that astronauts who may require anaesthesia in space may have to be managed by nonmedical personnel with limited training, in case the crew medical officer (CMO) is incapacitated or deceased. In addition, they maybe hypovolemic, deconditioned, at a risk for rhythm disturbances and gastric aspiration, and intolerant to succinylcholine [16].

Although ultrasound guided regional anaesthesia may be used safely and successfully, it requires considerable training [16, 38, 42, 43]. Spinal anaesthesia is not feasible in microgravity, its safety and efficacy is unpredictable because the heavy local anaesthetic solutions used depend on gravity. Epidural anaesthesia may be used, but it also requires considerable training and absolute asepsis, and therefore carries significant risks [16].

General anaesthesia with endotracheal intubation is suitable for all types of surgical conditions and is the recommended choice of anaesthetic technique.

Intubation, in general, is facilitated by use of general anaesthesia and muscle relaxants [16]. Furthermore, use of video laryngoscopy also increases the intubation success rate especially among new users [44, 45].

5.2 Technical considerations

5.2.1 Fluid generation and handling

Intravenous (IV) fluids have a limited shelf life and usually remain unused during that period. Shipping and storing them is expensive due to the added weight and the wastage of valuable storage volume. It is however expedient, and necessary to be able to generate IV fluids on demand using drinking water. This process was successfully tested on the ISS (project IVGEN) [16, 46].

Fluids and gases do not separate in space, owing to their different densities, which complicates fluid handling and drug preparation. Hence most drugs and intravenous fluids exist as a foamy liquid [16, 17].

It is advisable that injectable drugs be carried in prefilled syringes. Needleless vial adapters that allow direct drug aspiration into the syringe without the need for a needle to pierce the vial septum are also preferred. Experiments have been successfully conducted by NASA Scientific and Technical Information Program for removal of air bubbles [16, 17, 47].

Another important concern is that many medical devices such as anaesthetic vaporisers and suction equipment, that depend on gravity induced separation of fluids and gases, do not function properly in microgravity [17, 37].

5.2.2 Vascular access

During a medical emergency, vascular access may be difficult to obtain. In space, securing the body of the IV administrator as well as mastering fine motor skills to perform the required task can be a challenge. Also, microgravity causes small objects to float away. Sharp objects such as IV cannulae can present a potential hazard to the crew [47].

Using ultrasound to obtain central venous access either autonomously or robotically is currently under development. An intraosseous access kit has also been included in the ISS medical gear [16].

5.2.3 Closed cabin atmospheres

A spacecraft offers a tightly sealed environment. Medical procedures requiring the use of oxygen would risk oxygen enrichment in the closed cabin atmosphere, increasing the risk of explosion and fire. Use of a closed ventilation circuit limits the dumping of oxygen in the cabin. Volatile anaesthetics cannot be used in such an environment, as a gas leak can contaminate the on-board closed loop environment and therefore general anaesthesia must be provided by the technique of total intravenous anaesthesia (TIVA) [16, 17, 38]. Xenon may find utility as an anaesthetic gas in such space operations [17].

5.2.4 Medical equipment

Advanced medical care requires equipment such as a monitor, ventilator, suction equipment, and oxygen concentrator. Equipment carried to space must comply with specific spaceflight standards. There are a number of stipulations in terms of weight, size, and power consumption. For perspective: It costs about US$ 22,000 USD to transport one kilogramme of material into low Earth orbit. Drugs that do not need refrigeration and that which have a long shelf life are preferable, in this regard [16, 17].

5.2.5 Use of restraints

Airway management is made possible during spaceflight using restraints, allowing the operator's hands to be free to hold and guide the endotracheal tube into the airway [38]. Use of restraints are absolutely necessary to hold instruments, patients and personnel in place. For surgical procedures, it has been demonstrated that it is possible to restrain instruments in microgravity using various supplies ensuring sterility, operator accessibility, safe waste disposal, while maintaining ergonomic capability [48, 49].

5.2.6 Telemedicine and information technology

Telemedicine, as a medium for healthcare delivery, has tremendously improved and finds great many applications, in the current healthcare setting. Today, avail-ing a virtual opinion, of an expert, at a remote location, is fairly uncomplicated. This becomes very useful during spaceflight operations. However, a delay of about 20 minutes to receive one way communication is to be expected during a journey to Mars [4]. Since anaesthetic procedures and airway management require prompt and expedient responses, telemedicine—with its inherent latency issues—may not prove to be the most optimal solution. Hence, other advanced on-board information systems will need to supplement telemedicine technology in space [17].

5.3 Human factors

5.3.1 Medical skills and training

It must be borne in mind that the crew is unlikely to have a trained physician on board. Therefore anaesthetic procedures and airway management may have to be carried out by non-anaesthetists and non-physicians [35]. At present the Crew Medical Officer in each mission receives a training of about 80 hours [17]. Fading of skills during flight is an important concern and continuous training of the crew members is essential [1]. Fatigue and sleep debt during long duration space flight [2] can further affect performance of the medical officer during emergencies.

5.3.2 Incomplete knowledge about human physiology in partial gravity

At present we have very little information about human physiology in partial gravity. This knowledge is important in helping to plan the mission as well as preparing for medical contingencies. The moon has about one-sixth the Earth's gravity and Mars has about one-third. The Apollo moon missions did not include extensive physiological experiments unfortunately.

We have information about short term changes from transitioning from 1G to partial gravity levels. This has been obtained during parabolic flight, head-up tilt, lower body unweighting experiments. However, physiological impact of prolonged stay in reduced gravity is not available [16].

6. Airway management in space

An integrated space surgery research found that most procedures performed on Earth can be performed in microgravity with the right equipment and with the operator, subject and tools sufficiently restrained [38].

6.1 Training of crew

The medical team identified for the space exploration mission can be as lean as a single crew medical officer (CMO)—who is not necessarily a medical doctor [35].

On Earth, anaesthesia techniques, in high income countries, are performed only by experienced practitioners. In space, however, sophisticated medical expertise may be absent, or the CMO may have become injured, incapacitated or seriously ill, even requiring anaesthesia [35]. Since real-time telemedical support is not immediately forthcoming, the crew will have to be self-reliant. It may be imperative that lifesaving procedures may then have to be performed by personnel with limited training. Until now, neither an anaesthetic technique nor human surgery has been performed in space, except for local infiltration [50].

Simulation and ground research are important programmes in framing protocols since there is a poor knowledge base about managing medical events in space. Simulation tools and techniques are routinely used in the medical field for continuous training of doctors [51–53]. The benefits of such training on their performance has been well documented [51, 54]. Needless to say, the simulation setup must resemble the target environment as closely as possible [54, 55].

In low income countries, especially ones facing a shortage of trained medical professionals, non-physician medical professionals regularly provide anaes-thesia and perform surgical procedures. Many of them have a modest medical background [7]. They are trained mostly on the job and often work alone, even lacking recommended equipment and safety levels [35, 56–60] Ketamine based anaesthesia combined with the considerable skills that the providers acquire in a limited time could explain the perioperative mortality rate in these coun-tries— which is "only" about two to three times more— when compared to high resource ones [35]. However, a crucial difference between anaesthesia providers in these low income countries and future space exploration missions is that the former treat a high number of patients and therefore skill redundancy is not a factor [7].

Personnel with modest medical training may be able to perform invasive procedures safely which is being witnessed in austere environments in different parts of the world. Astronauts are perhaps best positioned to respond to such a challenge [35]. Apart from their multitude of skills, astronauts are also selected for their ability to tolerate extreme stress. They are unquestionably among the best candidates, besides healthcare providers, to be able to perform advanced and invasive medical procedures, in the remotest of settings [35].

Currently, the International Space Station (ISS) has an on-board crew medi-cal officer (CMO), who is not necessarily a trained medical doctor [7]. Given the uniqueness of future long distance space missions, the ideal profile of a crew physician is a subject for discussion [61, 62].

The CMO will need to possess a broad spectrum of knowledge, be competent in basic surgical skills and in the management of the critically ill [61–63]. He/She will have to be resourceful and flexible in his/her thinking and approach, and have the ability to improvise for unanticipated medical scenarios [62]. A single physician is required to manage both surgery and anaesthesia. If the CMO herself becomes ill, injured, incapacitated or dies, it is imperative that non-physicians take over. It is therefore prudent to train several of the crew members to manage at least the most common emergencies [7].

Recently, progress has been made in the field of artificial intelligence, especially in its application in medicine. It is now possible to achieve better monitoring, improve disease detection and bring in more efficient clinical decision support systems. Safety of crew members could be improved by autonomous diagnostic systems, closed-loop automated anaesthesia or other clinical decision support systems. Furthermore, these measures could also simplify the training programme [7].

6.2 Equipment

Devices for airway management that are brought on board the space shuttles, comprise of: a face mask, a pressure-cycled ventilator, a single-bladed laryngoscope, tracheal tubes, an introducer, a capnograph, and a tracheostomy kit [64].

Care has to be taken to ensure that the airway equipment carried on board is adequately restrained and made conveniently accessible to the person performing the procedure.

Cuffed endotracheal tube is the recommended device in view of the changes in gastric motility and reflux. However several studies have shown that non-anaesthesiologists can secure the airway more easily with supraglottic airways than with endotracheal intubation [65–67]. A laryngoscope is not required to be used in order to insert these devices and hence one hand is now available to stabilise the head as well as the neck [64]. The second generation supraglottic airways provides a better seal and also allow gastric drainage making it a good fit for emergencies, in microgravity [1].

Videolaryngoscopy may also be used with increased success in these scenarios since they have better glottic visualisation and a higher success rate with less experi-enced clinicians [68–70].

6.3 Technique

Airway management in space, with all its challenges, is amplified for non-medical personnel. Checklists and other such simple and minimal protocols will immensely help to streamline the process [7].

A mandatory pre-anaesthetic evaluation of all space travellers, before their departure from Earth, is recommended. A thorough airway assessment and detailed clinical documentation for every space traveller would prove to be valuable in case of emergencies. Clinical documentation may include details like the size of the endotracheal tube, requirement of additional intubation equipment, etc. Checklists for airway management and anaesthesia may include details of that equipment that must be available conveniently with the corresponding drug doses calculated and made available.

Conventional laryngoscopy and intubation with the patient placed in the sniffing position, without the use of restraints, is likely to have a high failure rate in microgravity. The head and neck move out of the field of vision during direct laryngoscopy, due to the anterior force exerted. It is not possible for the hand—the one not holding the laryngoscope—to stabilise the head–neck as well as direct the endotracheal tube toward the glottic opening simultaneously [64]. Anaesthesiologists exert a force of about 40 N, lasting for about 10–20 s during direct laryngoscopy. This force is sufficient for a 70 kg human to move about 0.3 m in 1 second in microgravity. Use of a restraint (**Figure 1**) will allow the stabilisation of the head and neck, so that the hand not holding the laryngoscope is free to direct the endotracheal tube toward the glottic opening [64]. However, there are some limitations in applying them during a medical emergency. Data indicates that it takes 5 to 10 seconds for strap application [7 1].

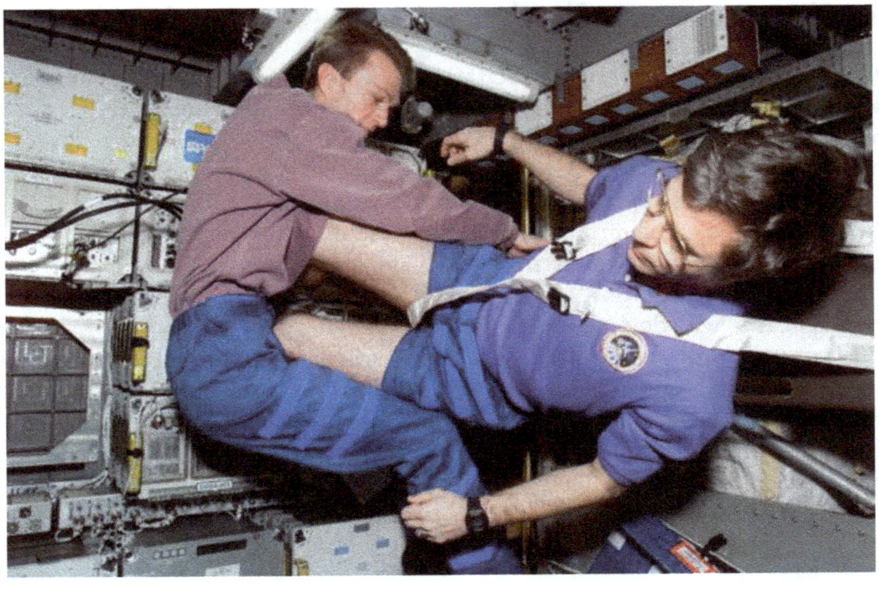

Figure 1.

*Crew medical restraint system used during the space shuttle missions. Restraints hold the patient in place and allow the operator's hands to be free. Source: Photograph S81E5933 - STS-081 - RME 1327 - Crew Medical Restraint System (**CMRS**); "STS-81 pilot Brent Jett straps mission specialist John Blaha into the Crew Medical Restraint System (**CMRS**) in the Spacehab module." January 1997; File Unit: STS-81, 4/12/1981 - 7/21/2011; Series: Mission photographs taken during the space shuttle program, 4/12/1981–7/21/2011; Record Group 255: Records of the National Aeronautics and Space Administration, 1903–2006; National archives at College Park, Adelphi road College Park (MD). Accessed on 30th April, 2021.*

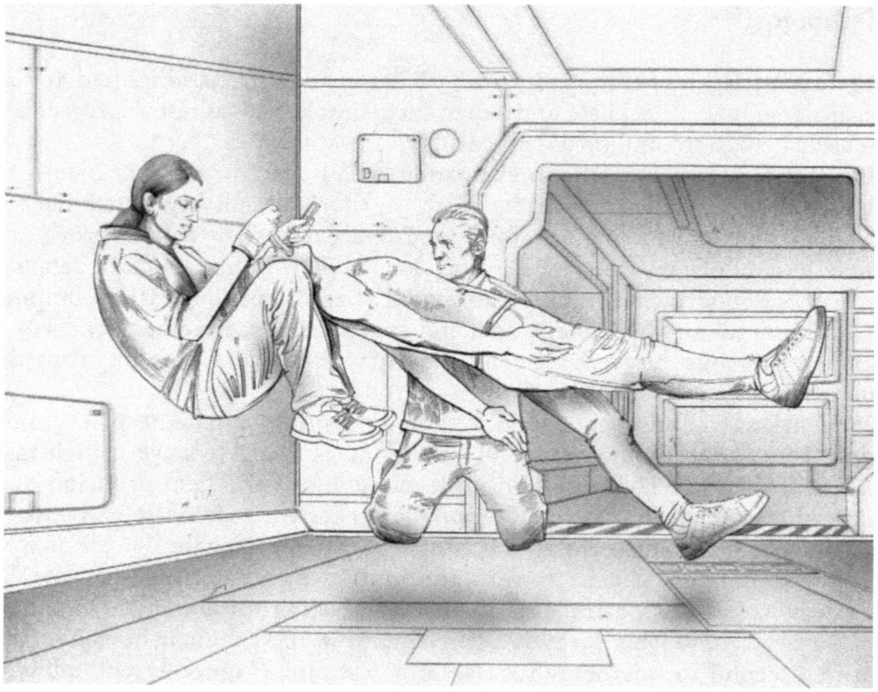

Figure 2.
An artist's representation of the sit down-lean back technique. (Image provided by the author.)

A self-retaining, bivalved laryngoscope may allow the hand that is holding the laryngoscope to be free in order to help stabilise the head and neck. The head and neck of the patient may also be stabilised between the knees of the person intubating the patient [64, 71].

During cardiopulmonary resuscitation, stabilising the head by gripping it between the knees (**Figure 2**) is recommended [71]. Even though this technique provides a distant view of the glottic inlet, it is stable and saves time. This technique may be compared to the "sit down–lean back technique" used by paramedics to stabilise the victim's head [72].

In the case of elective procedures, it is advisable to use restraints. If exercise of restraints is not feasible then extra tracheal airway devices may prove to be useful. It is not essential to position oneself at the head end of the patient for use of the extra-tracheal airway devices, in the interest of saving time [64].

Restraints are not necessary for either the patient, the equipment or the operator—on Mars—since its gravity is approximately one-third of Earth's gravity. In microgravity, however, restraining is recommended [35].

Robotic intubation may not find an application, at least in the foreseeable future, as far as space is concerned, in view of the undue up-mass they constitute and given that the chances of using this technique is remote [1].

7. Conclusion

Recent technological advances and scientific discoveries will lead mankind to realise its aspirations as a spacefaring species. If we are to support an enduring human exploration of space, we would have to rapidly update our understanding of human physiology, as well as find better ways to manage medical and surgical events in space. Persons other than professional astronauts are unprepared for the rigours of space travel and are prone to medical complications and emergencies. In this regard, it is of great importance that astronauts manage medical and surgical events in space.

With increasing flight durations, as a result of advances made in spaceflight enabling technologies that would afford deeper exploration into space, the prospect of a medical emergency entailing airway management is increased. Medical evacua-tion is not an option in such cases and therefore becomes imperative that immediate care is provided, while on board. Such situations call for a crew equipped with emergency care skills and training. Airway management is an important skill that is required to manage a great many medical emergencies.

This chapter briefly overviews the physiology of the airway, anaesthesia delivery and airway management, on Earth, to compare with and highlight the challenges to human physiology in space given the unique nature of the environment, so as to understand the specific challenges to anaesthesia delivery and airway management, in space.

The enabling technology for space travel, powered by Artificial Intelligence and Machine Learning algorithms, is advancing exponentially. Medical science typically tends to follow a more measured trajectory and tends to trail relative to the strides made by technology. We will also likely travel to Mars in the near future and this particular mission will come under more scrutiny than any other previous space missions. The crew will be required to travel farther and longer than any other human being in the history of our world. Techniques for airway management will need to undergo and be better defined in the coming years. As it stands, an expand-ing body of research that is trying to provide a better understanding of many aspects of human health in space is already underway. It is imperative for medical science to rapidly expand its body of knowledge when it comes to space travel, if it is to support human quest for space exploration.

Conflict of interest

The author declares no conflict of interest.

Notes/Thanks/Other declarations

Looking up into the night sky is looking into infinity – distance is incomprehensible and therefore meaningless.

–Douglas Adams, A Hitch-hikers Guide to the Galaxy.

Of all milestones and achievements in medicine, conquering pain must be one of the very few that has potentially affected every human being in the world.

–DH Robinson.

Special thanks to Preetam Satish.

Author details

Preethi Reddy
Independent Scientist, Bangalore, India

*Address all correspondence to: preethireddy3108@gmail.com

References

[1] Warnecke T, Tochtermann F, Kerkhoff S, Komorowski M, Neuhaus C, Hinkelbein J. Airway management in microgravity : a systematic review. Acta Anaesthesiol Scand. 2019;63:2-7. DOI: 10.1111/aas.13251

[2] Williams D, Kuipers A, Mukai C, Thirsk R. Acclimation during space flight: effects on human physiology. CMAJ. 2009 Jun 23; 180(13): 1317-1323. DOI: 10.1503/cmaj.090628

[3] Abadie LJ, Lloyd CW, Shelhamer MJ. The human body in space [Internet]. Houston: Johnson Space Center (US), NASA Human Research Program; [cited 2020 June 20]. Available from: https://www.nasa.gov/hrp/bodyinspace

[4] Keller C, Brimacombe J, Giampalmo M, Kleinsasser A, Loeckinger A, Giampalmo G, Pühringer F. Airway management during spaceflight: A comparison of four airway devices in simulated microgravity. Anesthesiology. 2000 May;92(5):1237-1241. DOI: 10.1097/00000542-200005000-00010

[5] LeJeune FE Jr. Laryngeal Problems in Space Travel. Aviat Space Environ Med. 1978 Nov;49(11):1347-1349.

[6] Rabitsch W, Moser D, Inzunza MR, Niedermayr M, Kostler WJ, Staudinger T, Locker GJ, Schellongowski P, Wulkersdorfer B, Rich JM, Meyer B, Benumof JL, Frass M. Airway management with endotracheal tube versus combitube during parabolic flights. Anesthesiology 2006; 105(4): 696 –702. DOI: 10.1097/00000542-200610000-00014

[7] Komorowski M, Fleming S, Mawkin M Hinkelbein J. Anaesthesia in austere environments: literature review and considerations for future space exploration missions. NPJ Microgravity. 2018 Feb 23; 4:5. DOI: 10.1038/s41526-018-0039-y

[8] Hagberg CA, Artime CA. Airway management in the adult. In: Miller RD, Cohen NH, Eriksson LI, Fleisher LA, Wiener-Kronish JP, Young WL, editors. Miller's Anaesthesia. 8th ed. Philadelphia, PA: Elsevier/Saunders; 2015. p. 1647-1683.

[9] Practice guidelines for preoperative fasting and the use of pharmacologic agents to reduce the risk of pulmonary aspiration: application to healthy patients undergoing elective procedures: an updated report by the American Society of Anesthesiologists task force on preoperative fasting and the use of pharmacologic agents to reduce the risk of pulmonary aspiration. 2017 Mar;126(3):376-93. DOI: 10.1097/ALN.0000000000001452

[10] Combes X, Andriamifidy L, Dufresne E, Suen P, Sauvat S, Scherrer E, Feiss P, Marty J, Duvaldestin P. Comparison of two induction regimens using or not using muscle relaxant: impact on postoperative upper airway discomfort. Br J Anaesth 2007;99(2):276-281. DOI: 10.1093/bja/aem147

[11] Gore MS, Harnagale KD. Evaluation of Intubating Conditions with Varying Doses of Propofol without Muscle Relaxants. J Anaesthesiol Clin Pharmacol 2011; 27(1):27-30.

[12] Erhan E, Ugur G, Gunusen I, Alper I, Ozya B. Propofol - not thiopental or etomidate - with remifentanil provides adequate intubating conditions in the absence of neuromuscular blockade. Can J Anaesth 2003 Feb;50(2):108-115. DOI: 10.1007/BF03017840

[13] Apfelbaum JL, Hagberg CA, Caplan RA, Blitt CD, Connis RT,

Nickinovich DG, et al. Practice guidelines for management of the difficult airway: an updated report by the American Society of Anesthesiologists Task Force on Management of the Difficult Airway. Anesthesiology. 2013 Feb;118(2):251-270. DOI: 10.1097/ALN.0b013e31827773b2

[14] The respiratory pathway. In: Ellis H, Feldman S, Griffiths WH. Anatomy for anaesthetists. 8th ed. Massachusetts: Blackwell Publishing; 2004. p.1-69.

[15] Demontis GC, Germani MM, Caiani EG, Barravecchia I, Passino C, Angeloni D. Human pathophysiological adaptations to the space environment. Front. Physiol. 2017 Aug 2;8:547. DOI: 10.3389/fphys.2017.00547

[16] Komorowski M, Fleming S, Kirkpatrick AW. Fundamentals of Anesthesiology for Spaceflight. J Cardiothorac Vasc Anesth. 2016 Jun;30(3):781-790. DOI: 10.1053/j.jvca.2016.01.007

[17] Norfleet WT. Anesthetic Concerns of Spaceflight. Anesthesiology. 2000 May;92(5):1219-1222. DOI: 10.1097/00000542-200005000-00006

[18] Alfrey CP, Udden MM, Leach-Huntoon C, Driscoll T, Pickett MH. Control of red blood cell mass in spaceflight. J Appl Physiol (1985). 1996 Jul;81(1):98-104. DOI: 10.1152/jappl.1996.81.1.98

[19] Norsk P, Asmar A, Damgaard M, Christensen NJ. Fluid shifts, vasodilatation and ambulatory blood pressure reduction during long duration spaceflight. J Physiol. 2015 Feb;593(3):573-584. DOI: 10.1113/jphysiol.2014.284869

[20] Ade CJ, Bemben DA. Differential MicroRNA expression following head-down tilt bed rest: implications for cardiovascular responses to microgravity. Physiol Rep. 2019 May;7(9):e14061. DOI: 10.14814/phy2.14061

[21] Martin DS, South DA, Wood ML, Bungo MW, Meck JV. Comparison of echocardiographic changes after short- and long-duration spaceflight. Aviat Space Environ Med. 2002 Jun;73(6):532-536.

[22] Perhonen MA, Franco F, Lane LD, Buckey JC, Blomqvist CG, Zerwekh JE, Peshock RM, Weatherall PT, Levine BD. Cardiac atrophy after bed rest and spaceflight. J Appl Physiol (1985). 2001 Aug;91(2):645-653. DOI: 10.1152/jappl.2001.91.2.645

[23] Simanonok KE, Charles JB. Space sickness and fluid shifts: a hypothesis. J Clin Pharmacol. 1994 Jun; 34(6):652-663. DOI: 10.1002/j.1552-4604.1994.tb02020.x

[24] LeBlanc AD, Spector ER, Evans HJ, Sibonga JD. Skeletal responses to space flight and the bed rest analog: a review. J Musculoskelet Neuronal Interact. 2007 Jan-Mar;7(1):33-47.

[25] Smith SM, Wastney ME, Morukov BV, Larina IM, Nyquist LE, Abrams SA, et al. Calcium metabolism before, during, and after a 3-mo spaceflight: kinetic and biochemical changes. Am J Physiol. 1999 Jul;277 (1 Pt 2):R1-10. DOI: 10.1152/ajpregu.1999.277.1.r1

[26] Smith SM, Heer M, Shackelford LC, Sibonga JD, Spatz J, Pietrzyk RA, Hudson EK, Zwart SR. Bone metabolism and renal stone risk during International Space Station missions. Bone. 2015 Dec;81:712-720. DOI: 10.1016/j.bone.2015.10.002

[27] Whitson PA, Pietrzyk RA, Morukov BV, Sams CF. The risk of renal stone formation during and after long duration space flight. Nephron. 2001

Nov;89(3):264-270. DOI: 10.1159/000046083

[28] Wantier M, Estenne M, Verbanck S, Prisk GK, Paiva M. Chest wall mechanics in sustained microgravity. Journal of Applied Physiology (Bethesda, Md. : 1985). 1998 Jun;84(6):2060-2065. DOI: 10.1152/jappl.1998.84.6.2060

[29] Prisk GM. Microgravity and the respiratory system. Eur Respir J. 2014 May;43(5):1459-1471. DOI: 10.1183/09031936.00001414

[30] Elliott AR, Prisk GK, Guy HJ, West JB. Lung volumes during sustained microgravity on Spacelab SLS-1. J Appl Physiol (1985). 1994 Oct;77(4):2005-2014. DOI: 10.1152/jappl.1994.77.4.2005

[31] Pierson DL, Stowe RP, Phillips TM, Lugg DJ, Mehta SK. Epstein-Barr virus shedding by astronauts during space flight. Brain Behav Immun. 2005 May;19(3):235-242. DOI: 10.1016/j.bbi.2004.08.001

[32] Wilson JW, Ott CM, Höner zu Bentrup K, Ramamurthy R, Quick L, Porwollik S, et al. Space flight alters bacterial gene expression and virulence and reveals a role for global regulator Hfq. Proc Natl Acad Sci U S A. 2007 Oct 9;104(41):16299-16304. DOI: 10.1073/pnas.0707155104

[33] Heer M, De Santo NG, Cirillo M, Drummer C. Body mass changes, energy, and protein metabolism in space. Am J Kidney Dis. 2001 Sep;38(3):691-695. DOI: 10.1053/ajkd.2001.27767

[34] Smith SM, Zwart SR, Block G, Rice BL, Davis-Street JE. The nutritional status of astronauts is altered after long-term space flight aboard the International Space Station. J Nutr. 2005 Mar;135(3):437-443. DOI: 10.1093/jn/135.3.437

[35] Komorowski M, Fleming S. Intubation after rapid sequence induction performed by non-medical personnel during space exploration missions: a simulation pilot study in a Mars analogue environment. Extrem Physiol Med. 2015 Nov;4:19. DOI: 10.1186/s13728-015-0038-5

[36] Convertino VA, Cooke WH. Evaluation of cardiovascular risks of spaceflight does not support the NASA bioastronautics critical path roadmap. Aviat Space Environ Med. 2005 Sep;76(9):869-876

[37] Komorowski M, Watkins SD, Lebuffe G, Clark JB. Potential anesthesia protocols for space exploration missions. Aviat Space Environ Med. 2013 Mar;84(3):226-233. DOI: 10.3357/asem.3427.2013

[38] Drudi L, Ball CG, Kirkpatrick AW, Saary J, Grenon SM. Surgery in Space: Where are we at now? Acta Astronaut. 2012;79:61-66. DOI: 10.1016/j.actaastro.2012.04.014

[39] Pavy-Le Traon A, Saivin S, Soulez-LaRivière C, Pujos M, Güell A, Houin G. Pharmacology in space: pharmacotherapy. Adv Space Biol Med. 1997;6:93-105. DOI: 10.1016/s1569-2574(08)60079-9

[40] Graebe A, Schuck EL, Lensing P, Putcha L, Derendorf H. Physiological, pharmacokinetic, and pharmaco dynamic changes in space. J Clin Pharmacol. 2004 Aug;44(8):837-853. DOI: 10.1177/0091270004267193

[41] Sørensen MK, Bretlau C, Gätke MR, Sørensen AM, Rasmussen LS. Rapid sequence induction and intubation with rocuronium-sugammadex compared with succinylcholine: a randomized trial. Br J Anaesth. 2012 Apr;108(4): 682-689. DOI: 10.1093/bja/aer503

[42] Niazi AU, Haldipur N, Prasad AG, Chan VW. Ultrasound-guided regional

anesthesia performance in the early learning period: effect of simulation training. Reg Anesth Pain Med. 2012 Jan-Feb;37(1):51-54. DOI: 10.1097/AAP.0b013e31823dc340

[43] Salinas FV, Hanson NA. Evidence-based medicine for ultrasound-guided regional anesthesia. Anesthesiol Clin. 2014 Dec;32(4):771-787. DOI: 10.1016/j. anclin.2014.08.001

[44] Ambrosio A, Pfannenstiel T, Bach K, Cornelissen C, Gaconnet C, Brigger MT. Difficult airway management for novice physicians: a randomized trial comparing direct and video-assisted laryngoscopy. Otolaryngol Head Neck Surg. 2014 May;150(5):775-778. DOI: 10.1177/0194599814521380

[45] Maharaj CH, Ni Chonghaile M, Higgins BD, Harte BH, Laffey JG. Tracheal intubation by inexperienced medical residents using the Airtraq and Macintosh laryngoscopes--a manikin study. Am J Emerg Med. 2006 Nov;24(7):769-774. DOI: 10.1016/j. ajem.2006.03.014

[46] McQuillen JB, McKay TL, Griffin DW, Brown DF, Zoldak JT. Final Report for Intravenous Fluid Generation (IVGEN) Spaceflight Experiment. [Internet] Cleveland, Ohio: NASA STI Program; July 2011 [cited 2020 June 25]. 86 p. Report No.: NASA/TM—2011-217033. Available from: https://ntrs. nasa.gov/citations/20110014585

[47] C9 and other microgravity simulations summary report. Houston, Texas: Space Life Sciences Directorate, Human Adaptation and Countermeasures Division, NASA Johnson Space Center; Sep 2009 [cited 2021 Feb 10]. 150 p. Report No.: NASA/TM-2009-214794 Available from: https://www.nasa.gov/centers/johnson/pdf/505838main_ FY09_ TM-20 09- 214794.pdf

[48] Campbell MR, Williams DR, Buckey JC Jr, Kirkpatrick AW. Animal surgery during spaceflight on the Neurolab Shuttle mission. Aviat Space Environ Med. 2005 Jun;76(6):589-593.

[49] Campbell MR, Dawson DL, Melton S, Hooker D, Cantu H. Surgical instrument restraint in weightlessness. Aviat Space Environ Med. 2001 Oct;72(10):871-876.

[50] Kirkpatrick AW, LaPorta A, Brien S, Leslie T, Glassberg E, McKee J, et al. Technical innovations that may facilitate real-time telementoring of damage control surgery in austere environments: a proof of concept comparative evaluation of the importance of surgical experience, telepresence, gravity and mentoring in the conduct of damage control laparotomies. Can J Surg. 2015 Jun;58(3):S88-S90. DOI: 10.1503/cjs.014214

[51] Smith HM, Jacob AK, Segura LG, Dilger JA, Torsher LC. Simulation education in anesthesia training: a case report of successful resuscitation of bupivacaine-induced cardiac arrest linked to recent simulation training. Anesth Analg. 2008 May;106(5): 1581-1584. DOI: 10.1213/ane.0b013e 31816b9478

[52] Galvez JA, Rehman MA. Telemedicine in anesthesia: an update. Curr Opin Anaesthesiol. 2011 Aug;24(4):459-462. DOI: 10.1097/ACO.0b013e328348717b

[53] Murray AW, Beaman ST, Kampik CW, Quinlan JJ. Simulation in the operating room. Best Pract Res Clin Anaesthesiol. 2015 Mar;29(1):41-50. DOI: 10.1016/j.bpa.2015.02.005

[54] Schwid HA, Rooke GA, Carline J, Steadman RH, Murray WB, Olympio M, et al. Anesthesia Simulator Research Consortium. Evaluation of anesthesia residents using mannequin-based simulation: a multi institutional study.

Anesthesiology. 2002 Dec;97(6):1434-1444. DOI: 10.1097/00000542-200212000-00015

[55] Weller JM, Nestel D, Marshall SD, Brooks PM, Conn JJ. Simulation in clinical teaching and learning. Med J Aust 2012;196(9):594. DOI: 10.5694/mja10.11474

[56] Bösenberg AT. Pediatric anesthesia in developing countries. Curr Opin Anaesthesiol. 2007 Jun;20(3):204-210. DOI: 10.1097/ACO.0b013e3280c60c78

[57] Martin J, Tau G, Cherian MN, Vergel de Dios J, Mills D, Fitzpatrick J, et al. Survey of the capacity for essential surgery and anaesthesia services in Papua New Guinea. BMJ Open. 2015 Dec 16;5(12):e009841. DOI: 10.1136/bmjopen-2015-009841.

[58] McQueen K, Coonan T, Ottaway A, Hendel S, Bagutifils PR, Froese A, et al. The Bare Minimum: The Reality of Global Anaesthesia and Patient Safety. World J Surg. 2015 Sep;39(9):2153-2160. DOI: 10.1007/s00268-015-3101-x

[59] Jochberger S, Ismailova F, Lederer W, Mayr VD, Luckner G, Wenzel V, et al. Anesthesia and its allied disciplines in the developing world: a nationwide survey of the Republic of Zambia. Anesth Analg. 2008 Mar;106(3):942-948. DOI: 10.1213/ane.0b013e318166ecb8

[60] Epiu I, Tindimwebwa JV, Mijumbi C, Chokwe TM, Lugazia E, Ndarugirire F, et al. Challenges of Anesthesia in Low- and Middle-Income Countries: A Cross-Sectional Survey of Access to Safe Obstetric Anesthesia in East Africa. Anesth Analg. 2017 Jan;124(1):290-299. DOI: 10.1213/ANE.0000000000001690

[61] Saluja IS, Williams DR, Woodard D, Kaczorowski J, Douglas B, Scarpa PJ, et al. Survey of astronaut opinions on medical crewmembers for a mission to Mars. Acta Astronautica. 2008 Sep;63(5-6): 586-593. DOI: 10.1016/j.actaastro.2008.05.002.

[62] Kuypers MI. Emergency and wilderness medicine training for physician astronauts on exploration class missions. Wilderness Environ Med. 2013 Dec;24(4):445-449. DOI: 10.1016/j. wem.2013.03.024

[63] Stewart LH, Trunkey D, Rebagliati GS. Emergency medicine in space. J Emerg Med. 2007 Jan;32(1):45- 54.DOI: 10.1016/j.jemermed.2006.05.031

[64] Keller C, Brimacombe J, A FR, Giampalmo M, Kleinsasser A, Loeckinger A, et al. Airway management during spaceflight: A comparison of four airway devices in simulated microgravity. Anesthesiology. 2000 May;92(5):1237-1241. DOI: 10.1097/00000542-200005000-00010

[65] Deakin CD, Peters R, Tomlinson P, Cassidy M. Securing the prehospital airway: a comparison of laryngeal mask insertion and endotracheal intubation by UK paramedics. Emerg Med J. 2005 Jan;22(1):64-67. DOI: 10.1136/emj.2004.017178

[66] Brimacombe J. The advantages of the LMA over the tracheal tube or facemask: a meta-analysis. Can J Anaesth. 1995 Nov;42(11):1017-1023. DOI: 10.1007/BF03011075

[67] Davies PR, Tighe SQ, Greenslade GL, Evans GH. Laryngeal mask airway and tracheal tube insertion by unskilled personnel. Lancet. 1990 Oct 20;336(8721):977-979. DOI: 10.1016/0140-6736(90)92429-l

[68] Arulkumaran N, Lowe J, Ions R, Mendoza M, Bennett V, Dunser MW. Videolaryngoscopy versus direct laryngoscopy for emergency orotracheal intubation outside the operating room: a systematic review and meta-analysis. Br

J Anaesth. 2018 Apr;120(4):712-724. DOI: 10.1016/j.bja.2017.12.041

[69] Risse J, Volberg C, Kratz T, Plöger B, Jerrentrup A, Pabst D, et al. Comparison of videolaryngoscopy and direct laryngoscopy by German paramedics during out-of-hospital cardiopulmonary resuscitation; an observational prospective study. BMC Emerg Med. 2020 Mar 23;20(1):22. DOI: 10.1186/s12873-020-00316-z

[70] Liu DX, Ye Y, Zhu YH, Li J, He HY, Dong L, et al. Intubation of non-difficult airways using video laryngoscope versus direct laryngoscope: a randomized, parallel-group study. BMC Anesthesiol. 2019 May 15;19(1):75. DOI: 10.1186/s12871-019-0737-3

[71] Groemer GE, Brimacombe J, Haas T, de Negueruela C, Soucek A, Thomsen M, et al. The feasibility of laryngoscope-guided tracheal intubation in microgravity during parabolic flight: a comparison of two techniques. Anesth Analg. 2005 Nov;101(5):1533-1535. DOI: 10.1213/01.ANE.0000181001.25777.53

[72] Rabitsch W, Moser D, Inzunza MR, Niedermayr M, Köstler WJ, Staudinger T, et al. Airway management with endotracheal tube versus Combitube during parabolic flights. Anesthesiology. 2006 Oct;105(4):696-702. DOI: 10.1097/00000542-200610000-00014

Permissions

All chapters in this book were first published by InTech Open; hereby published with permission under the Creative Commons Attribution License or equivalent. Every chapter published in this book has been scrutinized by our experts. Their significance has been extensively debated. The topics covered herein carry significant findings which will fuel the growth of the discipline. They may even be implemented as practical applications or may be referred to as a beginning point for another development.

The contributors of this book come from diverse backgrounds, making this book a truly international effort. This book will bring forth new frontiers with its revolutionizing research information and detailed analysis of the nascent developments around the world.

We would like to thank all the contributing authors for lending their expertise to make the book truly unique. They have played a crucial role in the development of this book. Without their invaluable contributions this book wouldn't have been possible. They have made vital efforts to compile up to date information on the varied aspects of this subject to make this book a valuable addition to the collection of many professionals and students.

This book was conceptualized with the vision of imparting up-to-date information and advanced data in this field. To ensure the same, a matchless editorial board was set up. Every individual on the board went through rigorous rounds of assessment to prove their worth. After which they invested a large part of their time researching and compiling the most relevant data for our readers.

The editorial board has been involved in producing this book since its inception. They have spent rigorous hours researching and exploring the diverse topics which have resulted in the successful publishing of this book. They have passed on their knowledge of decades through this book. To expedite this challenging task, the publisher supported the team at every step. A small team of assistant editors was also appointed to further simplify the editing procedure and attain best results for the readers.

Apart from the editorial board, the designing team has also invested a significant amount of their time in understanding the subject and creating the most relevant covers. They scrutinized every image to scout for the most suitable representation of the subject and create an appropriate cover for the book.

The publishing team has been an ardent support to the editorial, designing and production team. Their endless efforts to recruit the best for this project, has resulted in the accomplishment of this book. They are a veteran in the field of academics and their pool of knowledge is as vast as their experience in printing. Their expertise and guidance has proved useful at every step. Their uncompromising quality standards have made this book an exceptional effort. Their encouragement from time to time has been an inspiration for everyone.

The publisher and the editorial board hope that this book will prove to be a valuable piece of knowledge for researchers, students, practitioners and scholars across the globe.

List of Contributors

Angelika Diefenbach
Microgravity User Support Center, Deutsches Zentrum für Luft- und Raumfahrt (DLR), Cologne, Germany

Stephan Schneider and Thomas Volkmann
Institut für Materialphysik im Weltraum, Deutsches Zentrum für Luft- und Raumfahrt (DLR), Cologne, Germany

Vladimir Pletser
Blue Abyss, Space Training Operations, Liverpool, United Kingdom
Chinese Academy of Sciences, Beijing, China
European Space Agency, Noordwijk, The Netherlands

Thais Russomano
InnovaSpace, United Kingdom
Centre for Human and Applied Physiological Sciences, School of Basic and Medical Biosciences Faculty of Life Sciences and Medicine, King's College London, United Kingdom

Nikolay Smirnov, Valeriy Nikitin and Evgeniya Kolenkina (Skryleva)
Moscow M.V. Lomonosov State University, Moscow, Russia
Scientific Research Institute for System Analysis of the Russian Academy of Sciences, Moscow, Russia

Mohamed Farhat
Ecole Polytechnique Fédérale de Lausanne, Switzerland

Gwennou Coupier
Laboratory for Interdisciplinary Physics (LIPhy), UMR5588 CNRS, Université Grenoble Alpes, Grenoble, France

Thomas Podgorski
Laboratory for Interdisciplinary Physics (LIPhy), UMR5588 CNRS, Université Grenoble Alpes, Grenoble, France
Laboratoire Rhéologie et Procédés (LRP) - UMR5520 CNRS, Université Grenoble Alpes, Grenoble-INP, Grenoble, France

Christophe Minetti
Microgravity Research Center, CP-165/62, Université Libre de Bruxelles (ULB), Brussels, Belgium

Valentina Shevtsova, Denis Melnikov, Yuri Gaponenko and Aliaksandr Mialdun
Microgravity Research Centre, Université libre de Bruxelles (ULB), Brussels, Belgium

Tricia L. Larose
Faculty of Medicine, Department of Community Medicine and Global Health, Institute of Health and Society, Oslo, Norway
Human Performance and Space Department, International Space University, Strasbourg, France

Jean-Louis Thonnard, Laurent Opsomer and Philippe Lefèvre
Institute of Neurosciences (IoNS), University of Louvain, Belgium
Institute of Information and Communication Technologies, Electronics and Applied Mathematics (ICTEAM), University of Louvain, Belgium

Joseph McIntyre
Centre National de la Recherche Scientifique, France
TECNALIA, Basque Research and Technology Alliance (BRTA), Spain
Ikerbasque Research Foundation, Spain

Peter Norsk
Center for Space Medicine and Department Molecular Physiology and Biophysics, Baylor College of Medicine, Houston, Texas, USA

Preethi Reddy
Independent Scientist, Bangalore, India

Index